FORTRAN 77
FOR ENGINEERS

Statement	Comment
SAVE [⟨name list⟩]	Used to retain values of local variables between calls.
(*)ENTRY ⟨name⟩[⟨arg. list⟩]	Not used in this text.
(*)EQUIVALENCE ⟨name list⟩	Not used in this text.
(*)BLOCK DATA ⟨name⟩	Not used in this text.

ASSIGNMENT AND PROGRAM CONTROL STATEMENTS (executable)

Statement	Comment
Assignment statement ⟨name⟩ = ⟨exp.⟩	⟨exp.⟩ may be any of the Fortran data-types depending on the type of ⟨name⟩.
STOP [tag]	[tag] is an INTEGER or CHARACTER constant.
(*)ASSIGN ⟨stmt. label⟩ TO ⟨name⟩	Not used in this text.
(*)PAUSE [tag]	Not used in this text.

FLOW-CONTROL STATEMENTS

Statement	Comment
RETURN [exp.]	Return from a subprogram. The optional [exp.] is used for alternate returns from a subroutine.
CALL ⟨subr. name⟩[⟨arg. list⟩]	Transfer of control to subroutine ⟨name⟩.
CONTINUE	Loop terminator or target of a GO TO.
GO TO ⟨stmt. label⟩	Unconditional GO TO.
GO TO (⟨stmt. label list⟩),⟨exp.⟩	Computed GO TO.
(*)GO TO ⟨assigned variable⟩	Assigned GO TO. Not used in this text.
IF(⟨arith. exp.⟩)sl_1,sl_2,sl_3	Arithmetic IF test.
IF(⟨logical exp.⟩)⟨execut. stmt.⟩	Logical IF test.
IF(⟨logical exp.⟩)THEN	Logical block IF.
ELSE	Optional ELSE block.
ELSE IF(⟨logical exp.⟩)THEN	Optional conditional block.
END IF	Block IF terminator.
DO ⟨stmt. label⟩ vn = $e_1,e_2[,e_3]$	DO loop. vn is a variable name, e_1,e_2,e_3 are expressions specifying index bounds.

FILE DIRECTIVE STATEMENTS

Statement	Comment
OPEN(⟨unit no.⟩,FILE = '⟨name⟩',[options])	Connects a file to an I/O unit.
CLOSE(⟨unit no.⟩)	Disconnects a file.
REWIND ⟨unit no.⟩	Positions a SAM file at the beginning.

Continued on back endsheets

FORTRAN 77
FOR ENGINEERS

G. J. BORSE

Lehigh University

 PWS ENGINEERING • BOSTON

PWS PUBLISHERS

Prindle, Weber & Schmidt • ☙ • Duxbury Press • ♠ • PWS Engineering • △ • Breton Publishers • ⚙
20 Park Plaza • Boston, Massachusetts 02116

PWS Publishers is a division of Wadsworth, Inc.

Printed in the United States of America

86 87 88 89—10 9 8 7 6 5 4

Library of Congress Cataloging in Publication Data

Borse, Garold J.
  Fortran 77 for engineers.

  Includes index.
  1. Engineering—Data processing.   2. FORTRAN
(Computer program language) I. Title.
TA345.B68   1985   620'.0028'5424   85–532
ISBN 0–534–04650–9

ISBN 0-534-04650-9

Cover photograph © David Wagner/Phototake
Sponsoring Editor—John E. Block
Signing Representative—Winston Beauchamp
Editorial Assistants—Gabriele Bert and Suzi Sheperd
Composition: Interactive Composition Corporation
Printing and Binding: Halliday Lithograph
Cover Printing: New England Book Components

This text is intended for use in a Fortran for Engineers course for beginning engineering and physical science students. In the last decade such a course has joined introductory calculus and physics to form the core curriculum of all engineering education. While the content and structure of calculus and physics courses have had many decades to evolve into a form specially suited to the needs of engineering and science students, Fortran programming has had to adapt more quickly. The two most significant recent changes have been (1) the inclusion of Fortran programming as a one- or two-semester course in the beginning engineering sequence and (2) the revision of Fortran (the form now used is called Fortran 77) to encourage program writing in a clear and structured style. This text encompasses the experiences and prejudices of the author gained over many years of designing and teaching such a course. The theme of this volume is that the two most important ingredients of a Fortran for Engineers course are as follows:

1. The course must be carefully matched to the abilities and interests of the students. In the present context, this means that full advantage must be taken of the superior mathematical abilities of engineering students. Moreover, there should be recognition of the increase in their mathematical skills as the semester progresses due to experiences in calculus and physics. Additionally, realistic examples gathered from a variety of engineering disciplines must be included to ensure that the student is convinced that his or her studies are related to real-world applications of the computer to engineering.
2. The course must instruct the student in the use of good programming style as an efficient means of problem solving. The step-wise refinement of rudimentary algorithms into clear pseudocode outlines and finally into working Fortran code is a skill developed only by working a large number of demanding problems under the guidance of an instructor. This text provides numerous examples of completely worked problems, from the design of the algorithm to the final printed output, which can be read and studied by the student. However, there is no substitute for intensive interaction with the computer by the student. For this reason, a large number of quite demanding programming assignments are inserted throughout the text and can be used as major bi-weekly projects in addition to the shorter programming problems appearing at the end of each chapter.

The most novel feature of the text is the collection of seven sets of programming problems. In addition to providing challenging tests of the student's programming and problem-solving skills, these problems are also used to introduce the various disciplines of engineering.

Organization of the Text

In the first three chapters the student is taught to read and write elementary Fortran code. A few simple computer problems are offered for immediate solution to familiarize the student with the details of submitting a program and the peculiarities of integer and real arithmetic on the computer. The ideas of a loop structure are explained and the first major programming problem is introduced.

The fundamental elements of mathematical computation are introduced and illustrated in Chapter 4. Some sections involving material of a more advanced mathematical nature (numerical integration and Newton's method, Sections 4.5 and 4.7) may be omitted without a loss of continuity. It is noteworthy that the statement function is introduced in this chapter in order to facilitate cleaner and more structured code and to anticipate the introduction of modular structures (subprograms) in Chapter 9.

Chapters 5 and 6 are concerned with the formatted input and output of data and the use of both sequential- and direct-access data files. Even the most elementary student programming projects will make use of limited data files, which justifies the relatively early presentation of this material.

Subscripted arrays and the DO loop structure are closely related concepts in Fortran and are introduced together in Chapter 7. The treatment of character variables and data in Chapter 8 is quite extensive. If the constraints of time do not permit such a broad coverage, Sections 8.3, 8.4, 8.6.2, and 8.7.2 may be omitted without loss of continuity.

Subprograms are covered in Chapter 9 with extensive examples. The use of Fortran EXTERNAL and COMMON statements is especially important in all modular codes written for engineering applications, and the description of these statements is detailed to an uncommon degree.

Many of the remaining features of Fortran that for pedagogic reasons have been omitted from the earlier chapters are collected together in Chapter 10. Depending on the nature of the class and the course, the instructor should feel free to present some of these concepts at earlier stages in the course. For example, many instructors prefer to use DATA statements from the very beginning of the course. (It has been my experience that the students find these statements somewhat confusing if presented at the same time as assignment statements and list-directed READ statements.)

An extensive Instructor's Manual containing suggested syllabi, transparency masters, suggested hand-outs for getting started, and complete solutions to all chapter problems and all programming assignments is available.

Acknowledgments

When designing a new introductory course of any type, one accumulates a great many debts to individuals, from deans to teaching assistants, who freely give their time and knowledge. In the construction and evolution of a course such as the one based on this text, the list is very long indeed. Colleagues in each of the engineering disciplines helped in the construction of problems illustrating engineering analysis. The computing center at Lehigh University was continually helpful. I would like to acknowledge the special assistance of my colleagues at Lehigh, John Karakash, Curt Clump, Robert Johnson, and Stephen K. Tarby, who permitted the use of a typewritten version of this text to be used in our Fortran for Engineers course.

My greatest debt is to the several thousand Lehigh engineering students who have read and used this text in its various preliminary forms and who have made innumerable suggestions for its improvement. Of course, my family, through their tolerance and encouragement, have contributed immeasurably to this work.

I am also grateful for the advice and suggestions of many reviewers of the original manuscript, in particular:

Linda Hayes, University of Texas at Austin; Glen Williams, Texas A & M University; William Kubitz, University of Illinois; Tom Boyle, Purdue University; Philip M. Wolfe, Oklahoma State University; John B. Crittenden, Virginia Polytechnic Institute; Allen R. Cook, University of Oklahoma; Terry Feagin, University of Tennessee; Betty Barr, University of Houston; Robert Good, Widener University; Bart Childs, Texas A & M University

The help and encouragement of John Block, engineering editor at PWS, and the many contributions of Deborah Schneider in the production of this book are cordially acknowledged.

CONTENTS

3 FLOW CONTROL STRUCTURES AND PROGRAM DESIGN

PROGRAMMING ASSIGNMENT I

4 ELEMENTARY PROGRAMMING TECHNIQUES

PROGRAMMING ASSIGNMENT II

5 USE OF DATA FILES

---

[1]Sections 4.5 and 4.7 contain somewhat more advanced material and may be omitted without loss of continuity.

6 ELEMENTARY FORMATTED INPUT-OUTPUT

PROGRAMMING ASSIGNMENT III

7 DIMENSIONED VARIABLES AND DO LOOPS

PROGRAMMING ASSIGNMENT IV

8 NON-NUMERICAL APPLICATIONS—CHARACTER VARIABLES

[2]Sections 8.3, 8.4, 8.6.2, and 8.7.2 may be omitted without loss of continuity.

10 ADDITIONAL FORTRAN FEATURES

PROGRAMMING ASSIGNMENT VII

APPENDIX   SUMMARY OF FORTRAN STATEMENTS AND GRAMMAR RULES

CHAPTER 1

THE OPERATIONS
OF A COMPUTER

The last century was the golden age of applied mathematics. A rather large fraction of the great minds of that era were interested in the solution of problems in the area of engineering and physical sciences. Men like Laplace, Gauss, Bessel, and others determined what problems could be solved and specified the form of the solutions in elegant mathematical analyses. These men shaped the course of this century far more than we realize. The method of approaching almost any problem in engineering is due to the mathematics developed more than 100 years ago. However, there is a revolution in applied mathematics going on right now, all around us, and it will most certainly affect the future in unforeseen but dramatic ways. Of course, I'm talking about the effect of computers on engineering mathematics. Problems that were impossible to solve by hand are now often trivial. Problems that previously required ingenious approximations and "tricks" are now solved with no approximations and no tricks, but on large computers. Problems that most thought were not worth the effort are now solved anyway.[1] The era of computers has come upon us so quickly and the capabilities of the machines

---

[1]Not long ago, π was computed to an accuracy of tens of thousands of digits.

are presently advancing so rapidly that mathematical techniques specially suited to large computers are currently being developed. These methods are generally not suitable for an elementary text. The numerical procedures discussed in this text were originally developed for hand calculations and have been adapted for use on large computers.

The uses of large computers can be divided into two classes:

1. Numerical computation
2. Information and data storage, retrieval, processing, and synthesis

The first classification is, of course, the reason for the invention of modern computers. However, it is the second classification that is experiencing the most rapid growth and already dominates all aspects of computer technology. For example, the billing and record keeping of all large companies are handled by computers. Most large newspapers use computers to edit the text so that the words fit neatly in columns. These are programs that can match a patient's symptoms to a list of diseases and assist in medical diagnosis. The evaluation of huge amounts of data is now universally done by machines. In fact, it was the Bureau of the Census that was among the first to make use of the early computers. And, of course, we are all aware of the enormous popularity of video games and other uses of small personal computers.

Today, the number and type of applications of computers in all aspects of a community are obviously too vast to attempt to catalog. Moreover, the growth in their use is expanding at an ever-increasing rate.

The variety of uses of computers in society is limited only by the number of people trained in their programming and who are simultaneously expert in some discipline of potential application.

In this text we shall concern ourself almost exclusively with numerical computation. There is an important reason for this. The first two years of an engineering or science program in college are among the most crucial periods in your career. You will be taking physics and calculus along with this programming course. The individual goals of each of these courses are to acquaint you with the important and useful topics that will ultimately become the basis upon which other engineering disciplines will build. Viewed collectively, however, these courses have an even more important function, that is, to develop an analytical sophistication. Each course is intended to reinforce the other and the total is then greater than the sum of its parts. The best programming of a problem is done by someone who understands the physical principles of the problem and the mathematics involved.

This text is concerned with developing fluency in the programming language called Fortran. There is a large variety of programming languages in addition to Fortran, and each has its hard core of enthusiastic supporters claiming superiority in some aspect of its use or its ease of learning. Many linguists contend that all human languages are of roughly the same difficulty when being learned as the first language, and I feel the same can be said of programming languages. As for the relative utility of programming languages, the analogy with spoken languages is again useful. It can be argued that French, for example, is more exact than English, or more poetic, or superior in some other sense. Regardless of the validity of arguments of this type, the essential point is that English is the international language of science and engineering, and every scientist or engineer, anywhere in the world, must have at least a reading knowledge of English. The same is true of Fortran. Regardless of its relative merit as compared with other programming languages, it is the universally accepted scientific programming language, and the overwhelming majority of scientific and engineering computer programs, both in the United States and elsewhere, are written in Fortran.

Since it was introduced in the mid-fifties, there have been many attempts to correct some of the failings and limitations of Fortran, which were generally perceived to be

1. Inadequate ability to read, write, and manipulate textual-type data.
2. Awkward program flow control commands that resulted in needlessly complex programs.
3. Limited ability to handle a variety of data-base types. The only way to store large blocks of data in Fortran is via a subscripted array like x_i. (We will encounter arrays in Chapter 6.) Newer languages are more versatile in this regard.
4. The lack of recursive-type procedures. Programs or subprograms in Fortran are not allowed to "call" themselves, while many numerical problems are most succinctly expressed in terms of a recursive-type relation as we shall see.

The introduction of the latest version of Fortran, Fortran 77, addresses the first two shortcomings rather nicely. The situations in which the latter two limitations become serious handicaps are ordinarily quite advanced, and by the time you encounter such problems you will, I hope, be fluent in at least one or two other computer languages and will be able to choose the one that best suits the overall situation.

Fortran is relatively easy to learn, much easier than, say, German or Spanish. However, it is essential for efficient and accurate programming to understand that the Fortran language is more than simply a mode of communication between humans and computers; it is a set of instructions for a series of operations of the computer. If you were to prescribe for a friend a set of procedures to follow to solve a mathematics problem,[2] a certain amount of ambiguity in the instructions can be tolerated. Depending on the sophistication of the person carrying out the instructions, grammatical errors or trivial assumptions will be overlooked and your friend will interpret what you intended. A computer is unable to do this and all instructions to the computer must be absolutely precise and complete. A misplaced comma will very likely cause the computer to not recognize or to misinterpret the instruction. Thus, it is important that we have at least a minimum understanding of what is going on inside the computer.

---

[2] A recipe for the solution of a problem is more formally called an *algorithm*.

1.2 OPERATING PRINCIPLES OF DIGITAL COMPUTERS

1.2.1 Binary Representation of Numbers and Information

The actual internal operations of a computer are not, in principle, very profound. Several centuries ago it was realized that since a switch has two states, open or closed, it can be used to represent the numbers 0 and 1. A large collection of switches may then be used to represent or store a great many 0s and 1s. Next, ordinary numbers (base ten) can be rewritten in terms of strictly 0s and 1s (i.e., base two) and then be represented by combinations of open and closed switches. For example, the 372 is base ten, which means:

$$372 = 3(10^2) + 7(10^1) + 2(10^0)$$

and can also be written as:

$$\begin{aligned}
372 &= 1(256) + 0(128) + 1(64) + 1(32) + 1(16) \\
&\quad + 0(8) + 1(4) + 0(2) + 0(1) \\
&= 1(2^8) + 0(2^7) + 1(2^6) + 1(2^5) + 1(2^4) \\
&\quad + 0(2^3) + 1(2^2) + 0(2^1) + 0(2^0)
\end{aligned}$$

or

$$(372)_{10} = (101110100)_2$$

Thus, nine switches would be required to represent the number 372.

The actual internal manipulations of the binary images of decimal numbers in the computer is not apparent to the user. Yet it is occasionally useful to have at least a cursory understanding of how binary arithmetic is effected. First of all, ordinary base-ten numbers without a decimal fraction can be converted to base two by successively dividing by two and keeping track of the respective remainders. Thus,

$$\begin{aligned}
(19)_{10} &= 2(9) + \underline{1} \\
&= 2(2(4) + \underline{1}) + 1 \\
&= 2(2(2(2) + \underline{0}) + 1) + 1 \\
&= 2(2(2(2(1) + \underline{0}) + 0) + 1) + 1 \\
&= 2(2(2(2(2(0) + \underline{1}) + 0) + 0) + 1) + 1 \\
&= \qquad\qquad 1 \times 2^4 + 0 \times 2^3 + 0 \times 2^2 + 1 \times 2^1 + 1 \times 2^0 \\
&= (10011)_2
\end{aligned}$$

or

$$19/2 = 9 \text{ R1} \qquad \langle = 9 \text{ with a remainder of } 1 \rangle$$
$$9/2 = 4 \text{ R1}$$

$$4/2 = 2 \text{ R0}$$
$$2/2 = 1 \text{ R0}$$
$$1/2 = 0 \text{ R1}$$

Thus

$$(19)_{10} = (10011)_2$$

Applying this procedure to the number used in the previous example, we obtain:

$$(372)_{10}/2 = 186 \text{ R1}$$
$$186/2 = 93 \text{ R0}$$
$$93/2 = 46 \text{ R1}$$
$$46/2 = 23 \text{ R0}$$
$$23/2 = 11 \text{ R1}$$
$$11/2 = 5 \text{ R1}$$
$$5/2 = 2 \text{ R1}$$
$$2/2 = 1 \text{ R0}$$
$$1/2 = 0 \text{ R1}$$
$$= (10111010)_2$$

To obtain the binary image of a decimal fraction, *multiply* the number repeatedly by two, keeping track of whether the successive results are or are not greater than one. For example,

$$0.3125 \times 2 = \underline{0} + 0.625$$
$$0.625 \times 2 = \underline{1} + 0.25$$
$$0.25 \times 2 = \underline{0} + 0.5$$
$$0.5 \times 2 = \underline{1} + 0$$

or

$$0.3125 = 2^{-1}(.625)$$
$$= 2^{-1}(0 + 2^{-1}(1 + 0.25))$$
$$= 2^{-1}(0 + 2^{-1}(1 + 2^{-1}(0 + 0.5)))$$
$$= 2^{-1}(0 + 2^{-1}(1 + 2^{-1}(0 + 2^{-1}(1 + 2^{-1}(0)))))$$
$$= 0 \times 2^{-1} + 1 \times 2^{-2} + 0 \times 2^{-3} + 1 \times 2^{-4}$$
$$= (0.0101)_2$$

Consider next the binary image of 1/10.

$$(0.1)_{10} \times 2 = \underline{0} + 0.2$$

$$0.2 \times 2 = \underline{0} + 0.4$$
$$0.4 \times 2 = \underline{0} + 0.8$$
$$0.8 \times 2 = \underline{1} + 0.6 \qquad \langle \textit{repeating binary fraction} \rangle$$
$$0.6 \times 2 = \underline{1} + 0.2$$
$$0.2 \times 2 = \underline{0} + 0.4$$
$$0.4 \times 2 = \underline{0} + 0.8$$

$$\cdots \quad \cdots \quad \cdots$$

so

·0 0 0 1 1 0 0 1 1 0 0 1 1

$$(0.1)_{10} = (0.001100110011 \ldots)_2$$

The arithmetic operations of binary long division and multiplication are constructed in the same manner as with decimal numbers, but you may find it amusing to try a few examples. Remember $(1 + 1)_2 = (10)_2$.

Example

$$\{ \ 0.1 * 10.0 \}_{10} \text{ converted to base two}$$

$$(0.1)_{10} = (0.001100110011 \ldots)_2$$

$$(10.0)_{10} = (1010)_2$$

so

$$(10.0) * (0.1) \rightarrow \quad 0.001100110011 \ldots$$
$$\times \ 1010.$$

$$0.001100110011 \ldots$$
$$000.110011001100 \ldots$$
$$\overline{0.111111111111 \ldots} \rightarrow 1.0$$

This is the binary equivalent of $(0.9999999 \ldots)_{10}$.

In addition to storing numbers, other forms of data can be converted to a binary code (i.e., written in terms of 1s and 0s) and stored and processed in a computer. For example, each of the letters of the alphabet can be assigned a coded sequence of 1s and 0s and ultimately words and sentences constructed and manipulated. The information content of a photograph can be (approximately) replaced by a dot matrix with the location, color, and density of each dot allocated a binary code and the resolution of the photograph may then be computer-enhanced or the entire contents of the picture transmitted to another site by phone lines or satellite relay.

1.2.2 Main Memory

Originally the actual elements that played the role of switches in a computer were electrical-mechanical relays. They were physically rather large, consumed significant electrical power, and were very noisy. Over the past several decades advances in microelectronics have resulted in the replacement of actual switches by small transistor circuits called *flip-flops*. Each flip-flop circuit can be in only one of two states, on (representing 1) and off (0). The state of each circuit can be sensed (information read from) or altered (stored into). Additionally, in the last two decades methods have been developed that permit the placement of several thousand discrete circuits on a single wafer or chip. The result is called a *very large scale integrated circuit* (VLSIC) although the size is actually about 1 centimeter square. (See Figure 1-1.) These technological advances have made possible modern computers that are much faster, more reliable, and considerably cheaper than those of an earlier generation.

The main memory of a large computer consists of a few million such storage elements for 1s and 0s (called *bits*). The bits are arranged in groups of eight (called *bytes*) in which may be stored the binary code for a letter, digit, special symbol such as a comma, or a control symbol such as a carriage return. The unit of information that is transferred to and from the main memory is called a *word*. A computer word may consist of from 4 to 16 bits for microcomputers like many of the popular personal computers; 12 to 32 bits for larger, so-called minicomputers; or 32 to 64 bits or more for large computers.

The details concerning the nature of a computer's main memory are usually of little consequence to a programmer except for the following two features that will be repeatedly referred to throughout this text.

1. Associated with each element or memory cell in main memory are two items:

    The contents of the memory cell. A unit of data or some type of instruction written in a binary-coded form.
    The address of the memory cell. The cells can be thought of as boxes in a row, each box labeled by its position in the row.

2. The size of each memory cell, i.e., a computer word, is finite. This means that for each number used in a computation only a finite number of digits can be stored. Thus numbers represented by nonterminating binary expressions must be truncated before they can be stored in main memory. Incidentally, this refers to nonterminating fractional numbers expressed in base two not base ten. For example, $(0.1)_{10}$ is a nonterminating fraction in base two. (See Problem 1-1.) The length of a computer word will then limit the accuracy of the numerical computations and is an important characteristic of each computer.

The access time (the time required to read the information stored in a

Figure 1-1 The central section of a 64K memory chip magnified about 400 times. The actual size of the chip is about 1/4 inch on a side. The access time for this chip is about 20×10^{-9} seconds and information is discharged in units of 16 bits at a time. The paired "L"-shaped patterns arrayed in each corner of the square are the actual memory cells. (Courtesy of IBM)

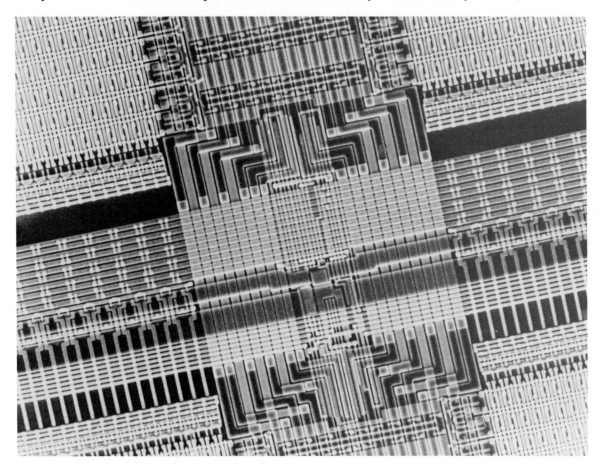

particular address) is of critical importance in the operation of the computer. A large computer can ordinarily add two numbers in less time than it takes to find them in main memory. The access time for a large semiconductor main memory consisting of 500,000 computer words is typically 5×10^{-7} seconds.[3] Memory of this type is called *random access memory* (RAM) meaning that the access time for all addresses is roughly the same. The information on a RAM unit is not stored sequentially, but for the time being we will find it simpler to think of all the information as being stored in order. The main memory of a large computer will typically have a storage capacity for 100,000 or more computer words.

---

[3] Even this incredibly short time can be reduced somewhat by temporarily storing the recently used data and instructions in a smaller, even faster memory device called *cache memory*. The idea is that the recently used items are likely to be used again shortly, and if so they can be more quickly accessed if grouped together.

In addition to main memory, there are several other possibilities for storing large amounts of data. These are called *secondary memory* and usually consist of magnetic tape or magnetic disk devices, although data can also be stored on cards or paper tape. The data in secondary memory are not directly accessible but must first be transferred to main memory for processing. Secondary memory storage is considerably less expensive than main memory, but the price is paid in significantly longer access times on the order of milliseconds or more.

1.2.3 The Central Processing Unit

The part of the computer that performs the operations on the data using the instructions stored in main memory is called the *central processing unit* (CPU). It is responsible for two distinctly different functions:

1. The CPU must monitor and control the entire system consisting of all the devices to get information in and out of the computer (I/O) and control the associated traffic of information flow between the various elements of the computer.
2. The CPU processes the binary-coded instructions transmitted to it from main memory.

The CPU may thus be thought of as two submodules to perform these operations. The *control unit* will execute the necessary control functions, whereas the data and instruction processing unit will be responsible for executing the elementary commands of a program. These consist primarily in performing the operations of arithmetic as well as the ability to compare two data items. This unit is thus called the *arithmetic-logic unit* (ALU). (See Figure 1-2.)

The control unit is responsible for fetching the next instruction and the

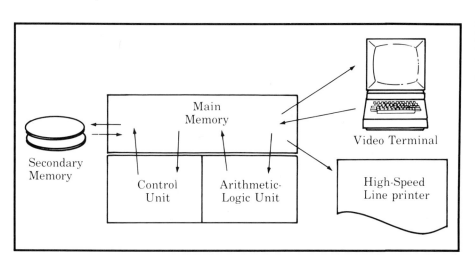

Figure 1-2
Block diagram of the functions of a computer.

address of the required data items from memory and temporarily storing them in local fast memory registers. The instructions are retrieved from memory in the order that they are stored in the program and the required data items are then likewise fetched from memory. The two data items are then transmitted to the ALU along with an instruction to add, subtract, compare, etc. The control unit is then responsible for moving the result back to main memory.

Recall that all of the items processed by the CPU, data and instructions, are written in a binary-coded fashion. This is the only form of communication that is directly understandable to a computer. Programs written in this manner are called machine language or microcode programs. Fortunately, it is unlikely that you will ever have to concern yourself with microcode programs.

1.2.4 Input/Output Devices

Two common and rather embarrassing errors in computer programming are to instruct the computer to execute some operation on data items that were never stored in memory, and to complete a computation and forget to print the result. The mechanism for getting data and the instructions for processing that data into the computer and then to have the results displayed in a useful form are obviously an important concern. The most common input devices include video terminals, punched card readers, and magnetic tape and magnetic disk reading devices. The results generated by the computer can then be displayed on the same video terminal, printed on a high-speed line printer, written on magnetic tape or disk, or punched on cards. We will be primarily concerned with input via the terminal and output at the terminal or on a line printer. Somewhat later (in Chapter 4) we will make use of the magnetic disk which will be described in more detail at that time.

All of the machinery comprising the CPU, the main and secondary memory units, and the various input/output devices are called computer *hardware*. The set of coded instructions or programs available on the computer is called *software*.

1.2.5 The Computing Speed of a Large Computer

In summary, then, what a computer is doing is rather trivial. It is simply manipulating 1s and 0s, nothing more. Its claim to fame rests on the fact that it can do the elementary operations on hundreds of thousands of stored numbers at an absolutely incredible speed. A typical time to add two 60-bit numbers (about 14 decimal digits) is about one-millionth of a second. To appreciate the enormous consequences of this fact, consider that it would take you about 10 seconds to add two 12-digit numbers, and there is a fair possibility that you would make a mistake. The computer adds the two numbers seven orders of magnitude faster, with the likelihood of an error being extremely small. Not only that, if you were highly paid to do this operation (perhaps because you are very accurate in addition and can do hundreds of thousands of these additions without becoming bored, in short have many of the attributes of a computer) and get say $10 per hour, this represents about three cents per addition. The large computer works at typically $500 per hour or 1.3×10^{-7} cents per addition. Thus the machine not only is faster by a factor of about 10^7, but cheaper by roughly the same factor. This is the main reason for all the excitement about computers. The tremendous advances in computing have taken place in a period of about 30 years. Consider for a moment the analogy with the advances over the years in transportation. For untold centuries, humans were limited by the fact that getting around meant walking, which meant about 5 mph. The discovery of the horse revolutionized civilization but was only a change of from 5 to about 20 mph, not even a factor of 10. Modern automobiles, which have a top legal speed of 55, represent a tremendous advancement (one factor of 10) and airplanes that travel at about 500 mph are a further advancement (two factors of 10). All of this in about one century. However, imagine what the consequences would be if the achievements in transportation could match those in computing (increase in speed by 10^7, decrease in cost by 10^{-7}, in 30 years). You could travel from New York to Los Angeles in less than 1 sec and the total cost of your car would be less than 1 cent. Clearly then, such dramatic changes in the field of computational mathematics will have very profound effects on society.

1.3 COMMUNICATING WITH THE COMPUTER

In the early days of computing, programming a computer to do calculations was extremely tedious. The difficulty was that the programmer was forced to communicate with the machine on its own terms. If a program required the product of two numbers represented by the symbols A and B, the programmer would have to code binary instructions to accomplish the following operations:

> Recall number A from address X
>   Transfer to arithmetic unit
> Recall number B from address Y
>   Transfer to arithmetic unit
> Multiply the two numbers
>   Transfer the result to address Z
> Etc.

All of this just to multiply two numbers. Also the machine language code written in this fashion would work only on the machine for which it was originally intended and no others. This meant that the use of the computer was limited to only those brave souls who had mastered this monotonous skill. The first significant step in broadening the pool of potential computer users occurred in 1955 at IBM when the first version of Fortran was developed. The idea was, once and for all, to write a program in machine language that would take a set of instructions written in a form resembling ordinary algebra and translate these instructions into all the transfer-here–store-there instructions required by the computer. Thus a statement like

$$X = 7.0 * A + 3.0 * B$$

which, of course,[4] represents

$$x = 7a + 3b$$

would be read off a punched card. A program permanently residing in the computer would then translate this into terms understandable to the machine, that is, the binary instructions for the operations:

> Fetch the number in the location allotted to symbol A
> > Transfer it to the arithmetic unit
> > Multiply by 7.0
> > Temporarily store the result
> Fetch the number in the location allotted to symbol B
> > Transfer it to the arithmetic unit
> > Multiply by 3.0
> Add the result to the previously stored multiplication
> Store the result in a location allotted to symbol X

The program that does the translation of the Fortran equation into machine language is called the *compiler*. The set of instructions written in Fortran is called the *source* code and the machine language translation is called the *object* code. Since machine code is written to conform to the specific features of an individual computer, an object code program will run only on the machine for which it was intended. The expression that is to be translated is, on the other hand, universal. That is,

$$X = C - D$$

would look the same in any program run on any machine. Fortran compilers were the first to appear and are still among the most common in use. Programs written in Fortran closely resemble ordinary algebra plus some simple English key words. (Fortran stands for FORmula TRANslator.) It is one

---

[4] The instructions to the computer to carry out the operations of multiplication and division are effected by the symbols * (asterisk, as in 7.0 * A) and / (slash, as in B/2.). Two side-by-side asterisks, **, are used to designate the operation of exponentiation as in X**2, which represents x^2. A complete description of how a sequence of arithmetic instructions is written in Fortran will be discussed in the next chapter.

of many so-called higher computer languages (i.e., above machine language) and has undergone several revisions. The form used in this book is Fortran 77.

Since the introduction of Fortran, numerous other programming languages have been developed and each of those listed below can very likely be compiled on every large computer in the world.

Common Higher-Level Computer Languages

ALGOL	
Pascal	General-purpose languages
PL/I	
C	
APL	
COBOL	Designed for applications in business
BASIC	A simplified language, similar to Fortran, and commonly used on microcomputers

The universal nature of these programming languages has resulted in very rapid growth in computer usage and also in collaboration between individuals working on different machines.

1.4 THE MECHANICS OF "GETTING ON" THE MACHINE

There are two modes of communication with a computer, *batch processing* and *interactive time sharing*. Batch processing generally refers to the execution of computer programs that have been previously punched on a deck of computer cards or that have been stored in a file on magnetic tape or disk. The instructions encoded on the cards or in the files will be carried out in sequence by the computer and ordinarily, once the computing job begins, the programmer cannot correct or alter the program until the job has passed through the machine. The programmer then checks the printed output for errors, corrects them, and resubmits the program. Thus batch processing is basically a "hands-off" computing mode.

Conversely, interactive computing is "hands-on" computing. The program is composed, edited, executed, and corrected at a video terminal connected to the computer. Both forms of computing have certain advantages; batch processing allows you to see the entire printed listing of the program and perhaps to make thoughtful changes at your leisure, whereas the terminal allows you to see only 20 lines or so at a time and program surgery at a terminal frequently tends to be somewhat hectic. In addition, the execution cost of a program run in a batch mode is considerably lower than the cost of the same job run interactively. In spite of these points, the last few years have seen a tremendous growth in interactive computing and consequent reduction in demands for batch processing. Card punches have all but disappeared from college campuses. Clearly this is due to the increased ease of computing via conveniently located terminals as compared with the repeated visits one was forced to make to the computing center in the past, lugging along boxes of computer cards. It is assumed throughout this text that you will be communicating with the computer by means of a cathode-ray tube (CRT) terminal. The response of the computer will usually be on the terminal screen, although computed results may also be sent to a high-speed line printer.

1.4.1 Interactive Processing

Interactive processing, also called time-sharing on large multiuser computers, allows you to monitor your program as it compiles and executes and to correct errors or alter the program immediately. The program is typed in at a terminal, the appropriate system commands are entered to execute the program, and the output is displayed on the terminal screen. The terminal may be wired directly to the computer or may use an audio link by means of a telephone. You may also have the computer produce a printed copy of the program and its results. There is a wide variety of terminals in use today, and the operational procedures for running Fortran programs vary considerably from one computer to the next; thus only general instructions can be given here. For the detailed instructions appropriate to your site you should obtain an "Interactive Computing User's Guide" or its equivalent from your computing center.

To execute a Fortran program at the terminal there are three levels of procedures that must be learned.

1. The system protocol or job control language (JCL) defined by your computing center. This will include a set of instructions for accessing the computer through your terminal and for having the computer translate (compile) the Fortran code into machine language object code, executing the program, saving the program, and getting results printed on a line printer. None of these instructions are Fortran and the precise forms of the JCL commands differ considerably from site to site. We will, however, discuss the general structure of some elementary JCL commands shortly.
2. Program composition and editing. In addition to translating and executing your Fortran program, the computer can be used to assist you in writing and correcting the program. Many computing centers provide a program that is accessed by entering a particular JCL command like EDIT or some variation that takes you into the next layer of commands called editing commands. These again are site-specific and at least a few of the elementary editing commands must be learned before you can enter and execute a Fortran program.
3. The third layer of procedures is the Fortran program itself.

1.4.2 The Format of Fortran Lines

When entering Fortran code at the terminal some general rules must be followed regarding the postitioning of various types of information on each line. The characteristics of a Fortran line are shown in Figure 1-3. The entire line is 80 columns wide. The Fortran statement must appear in columns 7 to 72. The compiler ignores blank columns, so the Fortran statements may appear anywhere in this field. Columns 1 to 5 are available for supplying an identifying statement number to a particular line of Fortran. Thus an instruction like

```
731        STOP
```

will result in termination of the program's execution. This Fortran statement can be referenced by means of its statement number, 731. Again blanks are ignored and the 731 could appear anywhere in the statement number field (columns 1 to 5). Column 6 is called the *continuation field*. If a Fortran statement is too long to fit on a single line it may be continued on the next line by including any symbol other than 0 in column 6. I would suggest using either a plus sign(+) or a dollar sign ($) to indicate that a line is a continuation of a previous line. The maximum length of a single Fortran expression is 20 lines (one line plus 19 continued lines). A line that is a continuation of a previous line may not have a statement number.

From the discussion thus far we would conclude that all of the Fortran statements at the top of page 16 would be translated identically.

Figure 1-3 The placement of information on a Fortran line.

Column No.					
	1	2	3	4	7	8

```
1....67..0.........0.........0.........0..      ..0.2.........0
        C
        O
Stmnt.  N     Fortran
        T                                          Identification
No.     I     Statements
        N
        U        Appear  in
        A                                          Field
        T        Columns
        I                                          73-80
        O        7 <==> 72
        N
```

```
              1             2
1,,,,67,,0,,,,,,,,,,0,,,
  731     STOP
  7 31     ST        OP

  731    S
        +T
        +0
        +P
```

Columns 73 to 80 on the Fortran line are called the *identification field*. Anything that is entered in this part of the line is ignored by the compiler but does appear on the listing of the program supplied by the compiler. Usually the identification field is used to add sequencing numbers to the Fortran lines, which are useful in reassembling a card deck that has been accidentally dropped. They are rarely used in interactive programming.

If a C or an asterisk(*) appears in column 1, the entire contents of the line are ignored by the compiler; but, as with the identification field, the information contained on the line is included with the listing. These lines are called *comment* lines and they constitute a very important part of a program's documentation. They are used to explain to someone reading the program what each element of the program is attempting to do. Even if you are the only one who will ever read a particular program, it is still a good idea to amply sprinkle comment lines throughout your program. Few tasks are more frustrating than trying to understand the operation of an undocumented code that you wrote six months ago. Some installations of Fortran permit the use of either lower- or uppercase letters within a program. If this is the case on your computer, I suggest that your comment lines be entered in normal lowercase English and that the Fortran be in uppercase only. An example of a possible use of comment lines is shown in Figure 1-4.

1.4.3 Logging On

The general form of the steps necessary to initiate a program at a terminal are as follows:

1. Log on to the machine. You will have to supply your name, a password, and perhaps additional information.
2. Enter the job control statements that specify that you will be writing a Fortran program. The Fortran code to follow will be stored on what is called a *file*, which has to be given a name.
3. Type in the Fortran program (See Section 1.4.4.)
4. Enter the job control statements that instruct the computer to compile and execute the program. If during execution the program is to read data,

Figure 1-4 An example of the use of comment lines.

```
            1          2          3          4          5
1234567     0          0          0          0          0
-----------------------------------------------------------
        PROGRAM ZERO
C
C   THIS IS A DEMONSTRATION PROGRAM THAT ILLUSTRATES THE
C   VARIOUS ARITHMETIC OPERATIONS IN FORTRAN AND SOME OF
C   THE ODD FEATURES OF INTEGER AND REAL (FLOATING POINT)
C   ARITHMETIC ON THE COMPUTER.  IT IS ALSO AN EXERCISE
C   IN LEARNING THE NECESSARY PROCEDURES TO ENTER AND
C   EXECUTE A FORTRAN PROGRAM INTERACTIVELY.
C      PROGRAMMED BY
C
C           ****************
C           *              *
C           * JOE STUDENT  *
C           * ROOM 6C      *
C           * NORTH QUAD   *
C           *              *
C           ****************
C
C   THE INPUT VARIABLES -- A,B -- ARE ASSIGNED VALUES BY
C   ENTERING THE NUMBERS, SEPARATED BY A COMMA, UPON
C   EXECUTION OF THE READ STATEMENT IN THE PROGRAM.
C
        READ *, A, B
```

the program will stop and wait for you to enter the data at the terminal. If there were either compilation or execution time errors, you can correct them and try again.

5. If all was successful, instruct the computer to produce a printed copy of both the program and its output.

In summary, before you commence writing Fortran programs you must learn the appropriate commands to handle the above steps. You should make a short list of these instructions for easy reference when you are working at a terminal. In addition, you should determine what some of the "panic" buttons are for your computer. That is, how to stop the program if it seems stuck in an endless loop, how to save a program for another try on another day. On many computers, simply entering the word "HELP" will call up a short tutorial session that may lessen some of your confusion. Also, do not be afraid of inadvertently entering some instuctions that will cause the machine to grind to a halt. For the novice, this is next to impossible to do.

1.4.4 Using an Editor Program To Enter the Fortran Code

The original versions of Fortran assumed that the code would be entered on cards. The Fortran statements would then be correctly positioned on each card and the entire card deck assembled and submitted. Even though the Fortran statements entered at a terminal are identical to the statements submitted on cards and each line on the screen can be viewed as a single card, the physical properties of terminals have resulted in a need to alter somewhat the rules described in Section 1.4.2 for a Fortran line. The description that follows gives some general characteristics of the use of an editor program to construct Fortran code. You should determine the specific features available in the editing program available on your computer.

Line Numbers

Each line of the Fortran code is assigned a sequencing line number. These line numbers usually begin at 100 and increase in equal steps of 10 or 100 to allow for later insertion of new lines between the old ones. Depending upon the operating system for the computer, these numbers may be displayed automatically for you, or you may have to type them in yourself. It is important to understand that these line numbers are *not* part of the Fortran program and will be useful primarily in editing the code later.

```
Line      Column  No.
                    1            2            3
          1234567  0            0            0
          - - - - - - - - - - - - - - - - - - - - - - - - -
00100  C     A COMMENT LINE
00110        A = 3.5
00120  651   X = 3.0 + A
00130  *     Another comment line
00140        Y = 2./X
00150        PRINT *,X,Y
```

Line 120 of the Fortran code bears the statement label 651. Elsewhere in the program this statement may be referenced by using this number. Statement numbers are most often used to designate particular Fortran statements as targets of various branchings in the program. Since the line numbers are not part of the Fortran program, they may not be referenced in the Fortran code itself.

A line may be inserted between line 120 and 130 for example by entering

```
00125    731   Z = A + X
```

or line 110 may be deleted by simply entering the line number again followed by a blank line (line number, RETURN). Of course line 110 could be altered by simply retyping the line number followed by the corrected Fortran statement.

Automatic Positioning of Fortran Code

The second major difference is a result of the difficulty in determining exactly at which column the terminal cursor is currently positioned. With some editing programs the TAB key on the terminal keyboard is used to move the cursor directly to column 7, whereas other editing programs will automatically reposition the information entered so that it satisfies the format rules of a valid Fortran line. The Fortran statements and their statement numbers are repositioned according to the following rules:

1. If the column immediately after the line number contains an asterisk, the entire line is interpreted as a comment. (An alternative is a C followed by a blank space.)
2. If the first column is not a C or an asterisk, then the line is interpreted as a Fortran statement. If a line begins with a number, it is viewed as a statement number which will then be followed by a blank and some Fortran statement. A line number followed by a blank space and any letter is interpreted as a Fortran statement without a statement number.
3. If the first column after the line number contains a plus sign,[5] it is considered to be a continuation of the previous Fortran statement line.

```
00100C   A COMMENT LINE
00110 A = 3.5
00120 651 X = 3.0 + A
00130*   Another comment line
00140 Y = 2./X
00150 PRINT *,X,
00160+ Y
```

If we next instruct the computer to list the Fortran code, we find that all the lines have been correctly sequenced by line number and that the Fortran has been automatically positioned to satisfy the rules we learned for the correct formatting of Fortran lines. (Section 1.4.2).

Correcting and Executing the Fortran Program

Typing errors are easily corrected at the terminal by backspacing or deleting before a RETURN is entered or by simply replacing the entire line. When you are somewhat more experienced, you will learn more powerful program editing procedures; but for now the simple ones will suffice.

---

[5] Some editor programs use a dollar sign for this purpose.

Even though we have not yet started the study of the Fortran language, I am sure you are anxious to try your hand at submitting and executing a simple program. Once you have determined the required log-on procedures, you should attempt to execute the simple programs below. I am sure that you will have no difficulty in reading and understanding the Fortran code.

1. Simple arithmetic

```
Column No.
         1               2               3
1....67..0...........0...........0...
        A = 2.5
        B = 3.5
        C = A + B
        PRINT *,A,B,C
        STOP
        END
```

The PRINT statement[6] instructs the computer to display the numbers A,B,C, which will then appear on the terminal screen.

2. Printing text

```
         1               2               3
1....67..0...........0...........0...
        A = 100.0
        B = A*A
        PRINT *,A,' SQUARED EQUALS ',B
        STOP
        END
```

The three items to be printed are separated by commas and in this case include the two-word phrase, SQUARED EQUALS, which must be enclosed in apostrophes. The symbols between the apostrophes are printed "as is", including blanks.

3. An infinite loop

```
         1               2               3
1....67..0...........0...........0..
C THIS PROGRAM CONTAINS AN INFINITE LOOP
        X = 100.
   12   PRINT *,' X = ',X
        X = X*X
        GO TO 12
        END
```

---

[6] The mechanism in Fortran for having a program display results obtained at any particular point in that program is quite easy. The word PRINT followed by an asterisk and a comma,

This program will first print the current value of X, i.e., 100., then replace the current value of X with the value of X squared. The GO TO 12 statement is a Fortran transfer statement that causes the program to branch directly to the Fortran line that bears the statement label 12. The GO TO statement will be discussed in more detail in Section 3.4.1, but for now its meaning is rather transparent and you should have no trouble in understanding the effects of GO TO statements. In this instance, the effect is to branch back to the print statement and continue with the calculation from that point. There is no way out of this loop and the values of X will grow without limit. Of course, there is a limit to the size of the numbers that can be stored in the computer and the program will terminate when this number is exceeded. This will also generate an error message on the terminal screen.

4. Reading data

```
          1              2              3
1....67..0...........0...........0...
        PRINT *,'INPUT X AND Y, SEPARATED BY A COMMA'
        READ *,X,Y
        Z = X * Y
        W = X/Y
        PRINT *,X,' TIMES ',Y,' EQUALS ',Z
        PRINT *,X,' DIVIDED BY ',Y,' EQUALS ',W
        STOP
        END
```

This program will execute the first PRINT statement and the program will then pause at the READ statement and wait for you to enter the values of X and Y at the terminal. The program will then resume after the values are entered (after the carriage return).

1.4.5 A Typical Terminal Session

In Figure 1-5 is the complete dialogue required with the computer to execute one of the above problems. This program was run on a CDC CYBER 730 computer and except for the Fortran itself you should expect all of the JCL and edit commands to differ from those on your computer. The general structure of the commands will be similar, however. In Figure 1-5 the output from the computer is in capital letters, while the response to the computer is in lowercase.

```
        PRINT *,
```

simply precedes the list of variables or constants to be printed or displayed on the terminal screen. These are in turn separated by commas. Upon execution of this line of the program, the *values* associated with the quantities in the list will be displayed. This manner of printing results will be discussed in more detail in Section 2.9.

Figure 1-5 An example of a dialogue with the computer while at a terminal.

```
LEHIGH UNIVERSITY CY170-730.
USER NAME: borse                  ⟨The response of the computer is in capital letters.
PASSWORD:                          My response is in lowercase.⟩

/senator                          ⟨The editor program on my computer is called
SENATOR VER 2.7 (84/05/09)         senator.⟩
*system,fortran                   ⟨The program will be written in Fortran.⟩
*new,test                         ⟨And will be stored on a file called TEST.⟩
*input
INPUT LINES

100    a = 100.0
110    b = a*a
120    print *,a,'squared equals ',b
130    stop
140    end

*list
                                  ⟨When the program is listed the editor will position
    100        A = 100.0           the Fortran correctly.⟩
    110        B = A*A
    120        PRINT *,A,'SQUARED EQUALS ',B
    130        STOP
    140        END

*run
                                  ⟨This is the local instruction to execute the
                                   program.⟩
   TEST         15:25      FORTRAN

   100.SQUARED EQUALS 10000.
*logout                           ⟨Here are the results of the program.⟩
```

1.5 PROGRAM ZERO

As a somewhat more demanding test of your abilities to execute a program, you should attempt to run the program listed in Figure 1-6 on your computer. Enter the Fortran code exactly as it appears, execute the program, and in addition obtain a printed listing of the program and its output. This program is more complicated than those described earlier and I do not expect you to be able to completely understand the Fortran. We will go over the code, line by line, in the next chapter.

Figure 1-6 A program to demonstrate arithmetic operations.

```
Column Number
         1            2          3          4          5
1234567  0            0          0          0          0
------------------------------------------------------------
       PROGRAM ZERO
       REAL A,B,C,D,E,F,G,H,P,Q,R,S,T,U,V
       INTEGER IA,IB,IC
       PRINT *,'INPUT VALUES FOR A,B SEPARATED BY A COMMA'
       READ *,A,B
       C = A + B
       D = A - B
       E = A/B
       F = A*B
       G = C*D/(E*F) + (B/A)**2 - 1.
C INTENTIONAL MIXED MODE FOLLOWS (6 LINES)
       IA = A
       IB = B
       IC = IA**IB
       H = IC
       P = LOG(H)/IB
       Q = EXP(P) - IA
       R = A/G
       S = B/Q
       V = 0.0
       PRINT *,'INPUT VALUES A = ',A,' B = ',B
       PRINT *,'COMPUTED VALUES G = ',G,' Q = ',Q
       PRINT *,'R = ',R,' S = ',S
       T = 1./V
       PRINT *,'THE RESULT OF 1 DIVIDED BY 0 IS ',T
       U = A*T
       STOP
       END
```

If you spot typing errors after you have the computer list the Fortran code, these should be corrected before you attempt to execute the program. If undetected typing errors remain, the computer will likely indicate these as errors in Fortran grammar and will give you some indication of the location and nature of the error. You should then have no trouble in correcting them.

You will notice that near the end of this code is a statement involving division by zero. Obviously this is going to cause problems during the execution of this program. Do not try to correct this line; let us just see what happens. If you find that this program will not execute at all, remove the line U = A*T and try again.

1.5.1 Error Diagnostics

Once you have typed the entire program in Figure 1-6 and have submitted the code for compilation and execution, the computer will first attempt to translate the Fortran into machine language. If there are typing errors, the compiler will endeavor to diagnose the nature of the error. For example, if the line C = A + B were typed C = A # B, the output would indicate:

```
                                           C  =  A  #  B
FATAL ERROR     *****   ILLEGAL USE OF OPERATOR   --   A #
```

This indicates that the compiler was unsuccessful in translating this statement. The symbol # is not an allowed symbol in Fortran.[7] This error has resulted in a line of code that does not satisfy the rigid rules of Fortran grammar that we will begin to learn shortly and results in a *compilation* time error. No object code is generated and there can be no attempt at execution.

But beware. Frequently, typing errors will result in a Fortran line that is grammatically correct but that has a meaning significantly different from that intended. For example, leaving out the addition operator in the line C = A + B results in C = A B, which is the same as C = AB. Since there has been no value assigned to a variable named AB, the value assigned to C will be meaningless and will likely cause the program to "die" during execution. (See Problem 1-8.)

Frequently the messages supplied by the compiler when a compilation error is encountered are extremely cryptic and they may not be readily understandable to the novice. In such cases you can consult the reference manual that describes the features of the Fortran compiler in use on your computer or seek help from your instructor. (These manuals are usually available in the computing center.) Ordinarily, merely knowing in which line the error is located is sufficient information to find a simple typing error.

Elimination of errors in programs is called *debugging,* and the correction of compilation time errors is the first and easiest stage of debugging a program. After you have all the Fortran corrected so that it now satisfies the grammar rules, the compilation of the Fortran can be completed and the program may then begin execution. However, the program may still die during execution, for example, by division by zero. An error such as this is called an *execution* time error. The second stage of program debugging then consists of eliminating all execution time errors. These are usually much more difficult to find and correct. Most modern compilers have available

---

[7] The only symbols allowed in Fortran are the letters of the alphabet, the integers, plus the special characters:

+	plus	.	period	,	comma
−	minus	'	apostrophe	=	equals
/	slash	(	left parenthesis	)	right parenthesis
$	dollar	*	asterisk		
:	colon				

special commands that will attempt to trace back an execution time error such as division by zero. These debugging programs will locate the general region of a program in which the error occurred and will print the values of the variables in the program at the time of the error. You should familiarize yourself with the instructions necessary to implement these features.

After these hurdles are passed and the program runs to completion, the final and most difficult phase of debugging a program begins; that is, to verify the validity of the results. For example, even a perfect Fortran code will not produce valid answers if the results depend on the value of π and π was incorrectly entered as PI = 3.4416. All three phases of program debugging will be dealt with in due course.

1.5.2 Additional Information Supplied by the Compiler

The example listing in Figure 1-7 is for a successful compilation and execution of program zero. The output was then sent to a line printer. The printed output you receive from your computing center may look quite different, but the overall features should be similar. First is a complete listing of the Fortran program you submitted. This is called the SOURCE code. The compiler will generally add sequenced line numbers off to the left and will insert error diagnositcs if required. Next comes the output of the compiler program, giving some details concerning the translation of the Fortran code. This is called the load map and may include such things as a list of all variables used in the program and all library functions required by your program (such as EXP). Additional information usually contained in a load map is the size of the program and the time for compilation.

Following the load map is the computed output of program zero. We will go over this in Chapter 2. The last item on a listing is usually the dayfile, which includes several bookkeeping items such as the time and date of program execution, a list of job control commands executed, central processor time (CP) used, main memory requirements, and of course the cost of the job. In addition, if execution time errors are encountered, they are listed here.

Figure 1-7 An example of the printed output of an interactive program.

```
1              PROGRAM ZERO
2              REAL A,B,C,D,E,F,G,H,P,Q,R,S,T,U,V
3              INTEGER IA,IB,IC
4              PRINT *,'INPUT VALUES FOR A,B SEPARATED BY A COMMA'
5              READ *,A,B
6              C = A + B
7              D = A - B
8              E = A/B
```

Continued

```
 9                    F = A * B
10                    G = C * D/(E * F) + (B/A)**2 - 1.
11          C   INTENTIONAL MIXED MODE FOLLOWS (6 LINES)
12                    IA = A
13                    IB = B
14                    IC = IA**IB
15                    H = IC
16                    P = LOG(H)/IB
17                    Q = EXP(P) - IA
18                    R = A/G
19                    S = B/Q
20                    V = 0.0
21                    PRINT *,'INPUT VALUES A = ',A,' B = ',B
22                    PRINT *,'COMPUTED VALUES G = ',G,' Q = ',Q
23                    PRINT *,'R = ',R,'S = ',S
24                    T = 1./V
25                    PRINT *,'THE RESULT OF 1 DIVIDED BY 0 IS ',T
26                    U = A*T
27                    STOP
28                    END
```

⟨A listing of the program⟩

⟨*A Site-Dependent LOAD MAP
usually appears here*⟩

```
INPUT VALUES FOR A,B SEPARATED BY A COMMA
INPUT VALUES A = 7. B = 13.
COMPUTED VALUES G = -1.42108547152E-14 Q = 0.
R = -4.925812092436E+14S = R
THE RESULT OF 1 DIVIDED BY 0 IS R
```

⟨*This is the output of
the program.*⟩

PROBLEMS

1. Rewrite the following base-ten numbers in binary (base two).
   a. 11   d. 2.5 ⟨use $.5 = 2^{-1}$⟩
   b. 33   e. 12.625
   c. 100   f. 0.1 ⟨this will be a continuing fraction⟩
2. Rewrite the following base-two numbers as base ten.
   a. 1011.0
   b. 110011.0
   c. 0.11
   d. 100000000.00001

3. Perform the following base-two arithmetic operations and verify by converting the problem to base ten.

    a. $\begin{array}{r} 1011 \\ +\ 11 \\ \hline \end{array}$ **b.** $\begin{array}{r} 1010 \\ -\ 11 \\ \hline \end{array}$

4. Convert the following base-ten arithmetic to binary and compute the result in binary arithmetic. Verify your answers with a base-ten calculation.

    a. $3 + 3 + 3$ ⟨or 3×3⟩

    b. 11×33

    c. 10×0.1

5. As a rare example of the utility of binary arithmetic that is not related to digital computers, consider the following description of the ancient oriental game of Nim: One player places any number of markers on the table arranged in rows. Any number of rows and any number of counters per row are allowed. The two players then alternately remove counters from the arrangement, taking any positive number from any *one* row. Either player may begin. The one removing the last counter is the winner. If you go first and if the initial arrangement is not a trivial one, you should be able to win almost every time. Consider the initial arrangement

$$
\begin{array}{cl}
\text{X X X X} & (4) \\
\text{X X X} & (3) \\
\text{X} & (1)
\end{array}
$$

The game-winning strategy is most easily explained by writing the number in each row in base two.

$$
\begin{array}{cl}
\text{X X X X} & (100) \\
\text{X X X} & (011) \\
\text{X} & (001)
\end{array}
$$

The correct next move is to remove enough counters from some row so that the base-two numbers of those remaining add up to even numbers when each column is added (base ten) separately. Thus the correct first move is to remove two markers from row one to obtain

$$
\begin{array}{cl}
\text{X X} & (0\quad 1\quad 0) \\
\text{X X X} & (0\quad 1\quad 1) \\
\text{X} & \underline{(0\quad 0\quad 1)} \\
& \text{even-even}
\end{array}
$$

Explain the reasoning behind the winning strategy for the simpler case of just two rows with any number in either row.

6. Explain the significance of the following columns on a Fortran statement line:
   a. column 1
   b. column 6
   c. columns 73 to 80
   d. How is a blank interpreted when it appears in columns 1 to 5?, column 6?, columns 7 to 72?

7. To answer the following questions you will have to consult the documentation supplied by your computing center that explains the necessary procedures for interactive computing.
   a. Explain the procedures for logging on at a terminal.
   b. What commands are necessary in order to initiate the writing of a Fortran program at the terminal? (e.g., How do you create a Fortran file? How is it sequenced? How do you position the Fortran statements to begin in column 7?)
   c. How do you erase a line, backspace or delete a character, insert a line, end the program at a terminal?
   d. How is the Fortran program then saved, executed, compiled? How do you get your results printed? How do you terminate the terminal session?

8. The following short program contains several grammatical errors. Enter and run the program as is. What error diagnostics are supplied by the compiler? Use these clues to correct the Fortran as best you can and try again. Were there any obvious errors that were missed by the compiler?

```
PROGRAM OPPS
PRINT *,'ENTER VALUES FOR A,B,C SEPARATED BY COMMAS'
READ *, A, B C
D = 2A + BC
E = D/-A
B = F
PRINT *,'A=',A,' B=',B,' C=',C
+' D=',D,' E=',E,' F=',F
STOP
END
```

9. Determine the answers to the following for your computing environment:
   a. What is the size of the central memory of your computer?
   b. What is the length of a computer word in bits? What is the corresponding number of decimal digits?
   c. Where are the public access terminals that are connected to this machine and what are the hours that they are available?
   d. Where is the reference manual for the Fortran compiler located?

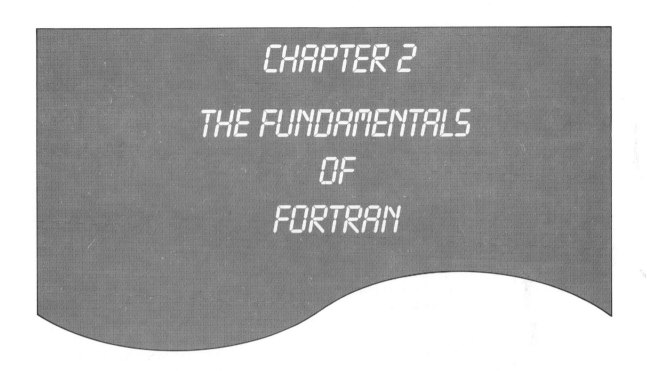

CHAPTER 2
THE FUNDAMENTALS
OF
FORTRAN

2.1 INTRODUCTION

The data types that can be processed by a Fortran program come in a variety of forms. The two most commonly used forms for numerical values are called *INTEGER* type and *REAL* type, while textual items (symbols) are of type *CHARACTER*. There are a few additional data types available and these will be described in Chapter 6. The three data types given above will be sufficient for most programs.

The arithmetic involving the numbers used in Fortran differs in many ways from the arithmetic you have previously used. Moreover, in the majority of cases in which a seemingly valid program gives incorrect results, the wrong answers are the result of the odd features of computer arithmetic. The purpose of this chapter is to enable you to translate algebraic expressions into Fortran code so that the computer will evaluate the expressions in precisely the manner intended.

2.2 CONSTANTS IN FORTRAN

A Fortran constant is a fixed quantity that may be either numerical or a specified collection of symbols. The numerical constants that we will use will be either of type INTEGER or of type REAL.

2.2.1 INTEGER Constants

An integer constant in Fortran is any number that does not possess a decimal point. It can be positive, negative, or zero. If it is positive, the plus sign can be omitted. For example:

```
        +3
        -17
        9999999999
```

Note that embedded commas are not permitted (e.g., 999,999,999), but since Fortran ignores blanks, large numbers could be written in groups of three digits as 999 999 999.

The computer stores each of these numbers in main memory in a computer word of from 8 to 64 bits in length. The largest allowed integer depends upon the word length, which varies from machine to machine but is, of course, finite in every case. Some examples of the maximum allowable integers for a variety of computers are listed in Table 2-1.

The finite nature of the number set in computer arithmetic has some odd consequences, as we shall see shortly.

2.2.2 REAL Constants

The numbers most often used in computation are numbers with decimal points. Numbers with a decimal point are called REAL.[1] These numbers can be written either with or without an exponent.

1. Reals without exponents. Again, if the number is positive the plus sign may be omitted.

---

[1] A somewhat older terminology calls numbers with a decimal point *floating-point* numbers, while numbers without a decimal point are called *fixed-point* numbers. Of course, the Fortran definition of the word "REAL" is somewhat different from and should not be confused with mathematical terminology wherein "real" is used to characterize all numbers that are not complex or imaginary.

Table 2-1 Maximum number of decimal digits permitted for integers for selected computers.

Type of Computer	Maximum Number of Decimal Digits
CDC-CYBER	14
IBM-4341	10
PDP-11	11
DEC-20	8
Typical Computer	10

Valid

    3.   ⟨This is not the same as integer 3⟩
    -3.14159259

Invalid

    999,888.0   ⟨Again, embedded commas are not allowed.⟩

2. Reals with exponents. The exponent consists of the letter E followed by a positive or negative integer and corresponds to the power of ten used when the number is written in scientific notation. If the integer exponent is positive, the plus sign may be omitted. The base of the number, that part of the number that precedes the exponent, should contain a decimal point.

Valid

    1234.56E-3   ⟨This is the same as 1234.56×10^{-3}; that is, 1.23456.⟩
    0.123456E+1
    1.E10

Invalid

    123.45E-5.5   ⟨A decimal point is not allowed in the exponent.⟩
    123E+45   ⟨The base should have a decimal point.⟩
    123.E-456   ⟨On most computers the exponent is limited to two
                     digits. See below.⟩

Once again the maximum allowable real number is determined by the word length of the particular computer. For example, on a computer with an exceptionally long word length of 64 bits, the limits are

$$10^{-294} \quad \text{to} \quad 10^{+322}$$

while on most other 32-bit machines the limits are typically

$$10^{-78} \quad \text{to} \quad 10^{+75}$$

Of course, you should determine what the precise limitations on the size of integers and reals are for your computer. (See Problem 2-1.)

2.2.3 CHARACTER-Type Constants

A CHARACTER-type constant is a string of any of the allowed symbols in Fortran (see footnote 7 in Chapter 1) that are enclosed in apostrophes. The string of characters is stored in memory exactly as it appears, with blanks included. The minimum number of symbols in a character constant is one, while the maximum is at least several thousand and depends on your local compiler. We will use character constants primarily to label the output of the program using the PRINT statement.

Examples of valid character constants:

```
'NOW IS THE TIME FOR ALL GOOD ...'
'X = '
' '          ⟨a single blank space⟩
'4.2'        ⟨the symbols for four–decimal point–two,
              not the numerical value⟩
```

Examples of incorrect character constants:

```
NOW IS THE TIME     ⟨apostrophes missing⟩
'END OF RUN''        ⟨too many terminating apostrophes²⟩
```

The values of constants are printed in a Fortran program by means of the PRINT statement. The form of the PRINT statement is[3]

```
PRINT *,      ⟨a list of constants or variables separated by commas⟩
```

This is an *executable* Fortran statement. All Fortran lines can be classified as either executable or nonexecutable. An executable Fortran line will ordinarily call for some action on the part of the computer, while nonexecutable Fortran lines will be used to preset various attributes in a program such as declaring which variables are of type INTEGER and which are of type REAL. (Type-declaration statements are covered in the next section.) Most nonexecutable statements are positioned at the very beginning of the program and are followed by the main body of the program, which consists of executable statements. The PRINT statement may be placed anywhere among the executable statements and may be used to display the values of integer, real, or character constants or variables that have been assigned values. For example, the three types of constants are all present in the

---

[2] If the character constant itself contains an apostrophe, the single apostrophe in the string is represented by two consecutive apostrophes in the character constant. Thus the expression ⟨don't quit⟩ would be written as a character constant as

```
'DON''T QUIT'
```

[3] In many of the examples of Fortran statements that will be given in this text, the correct form of the Fortran expression will be given in capital letters and a description of the remainder of the Fortran statement will be in lowercase and enclosed in angle brackets ⟨ . . . ⟩.

---

statement

```
PRINT *,100,'SQUARED IS ',1.E4
```

The output that results from this statement is:

```
100SQUARED IS 1.0E+04
```

The PRINT statement will be explained in more detail in Section 2.9.

2.3 FORTRAN VARIABLES

As in ordinary algebra, quantities may be associated with symbols or variable names and these may then be symbolically manipulated according to the rules of arithmetic. Each variable name identifies an address in memory, and whenever the variable name is referenced, the current value in that location is used in its place. Variable names in Fortran must satisfy two rules:

1. The variable name may be any combination of letters or numbers, but must begin with a letter. No special symbols are allowed in a variable name.
2. The number of symbols in a variable name is limited to six or fewer.

Variable names may be declared as containing a constant of type CHARACTER, REAL, or INTEGER.

2.3.1 Default Typing of Variable Names

Unless the Fortran code explicitly states otherwise (see the next section), the type of the number that is associated with a given variable name is determined by the following default typing rule:

All variable names that begin with the letters

I J K L M N

refer to integers. All others refer to reals.

A few examples are listed in Table 2-2.

Table 2-2 Examples of valid and invalid Fortran variable names

INTEGER Names			REAL Names		
Valid	I		Valid	X	
	K27B			XK57	
	INTGR			ANSWER	
	KOUNT				
Invalid	X	⟨real⟩	Invalid	IDIOT	⟨integer⟩
	K-8	⟨only letters and numerals allowed⟩		ENGINEER	⟨too many characters⟩
	INTEGER	⟨too many characters⟩			

2.3.2 Explicit Typing of Variable Names

A sound procedure in every Fortran program is to invent variable names that have a clear meaning when the code is read. For example, if your program is concerned with printing the current date, you will need IN-TEGER variable names to store the values of the month, day, and year. You could use IM, ID, IY; but the meaning of the Fortran code would be much more transparent if you used instead MONTH, DAY, YEAR. However, the variables DAY and YEAR would be default-typed as real and thus could not be used to store integer values. The method provided in Fortran to override default typing is called *explicit variable typing* and involves using the following Fortran statements.

> REAL   ⟨list of variable names⟩
> INTEGER   ⟨list of variable names⟩

These statements are called *specification statements* and should appear at the top of the Fortran code. They are nonexecutable statements in that they do not call for any action by the computer. The variables in the list are any valid Fortran names and are separated by commas. Of course, a variable name declared to be type REAL cannot also be declared to be type INTEGER or vice versa. Attempts to do so will result in a compilation time error.

Examples:

> REAL COST,LENGTH,MASS
> INTEGER DAY,YEAR,MONTH

The REAL and INTEGER statements can be in either order and one or both may be absent from the program. All variables that are not explicitly typed will follow the default-typing rule.

To avoid confusion, it is generally suggested that *all* variables that appear in a program be explicitly typed. This suggestion is often relaxed somewhat for simple variable names that are always used for the same function such as a subscript or counter (I or J), variables in an equation such as X, or common fixed constants such as PI.

Default typing can be used only for REAL and INTEGER variable names. To declare a variable to be of type CHARACTER an additional statement is required. This statement is slightly more complicated since the length of the character string that is to be stored in the variable must also be specified. The form of the CHARACTER specification statement is

```
CHARACTER name₁*<sl₁>,name₂*<sl₂>,...
```

where $name_1$, $name_2$, etc. are the names of the variables and $\langle sl_1 \rangle, \langle sl_2 \rangle, \ldots$ are positive integers corresponding to the length of the string stored in the corresponding name. For example,

```
CHARACTER NAME*10, STREET*12
```

If all of the variables in the CHARACTER statement are to be of the same length, the statement can be shortened by attaching the common string length to the word CHARACTER, as, for example,

```
CHARACTER*10 NAME,STREET,CITY
```

The two forms of the CHARACTER statement can be combined. For example, in the statement

```
CHARACTER*10 NAME, STREET*12, CITY
```

the variables NAME and CITY are of length 10 while STREET is of length 12.

2.4 ARITHMETIC EXPRESSIONS

An arithmetic expression in Fortran is an instruction to perform one or more arithmetic operations on constants or variables that have previously been assigned values. An expression can be quite simple, as A + B, or complicated, extending over several lines. The expression is evaluated and the entire expression is replaced by that value. The operations that are to be carried out in the expression are determined by the operators that it contains and by the sequence in which they appear.

2.4.1 Arithmetic Operators

The ordinary operations of arithmetic are effected in Fortran by means of the following symbols:

+   Addition
−   Subtraction
*   Multiplication
/   Division
**   Exponentiation

The symbols for multiplication, division, and exponentiation were chosen because they are available on every typewriter. Also, the exponentiation operator is defined to be a single symbol.

The rules for constructing arithmetic expressions are fairly simple:

1. No two operation symbols may occur side by side.
2. The multiplication, division, and exponentiation operators must appear in conjunction with two numbers or variables, e.g.,

```
A * B     D / 2.0     F ** 2
```

while the addition and subtraction operators may appear with a pair of numbers or variables,

```
A + B          D - 2.0
```

or with a single variable or number, as in negation.

```
+A     -B     -7.0
```

Note that in Fortran

1. Multiplication can never be implied as it often is in ordinary algebra. Thus, $a(b + c)$ must be written as A * (B + C), not A(B + C).
2. The rule about side-by-side operators must be carefully adhered to, even in cases where there appears to be no ambiguity. For example,

Incorrect Fortran	Valid Fortran	Algebraic Expression
X**−2	X**(−2)	x^{-2}
A * −3.0	A * (−3.0) or −3.0 * A	$-3a$

the entries in the first column will result in compilation time errors.
3. Note that, just as in ordinary algebra, terms may be grouped by using parentheses. The expression within a pair of parentheses is then evaluated before any operations outside the parentheses are executed.

2.4.2 Integer Arithmetic

Arithmetic expressions involving only integers will always result in a number that is an integer. This is especially important to remember when the expression involves division. If the division of two integers is not itself an integer, the computer automatically truncates the decimal fraction.

```
6/2 = 3   6/3 = 2   6/4 = 1   6/5 = 1   6/6 = 1   6/7 = 0
```

Because of this odd feature, integers should never be used in arithmetic expressions in which physical quantities are computed. They should be used exclusively as counters or indices.

2.4.3 Real Arithmetic

The actual computation in a Fortran program is done with real numbers and variables. The result of any arithmetic expression containing real numbers is a real number and so the above problem with integers does not occur.

```
6./2. = 3.     6./3. = 2.     6./4. = 1.5
6./5. = 1.2    6./6. = 1.     6./7. = 0.857142 . . .
```

However, real arithmetic has pecularities all its own that are a consequence of the finite word length of the computer. For example,

$$1./3. = 0.333333333 . . .$$

where . . . means that the decimal expression repeats indefinitely. The decimal answer stored in the computer, however, cannot repeat indefinitely but is limited by the number of significant figures in a computer word. A typical size for a computer word is about ten significant figures, so in real arithmetic

$$1./3. = 0.333333333 \qquad \langle no \ . . . \rangle$$

or, put another way, the result of 1./3. is slightly less than one-third. Thus

$$3.*(1./3.) - 1. \neq 0.0$$

while

$$2.*(1./2.) - 1. = 0.0$$

Of course, the numbers stored in the computer are stored in a base two notation. Thus, in base two

$$1/2 = (0.100000000 . . .)_2$$

$$1/3 = (0.010101010 . . .)_2$$

and thus

$$[2(1/2) - 1]_{10} = (10. \times 0.1 - 1.0)_2$$

$$= 1. - 1. = 0.0$$

$$[3(1/3) - 1]_{10} = (11. \times 0.0101010101 - 1.0)_2$$

$$= 0.1111111111 - 1.0$$

$$= -0.0000000001$$

This is true on most computing devices. (Try it on your pocket calculator.)

2.5   MORE COMPLICATED ARITHMETIC EXPRESSIONS

2.5.1   Hierarchy of Operations

The sequence in which the computer processes a series of mixed arithmetic operations is determined by a set of rules that have been formulated to remove potential ambiguities. The understanding of these rules is an essential element in the ability to readily read and program code in Fortran. This ordered sequence or hierarchy is listed below

The order in which the arithmetic operations in an expression are executed.

First: Clear all parentheses (innermost first).
Second: **
            Perform exponentiation.
Third: * or /
            Perform multiplication and/or division (equal priority).
Fourth: + or −
            Perform addition and/or subtraction (equal priority).

These rules are effected by successive scans of the expression, looking for each of the above in turn from *left* to *right*.[4] Some examples follow.

1. A + B + C is evaluated as $\underbrace{(A + B)}_{\text{first}} + C$

$\underbrace{}_{\text{second}}$

---

[4] Unfortunately, successive exponentiation is an exception to the left-to-right rule. Fortran 77 compilers are written so as to evaluate A**B**C as A**(B**C)—i.e., right to left. For operations of this type it is always best to avoid confusion and include parentheses to force the sequence of operations to be what you intended.

2. $4 * 3/2 \rightarrow 12/2 \rightarrow 6$
but
$3/2 * 4 \rightarrow 1 * 4 \rightarrow 4$
3. $2.**3 - 1. \rightarrow 8. - 1. \rightarrow 7.0$

Consider how the machine would process an expression like

$$A**B/C \;+\; D*E*(F-G)$$

Assume that the variables A through G have been previously assigned the values

Variable	A	B	C	D	E	F	G
Value	2.0	3.0	4.0	5.0	6.0	7.0	8.0

The first scan is to clear all parentheses and thus the first operation is to evaluate $(F - G)$ and temporarily store the result in say R_1. $\langle (7. - 8.) = -1. = R_1 \rangle$. The expression now reads

$$A**B/C \;+\; D*E*R_1 \qquad \langle R_1 = -1.0 \rangle$$

The next scan looks for exponentiation, replacing A**B with the temporarily stored value R_2 ($2.**3. = 8. = R_2$). So that we next have

$$R_2/C \;+\; D*E*R_1 \qquad \langle R_2 = 8.0 \rangle$$

The third scan carries out all the multiplication or division found, proceeding left to right

$$
\begin{aligned}
R_2/C &= R_3 & \langle 8./4. = 2. = R_3 \rangle \\
D*E &= R_4 & \langle 5. \times 6. = 30. = R_4 \rangle \\
R_4*R_1 &= R_5 & \langle 30. \times (-1.) = -30. = R_5 \rangle
\end{aligned}
$$

and we are left with

$$R_3 \;+\; R_5$$

The final scan executes all addition or subtraction proceeding left to right to obtain the final value of the expression

$$R_6 \;=\; (2. \;+\; (-30.)) \;=\; -28.0$$

Of course, additional parentheses could be inserted to alter the order of operations and perhaps the result. You should verify that the slightly changed expression

$$A**B/(C + D) \;*\; (E*(F - G))$$

has the value −5.333333333. A beginners' rule is, **When in doubt, always add parentheses.**

2.5.2 Mixed-Mode Expressions

All of the arithmetic expressions we have seen thus far have been carefully constructed to contain only elementary operations between the same types of numbers, integers added to integers, reals times reals, etc. The reason for this is that the ALU of the computer is set up to execute the operations of arithmetic or comparison *only* between numbers of the same type. It does not know how to multiply a real number times an integer. To carry out such an operation, the numbers must be first converted to the same type.

All modern compilers are written to handle an arithmetic operation between two numbers of different types by first converting the numbers to the same type and then carrying out the operation. To accomplish this in an unambiguous manner, levels of dominance are assigned to the number types REAL and INTEGER, with reals having dominance over integers. Thus an expression like

$$3.0*I$$

is evaluated by first converting the integer I to a real number and then multiplying by 3.0. The result of the operation is then real. An expression such as this is called a *mixed-mode expression*.

Mixed-mode expressions often cause considerable confusion among both beginning and more experienced programmers. This is especially true when the expression involves division. The presence or absence of a decimal point can dramatically alter the result.

Mixed-mode expression	is evaluated to be
1. + 1/2	1. ⟨1/2 → 0⟩
1 + 1./2	1.5 ⟨1./2 → 1./2. → .5
	1 + .5 → 1. + .5 → 1.5⟩

Mixed-mode expressions can serve very useful functions in programming; however, they should be avoided by beginners. Numerical computations should involve reals only, with integers being used primarily as counters. If you do find it necessary to use mixed-mode expressions, I suggest that, for a while, every Fortran statement you write that employs mixed-mode arithmetic be preceded by a comment line like

```
C   INTENTIONAL MIXED MODE FOLLOWS
```

which will serve both as a reminder and as an announcement to others.

The assignment operator = in Fortran bears a striking resemblance to the equal sign in ordinary algebra, but they have significantly different meanings. A Fortran expression like

```
X = 14. - 4.**.5
```

is a set of instructions to the computer to complete the arithmetic computation on the right and *to assign* that value to a variable called X. An important feature of higher computer languages like Fortran is that this statement will automatically determine and remember a storage location in main memory for this number. Subsequent access to this number is had by simply using the variable name, as in

```
Y = X**3
```

Since what was stored in X is 12.0 (i.e., 14. − 2.), the value stored in Y is 1728.0. With this understanding of what the = operator does, Fortran statements like

```
I = I + 1
X = X + 0.1
```

make sense, whereas in algebra they would be nonsense. That is, in Fortran the statement I = I + 1 means: Take the value already assigned to I, add 1 to it, and store the result in the location allotted to I.

2.6.1   Character Assignment Statements

Variables that have previously been declared as type CHARACTER may be assigned "values" by means of the assignment operator =. Of course, the value of a character variable is not numerical but is a string of symbols. For example,

```
CHARACTER NAME*6,STREET*6,CITY*7
NAME = 'MILLER'
STREET = 'E.MAIN'
CITY = 'CHICAGO'
```

In each case the character constant on the right of the assignment operator is assigned to the variable name on the left. Previously defined character variables may also appear on the right of the expression, as

```
NAME = 'MILLER'
STREET = NAME
CITY = 'CHICAGO'
```

The variables NAME and STREET now contain the same string of characters.

If the lengths of the string on the right of the expression are not the same as the specified length of the variable, the expression is altered to fit the length of the variable. If the expression is longer than the length of the variable, the expression is first truncated from the right until the lengths match.

```
CHARACTER NAME*6
NAME = 'WILLIAMS'        ⟨stored in NAME is the string |W|I|L|L|I|A|⟩
```

If the expression is shorter than the length of the variable, it is padded with blanks on the right.

```
NAME = 'DOE'       ⟨stored in NAME is the string |D|O|E|_ |_|_|⟩
```

2.6.2 Mixed-Mode Replacement

Next consider the consequences of statements like

```
I = 14./3.              ⟨the number stored in I is 4⟩
N = 3.*(1./3.) - 1.     ⟨the number stored in N is 0⟩
R = 4/3                 ⟨the number stored in R is 1.0⟩
```

In each case, the expression on the right of = is first evaluated as either a real or integer value and then the assignment is made to the variable on the left, which here requires the mode of the result (i.e., integer or real) be converted. Thus $14./3. = 4.66666667$ and the assignment to I automatically converts this to an integer by truncating the decimal part. The above statements are an illustration of what is called *mode conversion*, i.e., an integer (or real) is converted automatically into a real (or integer) by the assignment operator. The reason for this is that the number stored in the address allocated to I, for example, must be an integer and so the number must be converted to integer before it can be written in location I. One of the most common errors made by novice Fortran programmers is that of unintentional mode conversion, caused by using integer variable names for quantities that were intended to be real numbers. However, mode conversion can be an extremely useful feature of Fortran if used with care. As with mixed-mode expressions, I would strongly suggest that while you are learning Fortran every statement you write that employs mode conversion be preceded by a comment line like

```
C   INTENTIONAL MODE CONVERSION FOLLOWS
```

It should be obvious to you that expressions or assignment statements that mix character variables or constants with numerical values will always result in compilation errors. Thus, the code below

```
CHARACTER NAME*5
REAL X,C
X = 2.0
NAME = 'JONES'
B = X + NAME      ⟨Error—character and other type values may not be
                       mixed in an arithmetic operation.⟩
C = NAME          ⟨Error—character values cannot be converted to real.⟩
```

will result in two fatal compilation time errors.

2.7   PROGRAM ZERO

Fortran, like any skill, is best learned by doing and so it is probably best at this point to forestall any further exposition of rules and features and to carefully go over the program you executed earlier, program zero. You should have the complete output with you. The listing of the program given in Figure 1-6 is reproduced on the next page.

The first Fortran statement in the program is

```
PROGRAM ZERO
```

which simply gives this program a name of ZERO. This line is optional; if it is omitted, the compiler will assign a name. The next two lines of Fortran code

```
REAL A,B,C,D,E,F,G,H,P,Q,R,S,T,U,V
INTEGER IA,IB,IC
```

explicitly type the variables as either real or integer. If these two lines were omitted, the variables would have been default-typed exactly the same. However, it is usually good practice to explicitly type all variables that appear in a program.

The first executable statement in the program appears next.

```
PRINT *,'INPUT VALUES FOR A,B SEPARATED BY A COMMA'
```

This phrase will then be printed on the terminal screen and serves as a

Figure 1-6 A program to demonstrate arithmetic operations.

```
Column Number
          1              2              3              4              5
1234567   0              0              0              0              0
- - - - - - - - - - - - - - - - - - - - - - - - - - - - - - - - - - - - - - -
        PROGRAM ZERO
        REAL A,B,C,D,E,F,G,H,P,Q,R,S,T,U,V
        INTEGER IA,IB,IC
        PRINT *,'INPUT VALUES FOR A,B SEPARATED BY A COMMA'
        READ *,A,B
        C = A + B
        D = A - B
        E = A/B
        F = A*B
        G = C*D/(E*F) + (B/A)**2 - 1.
C INTENTIONAL MIXED MODE FOLLOWS (6 LINES)
        IA = A
        IB = B
        IC = IA**IB
        H = IC
        P = LOG(H)/IB
        Q = EXP(P) - IA
        R = A/G
        S = B/Q
        V = 0.0
        PRINT *,'INPUT VALUES A = ',A,' B  = ',B
        PRINT *,'COMPUTED VALUES G = ',G,' Q = ',Q
        PRINT *,'R = ',R,' S = ',S
        T = 1./V
        PRINT *,'THE RESULT OF 1 DIVIDED BY 0 IS ',T
        U = A*T
        STOP
        END
```

prompt for the READ statement that follows:

```
READ *,A,B
```

This is an instruction to read the values that will then be assigned to the variables A and B from a file called INPUT. When the program is run interactively at a terminal this will cause the program to pause and wait for you to enter the numbers. The values entered are separated by a comma and, since A and B are associated with REAL values, should contain a decimal point. Note, if we had neglected to insert the previous PRINT statement, the program would still stop at the READ and we could easily be confused as to why. Inserting this form of a prompt before each READ statement is essential if the program is to be run at a terminal. The READ * statement is an example of list-directed input and is discussed in more detail in Section 2.9.

The next PRINT statement,

```
PRINT *,'INPUT VALUES A = ',A,' B = ',B
```

is called an *echo print* and is an important part of every program. The echo print is a verification of the read statement. Typing errors when entering numbers are very common and this form of safeguard can save you considerable time in attempting to understand why an apparently valid program gives incorrect results.

When the program is executed assume that the following were entered as input:

$$7.0, \ 13.0 \quad \langle \textit{followed by a RETURN} \rangle$$

After execution of the READ line in the code, the variables A and B contain the values 7.0 and 13.0, respectively. Also, it is important that the type of the numbers entered as data (real or integer) agree with the type of the variable name in the READ statement.

The next four statements,

```
C = A + B
D = A - B
E = A/B
F = A*B
```

are trivial examples of real arithmetic operations and assignments. Notice that from what we know about the assignment operator, statements like,

```
A - B = D
   -D = B - A
```

would not make sense in Fortran although they would be perfectly valid in algebra.

The next statement is a bit more complicated.

```
G = C*D/(E*F) + (B/A)**2 - 1.
```

To untangle this, we need to use the hierarchy rules and successively scan the statement from left to right. Rewriting this expression in algebra, it would look like:

$$g = \frac{cd}{ef} + \left(\frac{b}{a}\right)^2 - 1$$

But recall

$$c = a + b$$

$$d = a - b$$

$$e = \frac{a}{b}$$

$$f = ab$$

so

$$
\begin{aligned}
g &= \frac{(a + b)(a - b)}{a^2} + \left(\frac{b^2}{a^2}\right) - 1 \\
&= \frac{a^2 - b^2}{a^2} + \left(\frac{b}{a}\right)^2 - 1 \\
&= 1 - \left(\frac{b}{a}\right)^2 + \left(\frac{b}{a}\right)^2 - 1 \\
&= 0.0
\end{aligned}
$$

Thus the value assigned to g is zero regardless of the values of a or b.

The statement IC = IA**IB appears simple enough but is deserving of a few moments consideration. To execute the exponentiation operation, the computer calls up a special program that is stored in main memory whose assignment is to take an integer base (IA) and raise it to an integer power (IB). The procedure used to calculate IA**IB is simply to multiply IA times IA, IB − 1 times, and if IB is negative, invert the result. Clearly, IA and/or IB may be either positive or negative.

The exponentiation in a previous statement, (B/A)**2, is of a different sort (real base, integer exponent) and a different subprogram is required to process the operation. Once again, to execute the exponentiation, if the exponent is INTEGER, the subprogram simply multiplies the base times itself the required number of times. The base and/or the exponent may be positive or negative. However, an operation like

$$(B/A)**2.3$$

requires special care. The subprogram that calculates this quantity must use logarithms, and since the logarithm of a negative number is undefined in our number system (it is actually an imaginary number), we must take care that the base is always positive. For example,

(−3.)**3	works	gives −27.
(−3.)**(−3)	works	gives −1./27.
(−3.)**(−3.)	will not work	

If the exponent is real—the base must be positive.

You have perhaps noticed that we have been cavalierly mixing modes without including the appropriate comment lines. The reason for this is that exponentiation is an exception to the previously stated mixed-mode rule. From the explanation given for the execution of the operation

$$X**7 \qquad \langle \textit{real base, integer exponent} \rangle$$

we can see that at no time are two quantities of differing mode involved in an elementary arithmetic operation (addition, subraction, multiplication, division). In fact, whenever possible, exponentiation should be of the form

$$X**I$$

rather than

$$X**R$$

since the former is much faster and safer.

The next several lines in the program employ intentional mixed-mode arithmetic or replacement.

```
C INTENTIONAL MIXED MODE FOLLOWS (6 LINES)

      H = IC
      P = LOG(H)/IB
      Q = EXP(P) - IA
```

We have no difficulty in predicting the result of the statement H = IC; however, the next two statements are probably confusing. They both make use of what are called *intrinsic functions*. That is, subprograms stored in main memory that can be used by a Fortran program to calculate several common mathematical functions. Thus, LOG(X) computes the natural logarithm of X and EXP(X) computes e^x. For both of these intrinsic functions, the argument (X) should be real. The operation of these functions is very similar to those in your pocket calculator when you push the appropriate key. (e.g. $\ln(x)$ or e^x). These are but two of a long list of intrinsic functions available to Fortran programs. A few of the more commonly used functions are listed in Table 2-3.

Returning to the assignment statements for P and Q; once the $\ln(h)$ and e^p have been calculated, the rest of the statements involve authentic mixed-mode arithmetic, which should be avoided. However, in this instance we can easily figure out how the machine handles it. As mentioned earlier, before the division and subtraction operations are executed, both numbers involved [i.e., LOG(H) and IB], are converted to the dominant mode, in this case real. Now to see the effect of all these statements, let us rewrite them in algebraic notation.

$$h = i_c = (i_a)^{i_b}$$

Table 2-3 Some of the intrinsic functions available in Fortran.

FORTRAN	Algebra	Description	Argument	Result	Example
LOG(X)	$\ln(x)$	Natural log	Real	Real	Y = LOG(3.1)
EXP(X)	e^x	$e = 2.71828\ldots$	Real	Real	P = EXP(1.5)
SQRT(X)	$\sqrt{x}$	Square root	Real	Real	R = SQRT(4./6.)
SIN(X)	$\sin(x)$	Trigonometric sine	Real $\langle$radians$\rangle$	Real	S = SIN(3.14)
COS(X)	$\cos(x)$	Trigonometric cosine	Real $\langle$radians$\rangle$	Real	T = COS(0.)
ABS(X)	$|x|$	Absolute value	Real	Real positive	W = ABS(-5.5)
ACOS(X)	$\cos^{-1}(x)$	Inverse cosine, if $x =$ $\cos(\theta)$ then $\theta =$ $\cos^{-1}(x)$	Real	Real $\langle$radians$\rangle$	PI = ACOS(-1.)

$$p = \frac{\ln(h)}{i_b} = \frac{\ln[(i_a)^{i_b}]}{i_b}$$

$$= \frac{i_b \ln(i_a)}{i_b} = \ln(i_a) \qquad \langle \textit{Note: } \ln(x^y) = y \ln(x) \rangle$$

$$q = e^p - i_a$$

$$= e^{\ln(i_a)} - i_a$$

$$= i_a - i_a = 0.0 \qquad \langle \textit{Note: } e^{\ln(x)} = x \rangle$$

Thus, q is zero regardless of the values of i_a and i_b. Also recall that we determined that g is identically zero as well.

The next two assignment statements then do forbidden things:

```
R = A/G
S = B/Q
```

To see the results of the calculation thus far, the program next prints the current values of some of the variables.

```
PRINT *,'COMPUTED VALUES G = ',G,' Q = ',Q
PRINT *,'R = ',R,' S = ',S
```

We expect the computed values for both G and Q to be 0 and thus the values for R and S will be undefined (i.e., infinity). On your job run find the numbers printed for G and Q. The results are perhaps something like

```
INPUT VALUES A = 7.00000   B = 13.0000
COMPUTED VALUES G = -2.8422E-14   Q = -1.1366E-13
R = -2.4629E+14   S = -1.1437E+14
```

Both G and Q are very small numbers, but not zero. Why not? The answer once again is due to the fact that the machine, when working with real numbers only carries about ten significant figures, so

$$B/A = 13.0/7.0 = 1.85714285714218 \ldots$$

Thus the statements we thought were errors (division by zero) were not caught by the machine. The computer does, however, catch division by precisely zero, as in the later statement

$$T = 1./V \quad \langle V \text{ has been assigned the value } 0.0 \rangle$$

This is genuinely bad. However, the computer assumes you know what you are doing and dutifully assigns positive indefinite (i.e., + infinity) to T. If you attempt to print T, some compilers will print an "R" (meaning indefinite) or some other symbol; others will terminate the program at this point with an execution time error. If your program has made it this far, it will not get past the next statement:

$$U = A*T$$

At this point the machine is forced to conclude that all the faith it had in you was not well founded. You do not know what you are doing. Since no definite value has been assigned to T, this operation cannot be processed. The program dies at this point, not at the point where division by zero occurred but where an attempt was made to use the result of division by zero. This is an illustration of a common execution time error. When a program dies by an execution time error, the computer will inform you of the mode of the program's death and the approximate location of the error. In this example, the fatal execution time error is in the statement $U = A * T$.

The last two lines of the code are essential to the execution of a Fortran program. The statement

$$STOP$$

will cause termination of the program. This is an *executable* statement because it calls for some action on the part of the computer. Other examples of executable statements that we have seen so far are the simple arithmetic assignment statements.

The last line in the program,

$$END$$

is also executable and can be used to terminate the program. However, the principal use of END statements is as a marker to inform the compiler where the program ends and where to stop the translation into machine language. It is suggested that you use the END statement only for this purpose. Every Fortran program or subprogram must have END as its last line.

2.8 TRANSLATING ALGEBRA INTO FORTRAN

In order to develop a facility in the use of the hierarchy rules, you should next translate several moderately complicated algebraic expressions into Fortran and vice versa. A few examples follow:

1. Translate the following into Fortran (a) using parentheses and (b) totally without parentheses.

$$x = \frac{rP}{1 + (1 + r)^{-n}}$$

a. `X = R*P/(1, + ((1, + R)**(-N)))`
b. `TERM = 1, + R`
   `TERM = TERM**N`
   `BOTTOM = 1, + 1,/TERM`
   `X = R*P/BOTTOM`

2. Write a single algebraic formula for the following Fortran statements.

```
TERM1 = C + 1,
TERM1 = A*B/TERM1
TERM2 = 1, + R
TERM2 = TERM2**N
TERM2 = TERM2 - 1,
TERM2 = TERM2/R
T = X*TERM1 - P/TERM2
```

Transcribing this line by line into algebraic notation, we obtain

$$t_1 = c + 1$$
$$t_1 = ab/(c + 1)$$
$$t_2 = 1 + r$$
$$t_2 = (1 + r)^n$$
$$t_2 = (1 + r)^n - 1$$
$$t_2 = \frac{(1 + r)^n - 1}{r}$$

Thus the final result can be written as

$$t = \frac{ab}{1 + c}x - \frac{rP}{(1 + r)^n - 1}$$

3. The velocity of very small water waves is given by the formula

$$v = \sqrt{\frac{2\pi t}{\lambda d} + \frac{g\lambda}{2\pi}}$$

where t is the surface tension (N/m), d is water density (kg/m^3), g is the gravitational acceleration (m/sec^2), and λ the wavelength (m) of the wave. The computed velocity will be in meters per second.

Write a complete program to read t and λ from a terminal and compute the wave velocity. The density d and the gravitational constant g should be assigned values of 1000. and 9.8, respectively, in the program.

You should have no difficulty in writing this program. Your result should resemble the code given below:

```
PROGRAM WAVES
D = 1000.
G = 9.8
PI = 3.14159265
READ *,T,WAVLTH
V = 2.*PI*T/(WAVLTH*D) + G*WAVLTH/2./PI
V = SQRT(V)
PRINT *,V
STOP
END
```

Note:

Since the variable names were not explicitly typed as INTEGER or REAL, using the names L or LAMBDA for the wavelength would have been incorrect and might have resulted in errors. Unlike most pocket calculators, the computer has no stored value for π. A somewhat more accurate assignment is obtained by

```
PI = ACOS(-1.)
```

The successive divisions at the end of the sixth line of code are equivalent to

```
G*WAVLTH/(2.*PI)
```

2.9   LIST-DIRECTED INPUT AND OUTPUT

The Fortran programs we have seen up to this point have made use of list-directed input statements of the form

```
READ *,    ⟨a list of variables⟩
```

The values entered at the terminal must agree in number and in type with the variable names in the READ statement. The values should be separated by commas.[5] For example, the statement

```
READ *,X,IA,Y,KOUNT
```

when used to enter the following values

```
5.72,4,3.E6 ,10002
```

is equivalent to the assignments,

```
X       = 5.72
IA      = 4
Y       = 3000000.
KOUNT   = 10002
```

If the data were entered as

```
5.72, 4,    ⟨CR⟩
    3.E6,10002
```

where ⟨CR⟩ stands for carriage return, exactly the same assignments would be made. The input

```
5.72,  ,  ,10002
```

would read zero(0) for IA and zero(0.0) for Y.

The list-directed read statement in the form

```
READ *,    ⟨variable list⟩
```

can also be used to read values for variables of type CHARACTER, provided the character strings that are read in are enclosed in apostrophes.

The list-directed output statement

```
PRINT *,    ⟨a list of variables, arithmetic expressions,
               or character strings⟩
```

operates in a similar fashion, with the additional features that arithmetic expressions may be included in the list and also strings of characters may be printed if enclosed by apostrophes. A few examples are listed in Table 2-4. The ability to print character strings as well as numerical values can be used to facilitate data input at a terminal. As suggested earlier, including an explanatory PRINT before each READ will eliminate a great many mistaken variable assignments. An additional reminder: It is a good idea to immediately print the values just read to verify they are indeed what you intended. This is called an echo print. For example, the water wave velocity

---

[5] You may alternatively separate the numbers by blanks, but this is not recommended.

Table 2-4 Examples of list-directed output.

Fortran	Output
| ```
X = 5.
Y = 8.
I = 6
J = 12
PRINT *,X,Y,I,J
PRINT *,Y/X,I+J
PRINT *,I+SQRT(X/Y)
PRINT *,'X = ',X,'I-J =',I-J
``` | ```


          5.0000    8.0000   6    12
          1.6000      18
          6.79057
X = 5.0000    I-J =    -6
``` |

problem in the previous section would be incomplete without the addition of
the PRINT/READ lines:

```
PRINT *, 'ENTER T(SURFACE TENSION) AND THE WAVELENGTH'
READ *,T,WAVLTH
PRINT *,'T = ',T,'WAVELENGTH = ',WAVLTH
```

2.10 SUMMARY

As a review of some of the concepts covered in this chapter, Table 2-5 lists
several common examples of incorrect transcriptions of algebra into Fortran
along with the corrected versions.

Table 2-5 Common errors in translating algebra into Fortran.

Algebra	Incorrect Fortran	Correct Fortran	Comments
$a(b + c)$	A(B + C)	A*(B + C)	
$2a + 4$	2*A + 4	2.*A + 4.	Legal but mixed mode not appropriate for beginners
a^{n+1}	A**N + 1	A**(N + 1)	
$a^{(1/n)}$	A**(1/N)	A**(1./N)	Legal; however (1/N) is likely zero (integer arithmetic)
$\dfrac{ab}{cd}$	A*B/C*D	A*B/(C*D) or A*B/C/D	
$(-x)^n$	−X**N	(−X)**N	In Fortran 77 exponentiation precedes negation. −X**N is the same as −(X**N)
$x = 3. \times 10^6$	X = 3.*10.**6	X = 3.E+6	Note the correct version requires no arithmetic operations

PROBLEMS

1. What are the limits for numerical values on your computer? That is, find out what the maximum number of digits are for an integer, for a real number.
   a. Evaluate $[(\sqrt{2}.)^2 - 2.]$ directly on your computer (or calculator) and from the result determine the number of significant figures in a computer word.
   b. Execute the following code on your computer

```
        I = 1000
   1    PRINT *,I
        I = I*10
        GO TO 1
        END
```

and from the results determine the maximum allowable integer. (You cannot use your calculator for this part; calculators do not have integer arithmetic.)

2. From the definition of the assignment operator, explain why the following Fortran statements are in error.
   a. `3. = K`
   b. `A + B = C`   *(Assume that A and B have been assigned values.)*
   c. `X = 1.0`
      `X = Y`
      `PRINT *,X,Y`

3. For every computer the following arithmetic is true:

$$1000. \ + \ EPS = 1000.$$

provided EPS is chosen small enough.
   a. If the maximum number of digits on your computer is six, what is the largest value of EPS for which the above is true?
   b. If EPS = 0.0001, what is the result of

   `EPS + EPS + EPS + · · · · + EPS`   *(100 million terms)*

   c. Is this the same as 100 million times EPS? (i.e., 1.E8 * EPS)

4. Identify any compilation errors in the following. If none write "OK."
   a. `REAL IJKLMN`
   b. `INTEGER REAL`
      `REAL INTEGER`
   c. `INTEGER*5 X,Y`
   d. `REAL X,Y,Z`
      `INTEGER IX,Y`
   e. `INTEGER IX,IY`
      `REAL IX IY`
   f. `INTEGER X`
      `X = 17`
      `REAL Y`
      `Y = 4.`

g. CHARACTER TWO*2,ZERO*0   **i.** INTEGER X, REAL Y

h. N = 7

    M = 6

    CHARACTER NAME*N,SEX*M

5. Identify the compilation errors in the following. If none, write "OK."

  a. X = 1,000,001.2   **g.** I = 7./7.1

  b. Y = -1.73E-6.2   **h.** S = X**-2.

  c. Z = 61E-06   **i.** U = (-3.)**.5

  d. W = 7   **j.** V = 16./2./2./2./2.

  e. R = 2.E2   **k.** A = I

  f. 6XA = 12.   **l.** X = X

6. Determine the output of the following programs.

<table>
<tr><td>

a. I = 2
   J = 3
   K = I + J
   L = K + I
   I = I + L + 1
   L = K/J
   PRINT *,I,J,K,L
   STOP
   END

</td><td>

c. I = 2
   J = 3
   K = 4
   L = I**J
   M = I**(-J)
   N = (J**I)**K
   NN = J**(I*K)
   MM = J**(I**K)
   PRINT *,L,M,N,NN,MM
   STOP
   END

</td></tr>
<tr><td>

b. R = 0.07
   P = 2000.
   N = 20
   T = 1. + R
   T = T**N
   X = R*P/T
   PRINT *,P,R,N,X
   STOP
   END

</td><td>

d.   X = 1.
  1 Y = 1. + X
    Z = X/Y
    W = (Y + X)*(Y - X)/Y/Y
    W = W + Z**2 - 1.
    C = SQRT(W)
    PRINT *,C
    X = X + 1.
    GO TO 1
    END

</td></tr>
</table>

7. Assuming mixed-mode arithmetic is, for the moment, acceptable, determine the value of the following arithmetic expressions:

  a. 1 + 1./.5   **e.** 28/3/2/3   **i.** 9.**1./2.

  b. 5*4/5   **f.** 28/3/2./3   **j.** 3.**9**.5

  c. 4/5*5   **g.** 28/(3/2)/3   **k.** 27/3**3

  d. 4./5*5   **h.** 4/1+1

8. Translate the following algebraic expressions into Fortran expressions. Use Fortran intrinsic functions where indicated.

  a. $\sin[\cos^{-1}(\beta)]$ **e.** $\tan^2(x/\pi + y)$

  b. $e^{\alpha+\beta} - \sin(\alpha - \beta)$ **f.** $\cos^{-1}(x + |\ln(y)|)$

  c. $\dfrac{1}{|a + b|}c + d$ **g.** $\left(\dfrac{x}{y}\right)^{n+1}$

  d. $\dfrac{x/y + \pi}{\pi - y/x}$

9. The following translations of algebra into Fortran are incorrect. Rewrite the correct Fortran expressions.

a. $\dfrac{xy}{z+1}$ →   XY/Z+1

b. x^{n+1} →   X**N+1

c. $x^{1/2}$ →   X**1./2.

d. $\cos^{-1}(|\ln(x)|)$ →   ACOS(LOG(ABS(X)))

e. $(x^a)^b$ →   X**A**B

10. The equation for the height of a falling object is

$$y = y_0 + v_0 t - (g/2)t^2$$

where y_0 is the starting height at $t = 0$, v_0 is the starting velocity, and g is the gravitational acceleration (9.8 m/sec^2). Write a program to read y_0, v_0, and a value of y that is less than y_0 and return a value of t that satisfies this equation. (You will have to use the quadratic formula.)

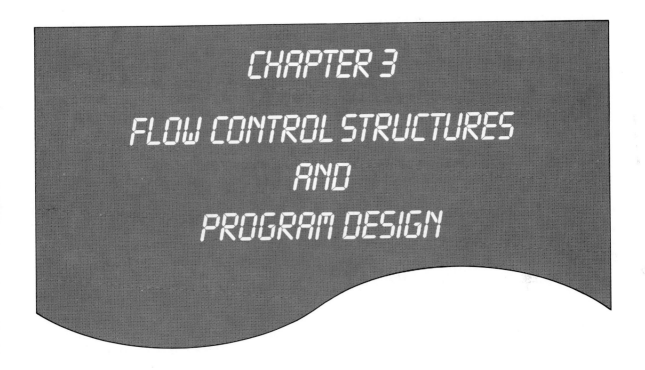

CHAPTER 3
FLOW CONTROL STRUCTURES
AND
PROGRAM DESIGN

3.1 INTRODUCTION

A program normally proceeds from one statement to the next. Fortran flow
control structures may be used to alter this in a number of ways, such as

Decision structures: Compare the values of two quantities and,
based on the result, branch to a variety of points in the program.
Loop structures: Return to a previous statement and repeat the
calculation using different numbers.

It is mainly this ability to follow diverse paths through a code that
makes the computer the useful computational tool that it is. If the program
only requires one straight-through pass, it is merely duplicating the operation of a simple calculator and in almost all cases it would be more efficient
to do the calculations by hand on your calculator.

In this chapter we will discuss a few of the control statements available
in Fortran that make it possible to create interesting repetitive programs. In
addition, since these programs will execute via numerous alternate paths,
the structure of the program's flow can be quite complicated and so some
systematic procedures for designing efficient programs will be discussed.

57

The step-by-step recipe for the solution of a problem is called an *algorithm,* and the design of clear, concise, and effective algorithms is the keystone of computer programming. Once an algorithm has been designed, the construction of a Fortran code to implement the algorithm is usually straightforward. There are a variety of mechanisms for preparing computational algorithms. The two most common schemes presently in use are called *flowcharts* and *pseudocode.*

3.2   THE USE OF FLOWCHARTS AND PSEUDOCODE IN PROGRAM DESIGN

The construction of programs that employ complicated branchings requires significantly more forethought than was indicated in the program examples to this point. Frequently both the reading and the writing of such code can create considerable confusion in even the best of programmers. The difficulty is that in such a project you can no longer trace the computation straight down from the first line to the END statement, but are forced to consider two or more possible alternatives simultaneously. Facing similar problems in, say, developing a complicated essay, you would resort to an outline. The outline of a Fortran program has been standardized and is called a flowchart. Before you begin to write any moderately complicated program, some sort of flowchart is essential as a guide to the construction of the code. Flowcharts themselves are not always easy to read and are often difficult to alter, especially if the program is quite long; and their popularity among professional programmers and engineers has diminished considerably in recent times. An alternative to flowcharts will be discussed in Section 3.2.2. Nonetheless, preparing a neat flowchart is an excellent means of accurately organizing your thoughts on a computational algorithm. Additionally, the flowchart can be an important part of a program's documentation and is especially useful if you intend to discuss your code with colleagues or your instructor. It is usually much easier to read someone else's flowchart than their Fortran.

3.2.1   Flowcharts

A flowchart is a method of diagramming the logic of an algorithm using a standardized set of symbols to indicate the various elements of the program. The most common symbols that are used in flowcharts are shown in Table 3-1. In a complete flowchart, short messages are ordinarily written within each symbol to explain the current activity.

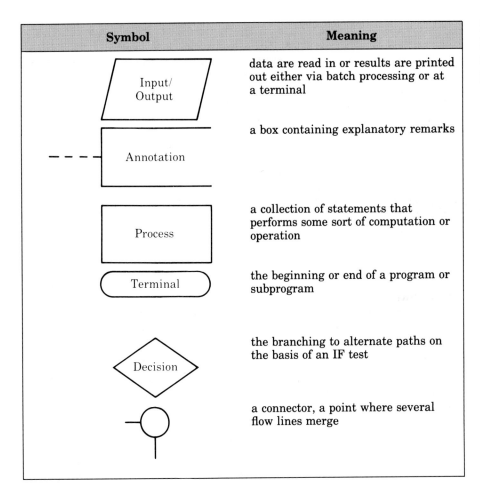

Symbol	Meaning
Input/Output	data are read in or results are printed out either via batch processing or at a terminal
Annotation	a box containing explanatory remarks
Process	a collection of statements that performs some sort of computation or operation
Terminal	the beginning or end of a program or subprogram
Decision	the branching to alternate paths on the basis of an IF test
(connector)	a connector, a point where several flow lines merge

Table 3-1 Symbols used in flowcharting fortran programs.

A flowchart should have one *start* and one or more *stops*. The logical flow of the algorithm is from top to bottom, and alternative paths are indicated by flow lines with arrowheads to indicate the direction of the calculation. Flow lines may cross, but the merging of two or more computational paths at a point in a program is indicated by means of the connector symbol (circle) which may also include the statement number of the junction point of the lines in the program.

As an example of a flowchart representation of a simple program, consider the problem of computing the wages for a worker who has worked a given number of hours this week and who is paid at a rate of PAYRAT in dollars per hour for the first 40 hours and at a higher rate of OVTRAT in dollars per hour for time exceeding 40 hours. The flowchart for this rather simple problem is shown in Figure 3-1. The key element of the algorithm is the decision structure represented by the diamond-shaped symbol. At this point, the program is expected to compare the value read for hours worked with the number 40. If the hours are more than 40, a calculation of the

Figure 3-1
Illustration of a
flowchart for a
simple program.

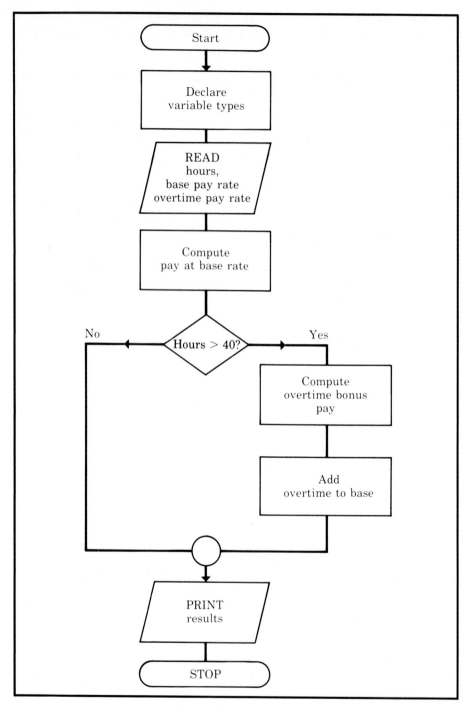

overtime pay is executed and this result is added to the base pay; otherwise
the program skips the overtime computation. The Fortran implementation
of decision structures is by means of the IF(...)THEN statements which
will be discussed in Section 3.3.

More complicated flowcharts will be illustrated in subsequent examples, but first we will consider a competing method of outlining an algorithm.

3.2.2 Pseudocode Outlines of Computational Algorithms

Before a program can be written, the problem to be solved must be mapped out and some form of outline constructed. One procedure for doing this is the flowchart described above. However, a great many scientists and engineers who write complicated programs feel that the flowchart is too formal and artificial a device for a preliminary draft and resort instead to a highly informal procedure called pseudocode. The idea is to describe the operation of the program using a simplified mix of Fortran and English. Since pseudocode is a response to the rigidity of flowcharts, there are very few rules. Basically you write out the operation of the program using minimal English, capitalizing Fortran phrases, using one separate line for each distinct segment of the code. It is also a good idea to indent subsegments as in an ordinary outline. A pseudocode outline of the salary problem follows:

```
READ in hours worked, base pay rate, and overtime rate
Echo PRINT

compute pay using base rate
     PAY = PAYRAT*HOURS

IF Hours worked > 40 THEN
   multiply excess of 40 by overtime bonus rate
       (HOURS - 40)*(OVTRAT - PAYRAT)
   add to base pay
ELSE
   skip overtime computation

PRINT pay
STOP
```

In this form of the outline, the decision structure in the program is indicated by the English words IF and ELSE. This is very close to the form of the actual Fortran implementation of the algorithm as described in the next section.

3.3 DECISION STRUCTURES IN FORTRAN

In addition to the elementary arithmetic operations, the ALU in the computer is designed to compare two items and execute instructions based on that comparison. In the most elementary type of comparison, the two items are either the same or not the same. In order to facilitate the construction of comparison tests that are easy to read and understand, a new form of expression, in addition to arithmetic and character expressions, has been added to Fortran. This is called the *logical expression* and it constitutes the central ingredient in decision structures. Logical expressions are employed in logical IF tests and logical block IF structures. The logical IF test is a short form of the logical block IF structure and will be discussed in Section 3.6.1.

3.3.1   The Logical Block IF Structure

The Fortran structure which is the keystone of all decision procedures is the logical block IF structure which is of the following form.

> **IF(⟨logical expression⟩)THEN**
>
>> Set or block of Fortran statements that will be executed only if the logical expression is evaluated as TRUE.
>
> **END IF**
>
> **If the logical expression is TRUE—**
> **The block of Fortran statements is executed.**
> **If the logical expression is FALSE—**
> **The block of statements is ignored and the program proceeds to the next statement after the END IF.**

The IF(. . .)THEN occupies one line and the Fortran statements that constitute the execution block appear on subsequent lines. The block of statements to be executed conditionally *must* be followed by the statement END IF. The END IF statement *should not* have a statement number. Also, for every IF(. . .)THEN there must be a corresponding END IF. To improve the readability of the code, the block of Fortran statements is usually indented with respect to the IF(. . .)THEN and the END IF.

Logical Expressions   The action of the IF statement of course depends on the definition of the logical expression. A logical expression is built up from combinations of one or more *relational expressions* of the form

$$a_1 \text{ op } a_2$$

where a_1, a_2 are arithmetic expressions, variables, constants, or character strings; in short, things that have values that can be compared. By "op" is meant a "relational logic operator" belonging to the following set:

Relational logic operator	Meaning
.EQ.	Equal to
.NE.	Not equal to
.GT.	Greater than
.GE.	Greater than or equal to
.LT.	Less than
.LE.	Less than or equal to

Note: The periods are part of the operator and must always be present. A relational expression must have a value[1] of ⟨true⟩ or ⟨false⟩. The simplest logical expression consists of a single relational expression like

```
12 .GT. 6     This has a value of (true).
```

A logical expression may then be incorporated in an IF block structure, for example,

```
IF(TEMP .GT. 450.)THEN
     PRINT *,'STEAM TEMPERATURE DANGEROUSLY HIGH'
     STOP
END IF
```

More complex logical expressions can be built up by combining two or more relational expressions by means of the following combinational operators:

Combinational logic operator	Meaning	
.OR.	Or	
.AND.	And	
.NOT.	Not	(Changes a value ⟨true⟩ into a value ⟨false⟩ and vice-versa)

The evaluation of logical expressions is fairly transparent, as can be seen from the following examples.

1. If the variable SIZE has previously been assigned the value 12.0, then the expression

---

[1] In addition to the data types REAL, INTEGER, and CHARACTER, Fortran permits values of type LOGICAL. Logical variables may only have a value of .TRUE. or .FALSE. (Again, the periods are part of the expression.) The Fortran values are indicated in the text as ⟨true⟩, ⟨false⟩. LOGICAL variables are described in more detail in Section 10.1.3.

$$\text{(SIZE .LT. 100.0)}$$

Has a value of ⟨true⟩.

2. All arithmetic operations are processed before the logical expression is evaluated. Thus

$$\text{(SIZE .LT. 10. * SQRT(100.))}$$

has the same value as the previous expression.

3. Parentheses may be added for clarity or to alter the value of the expression.

$$\text{((SIZE - 6.) .LT. (10. * SQRT(100.) - 6.))}$$

4. Logical subexpressions may be combined by using the operators .AND. and .OR. and are then evaluated according to the following rules.

$$⟨true⟩ \text{ .AND. } ⟨true⟩ = ⟨true⟩$$

$$⟨true⟩ \text{ .AND. } ⟨false⟩ = ⟨false⟩$$

That is, the entire expression is ⟨false⟩ if either side of the .AND. is ⟨false⟩.

$$⟨true⟩ \text{ .OR. } ⟨false⟩ = ⟨true⟩$$

$$⟨false⟩ \text{ .OR. } ⟨false⟩ = ⟨false⟩$$

The entire expression is ⟨true⟩ if either side of the .OR. is ⟨true⟩. For example, if the variables A and B have values 2. and 8., respectively, then

`(A .GT. 6. .AND. 2. * B .LT. 20.)` *Is evaluated as ⟨false⟩ .AND. ⟨true⟩ which is then ⟨false⟩*

`(A * B .EQ. 0. .OR. A .LT. 10.)` *Is evaluated as ⟨false⟩ .OR. ⟨true⟩ which is then ⟨true⟩*

A New Hierarchy Rule The processing of complicated logical expressions can lead to ambiguities, and so an additional hierarchy rule is required over those used in ordinary arithmetic (see Section 2.5.1). It reads

A logical expression is evaluated by first processing all arithmetic expressions according to the hierarchy rules pertaining to arithmetic expressions. Then the logic operators are processed scanning from left to right. The subexpressions are combined (the .AND. .OR. operators processed) from left to right with the .AND.'s processed before the .OR.'s.

Consider the meaning of the following rather complicated logical statement:

```
(I .EQ. 10 .OR. X .LT. 1. .AND. Z .GE. 0.0)
```

This could also be written as

```
((I .EQ. 10) .OR. (X .LT. 1.) .AND. (Z .GE. 0.0))
```

and has the same meaning as

```
((I .EQ. 10) .OR. ((X .LT. 1.) .AND. (Z .GE. 0.0)))
```

That is, .AND. is done before .OR. If the values assigned to the variables are

$$I = 10$$
$$X = 0.$$
$$Z = -1.$$

The expression reads

$$(\langle\text{true}\rangle \text{ .OR. } (\langle\text{true}\rangle \text{ .AND. } \langle\text{false}\rangle))$$

Since ⟨true⟩ .AND. ⟨false⟩ = ⟨false⟩, this expression is equivalent to

$$(\langle\text{true}\rangle \text{ .OR. } \langle\text{false}\rangle)$$

which has a value of ⟨true⟩. The expression forcing .OR. before .AND. has a quite different meaning.

```
( ((I .EQ. 10) .OR. (X .LT. 1.)) .AND. (Z .GE. 0.) )
```

Using the same values for the variables, this is equivalent to

$$((\langle\text{true}\rangle \text{ .OR. } \langle\text{true}\rangle) \text{ .AND. } \langle\text{false}\rangle)$$

or

$$(\langle\text{true}\rangle \text{ .AND. } \langle\text{false}\rangle)$$

and in this case the expression has a value of ⟨false⟩.

A Few Potential Pitfalls   There are several very common errors made when using a logical IF test.

1. *Never* test for equality of reals obtained from computation. The reason is of course due to the approximate nature of the arithmetic operations involving finite word length representations of real numbers. For example, if

---

```
A = 2.
B = (SQRT(2.)**2)
IF(A .EQ. B)THEN
    STOP
END IF
```

the test will possibly fail since B may have been assigned the value
1.99999999.

2. If it is necessary to test whether a quantity is smaller than some very
small number, say EPS, the form of the test should never be

```
IF(X .LT. EPS)THEN       (Incorrect test for smallness)
```

but rather

```
IF(ABS(X) .LT. EPS)THEN
```

The point being that X might possibly be negative, and any negative
number, regardless of size, would satisfy the first IF test. The comparison
of the two reals, A and B, should then possibly read

```
IF(ABS(A - B) .LT. EPS)THEN
```

There are several other forms of IF statements which will be discussed
in Section 3.6. However, all of the additional features they present can be
duplicated by combinations of block IF structures and, more importantly,
these alternate forms of the IF test date from earlier versions of Fortran and
are partly responsible for generating Fortran code that is difficult to read
and even more difficult to change or correct. Whenever possible, the decision
structures in a program should employ the IF(. . .)THEN statements.

3.3.2 Examples of IF(. . .)THEN–END IF Structures

A very important part of any program is the error diagnostic. As we shall see
in Section 3.4.3, a great many programs are designed to monitor the behavior of some function, and if the computation proceeds as anticipated, print
the result. If problems develop during the calculation, the program should be
written so that it will flag the error and take some action. For example,

```
IF(VOLTGE .GT. 125. .OR. VOLTGE .LT. 105.)THEN
      PRINT *,'DANGER WARNING'
      PRINT *,'VOLTAGE OUTSIDE ACCEPTABLE LIMITS'
      STOP
END IF
```

Note that this test could not be written as

```
IF(VOLTGE .GT. 125. .OR. .LT. 105.)THEN      ⟨Incorrect⟩
```

Since the logical expression has two operators side by side (.OR. .LT.) the
statement will lead to a compilation time error. An analogous statement
that will not lead to compilation errors but which is also incorrect is to write
the test for I = 0 or I = 10 as

```
              IF(I .EQ. 0 .OR. 10)THEN      ⟨Incorrect⟩
```

Both sides of the .OR. operator must have a value of either ⟨true⟩ or ⟨false⟩
and in this case the value of the expression on the right is 10.[2]
   Another example is to simply flag the errant condition and continue.

```
              REAL BALANC ,WITHDR
              INTEGER FLAG
              FLAG = 1
              IF(BALANC - WITHDR .LT. 0.0)THEN
                    FLAG = 0
                    WITHDR = 0.0
              END IF
                 ...          ...
                 ...          ...
              IF(FLAG .EQ. 0)THEN
                    PRINT *,'INSUFFICIENT FUNDS'
                    STOP
              END IF
```

   The Fortran code for the calculation of overtime pay discussed in Section
3.2.1 may now be constructed by using the IF block as

```
          PROGRAM PAY
          REAL PAY ,PAYRAT ,OVTRAT ,OVRTYM ,HOURS
          READ * ,HOURS ,PAYRAT ,OVTRAT
    *
    *                               FIRST COMPUTE THE PAY BASED ON
    *                               THE BASE HOURLY PAY RATE
    *
          PAY = PAYRAT * HOURS
    *                               IF MORE THAN 40 HOURS , COMPUTE
    *                               OVERTIME BONUS AT HIGHER RATE FOR
    *                               EXCESS HOURS
```

Continued

---

[2] The results of this statement are unpredictable. Some compilers will check one side of an
.OR. and if it is ⟨true⟩ ignore the other side. (If either side is ⟨true⟩, the expression is ⟨true⟩.) A
similar operation will occur if either side of an .AND. is ⟨false⟩.

```
*
      IF(HOURS .GT. 40.0)THEN
          OVRTYM = (HOURS - 40.) * (OVTRAT - PAYRAT)
*
*                            ADD OVERTIME PAY TO BASE PAY
*
          PAY = PAY + OVRTYM
      END IF
      PRINT *,'HOURS WORKED = ',HOURS
      PRINT *,'BASE PAY RATE = ',PAYRAT
      PRINT *,'OVERTIME PAY RATE = ',OVTRAT
      PRINT *, 'PAY = ',PAY,'DOLLARS'
      PRINT *,'OF WHICH ',OVRTYM,'DOLLARS WAS OVERTIME PAY'
      STOP
      END
```

Note the similarity between the actual Fortran and the pseudocode version of the program.

The block IF could also be used to convert a number in radian measure to an angle θ between 0 and 360°. The number must first be scaled so that it is between 0 and 2π by subtracting integer multiples of 2π.

```
      REAL THETA,RADIAN,PI
      INTEGER MULTPL
      PI = ACOS(-1.)
      READ *,RADIAN
*
*                            MULTPL IS THE INTEGER NUMBER
*                            OF 2 PI MULTIPLES IN RADIANS.
*                            THIS STATEMENT INVOLVES MIXED-
*                            MODE REPLACEMENT.
*
      MULTPL = RADIAN/(2. * PI)
      IF(MULTPL .GT. 0)THEN
*
*                            THIS STATEMENT EMPLOYS MIXED-
*                            MODE ARITHMETIC.
*
          RADIAN = RADIAN - MULTPL * 2. * PI
      END IF
      THETA = RADIAN * 360.0/(2. * PI)
      PRINT *,RADIAN,'RADIANS = ',THETA,'DEGREES'
      STOP
      END
```

3.3.3 The IF(. . .)THEN–ELSE Structure

Frequently an algorithm will have two computational branches as a result of a logical IF test. If the condition is ⟨true⟩, a complete block of statements is to be executed, while if ⟨false⟩, an alternate set of statements is to be executed. You could easily accomplish this using two block IF structures. However, an additional option in the block IF structure enables you to construct a code that is easier to read and, more importantly, ties together the two related branches into one structure. This is the ELSE statement, which is placed between the IF(. . .)THEN and END IF.

> **IF(logical expression)THEN**
>   Block of statements to be executed only if the expression is
>   ⟨true⟩
> **ELSE**
>   Block of statements to be executed only if the expression is
>   ⟨false⟩
> **END IF**

As with the END IF, the ELSE statement occupies a line all by itself and it should not have a statement number.

Consider the problem of writing an algorithm to find the smallest-magnitude real root (if any) of the quadratic equation

$$ax^2 + bx + c$$

The nature of the roots depends on the value of the discriminant, $\Delta = b^2 - 4ac$

If	Then
$\Delta > 0$	two real and distinct roots
	$x_+ = \dfrac{1}{2a}(-b + \Delta^{1/2})$
	$x_- = \dfrac{1}{2a}(-b - \Delta^{1/2})$
$\Delta = 0$	two real roots, both identical
	$x_+ = x_- = -\dfrac{b}{2a}$

$$\Delta < 0 \qquad \text{two complex and distinct roots}$$

$$x_+ = \frac{1}{2a}[-b + i(-\Delta)^{1/2}]$$

$$i = (-1)^{1/2}$$

$$x_- = \frac{1}{2a}[-b - i(-\Delta)^{1/2}]$$

If it is desired to compute the smallest-*magnitude* real root of a quadratic with coefficients a, b, c, the program will have to first compute the discriminant Δ. If Δ is negative, the program will print a message (complex roots) and stop. If Δ is not negative (including the case $\Delta = 0$) the smallest-magnitude real root is

$$\frac{1}{2a}(-b + \Delta^{1/2}) \qquad \textit{if b is positive}$$

$$\frac{1}{2a}(-b - \Delta^{1/2}) \qquad \textit{if b is negative}$$

The flowchart for this problem is given in Figure 3-2 and the Fortran program is given in Figure 3-3. In this code there are two nested IF blocks. The inner IF block is completely contained within the outer IF block, and for each IF(...)THEN there is one corresponding END IF. IF blocks may be nested in this manner, one inside the other, but they must never overlap.

```
IF(A .LT. 0.)THEN
          ...
          ...
          IF(B .GT. 0.)THEN
                  ...
                  ...
     END IF
               ...
               ...
          END IF
```

In spite of the suggestive indentations, the first END IF is paired with the inner IF block and thus the code will not execute in the manner that was probably intended.

The algorithm used to determine the roots of a quadratic has three natural branches depending on the value of the discriminant $b^2 - 4ac$, and even though the program will correctly handle the third possibility, that of $\Delta = 0$, to avoid confusion, it is best to use a decision structure better suited to a situation with more than two alternatives. There is a further option available in the block IF structure that is designed to handle multiple paths as the result of an IF test: the ELSE IF structure.

Figure 3-2 The flowchart for the smallest-magnitude real root of a quadratic.

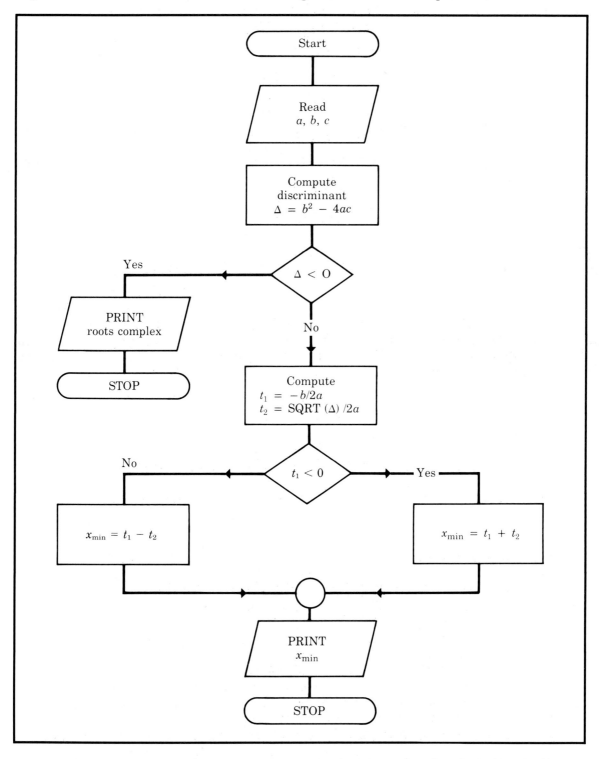

Figure 3-3 The Fortran program to compute the smallest-magnitude real root of a quadratic.

```
PROGRAM QUAD
REAL A,B,C,DISCR,T1,T2,XMIN
READ *,A,B,C
DISCR = B * B - 4. * A * C
IF(DISCR .LT. 0.)THEN
      PRINT *,'BOTH ROOTS ARE COMPLEX'
      STOP
ELSE
*
*                            DEFINE THE TWO TERMS T1 AND T2
*
      DENOM = 2.*A
      T1 = -B/DENOM
      T2 = SQRT(DISCR)/DENOM
      IF(B.GE.0.)THEN
            XMIN = T1 + T2
      ELSE
            XMIN = T1 - T2
      END IF
*
END IF
PRINT *,'SMALLEST-MAGNITUDE REAL ROOT = ',XMIN
STOP
END
```

3.3.4 The ELSE IF Statement

Frequently the possible branches of a computational algorithm are more numerous than the two permitted in a true-false test. To accommodate these cases the ELSE IF structure is used and is best explained by an example. The code for the roots of the quadratic may be written using this structure in the following way:

```
READ(*,*)A,B,C
DELTA = B * B - 4. * A * C
IF(DELTA .GT. 0.)THEN
      . . .
      . . .
      . . .
ELSE IF(DELTA .EQ. 0.)THEN
      . . .
      . . .
      . . .
ELSE
```

```
    • • •        ⟨These statements are executed
    • • •         only if both IF tests fail.⟩
    • • •
    END IF
```

Notice that the ELSE IF(. . .)THEN is *not* paired with an END IF. Also, as with the simple ELSE, the ELSE IF(. . .)THEN occupies a single line by itself. An unlimited number of ELSE IFs may be placed within the block IF structure.

As a second example of nested block IFs with multiple alternatives, consider the problem of determining whether three lengths, a, b, c, can form a triangle, and if so, whether the triangle is isosceles or equilateral.

Three lengths a, b, c can form a triangle if

$$|a - b| < c < a + b$$

The triangle is isosceles if

$$a = b \quad \text{or} \quad a = c \quad \text{or} \quad b = c$$

The triangle is equilateral if

$$a = b \quad \text{and} \quad a = c \quad \text{and} \quad b = c$$

The program for this problem will have to make multiple comparisons and account for several possibilities. The flowchart for the solution is shown in Figure 3-4 and the Fortran program is given in Figure 3-5.

3.4 LOOP STRUCTURES IN FORTRAN

Perhaps the most common and useful computational structure in programming is the loop structure, wherein a block of statements is executed and the block is then simply repeated with some of the parameters changed. The formal construction for operations of this type is a loop structure, called a *DoWhile* loop structure. The characteristics of a DoWhile loop are

1. An entry point (the top) labeled as DoWhile(. . .)
2. An execution block. The body of the loop containing the block of Fortran statements to be conditionally repeated.
3. A normal exit point (the bottom) labeled as EndDo.
4. Optional abnormal exit points (jump out of the loop from within)
5. Loop control specifications. Conditions on a parameter that determine when the cycling of the loop is to be terminated. The conditions are placed in the parentheses in the DoWhile entry point labels.

Figure 3-4 A flowchart for determining whether a triangle is isosceles or equilateral.

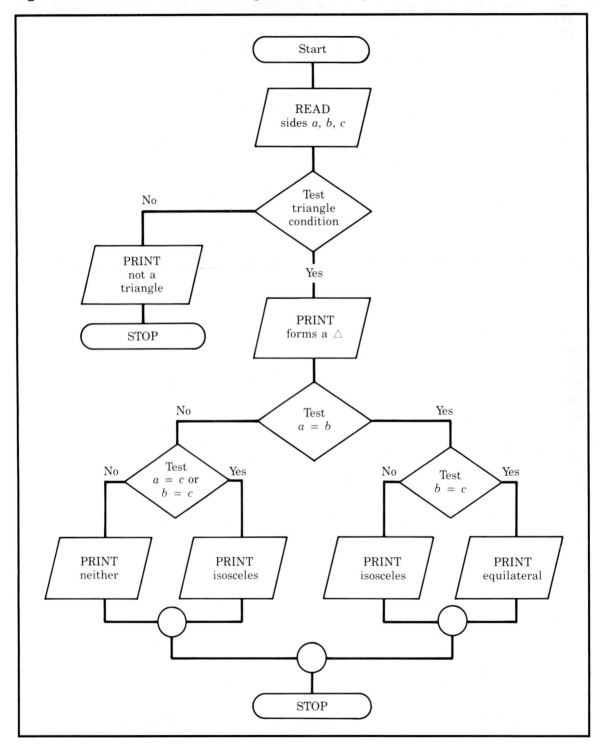

Figure 3-5 A program to determine if a triangle is isosceles or equilateral.

```
READ *,A,B,C
IF(ABS(A - B) .LE. C .AND. C .LE. A + B)THEN
      PRINT *,'SIDES ',A,B,C,' DO FORM A TRIANGLE'
      IF(A .EQ. B)THEN
            PRINT *,'THAT IS ISOSCELES'
            IF(B .EQ. C)THEN
                  PRINT *,'AND EQUILATERAL'
            END IF
      ELSE IF(A .EQ. C .OR. B .EQ. C)THEN
            PRINT *, 'THAT IS ISOSCELES'
      ELSE
            PRINT *,'THAT IS NOT ISOSCELES OR EQUILATERAL'
      END IF
ELSE
      PRINT *,'SIDES ',A,B,C,'DO NOT FORM A TRIANGLE'
END IF
```

I must point out at the very beginning of the discussion that at present *Fortran 77 does not contain specific DoWhile or EndDo statements for constructing loop structures*. However, the loop structures are more easily understood when expressed in terms of constructions analogous to the statements used in decision structures, and for this reason these statements have become more or less standard in pseudocode outlines of a program even though they *do not exist* as actual Fortran statements. Fortunately, it is quite easy to combine existing Fortran statements to accomplish the objectives of the missing statements. Future revisions of Fortran will almost certainly add the DoWhile and the EndDo to the vocabulary. A flowchart of the operation of the loop structure is shown in Figure 3-6.

As an example of a problem employing loop structures, consider the task of reading a list of student exam scores and simply counting the number who passed (score $\geq$ 60). The program could be arranged in a DoWhile structure as

```
NPASS = 0
DoWhile (<there is input left>)
   <Test: Any more input? If so, proceed; if not, quit>
   READ *,SCORE
   IF(SCORE .GE. 60.)THEN
       NPASS = NPASS + 1
   END IF
EndDo <Return to top of execution block.>
```

The only significant difference between the two structures shown in Figure 3.6 is the placement of the test for completion of the loop.

Figure 3-6 Loop structures in Fortran.

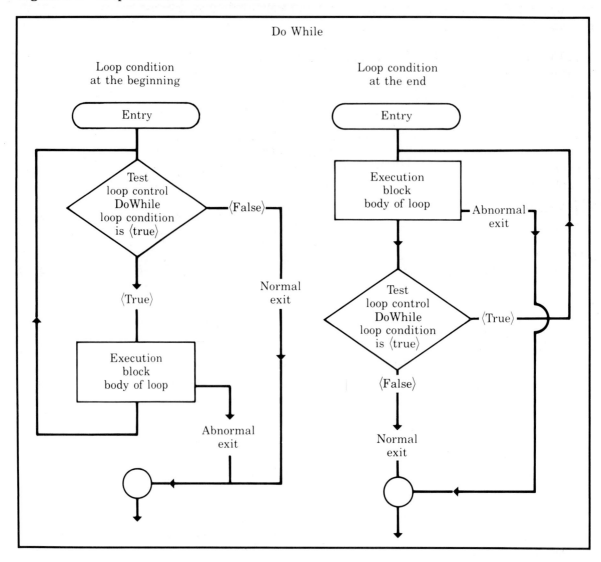

The above examples are, of course, not valid Fortran programs but are merely pseudocode versions of the algorithm, as discussed in Section 3.2.2. The construction of loop structures using valid Fortran statements is the subject of the next section.

3.4.1 Fortran Implementation of DoWhile Structures

The basic ingredients of a loop structure are the test for completion and the instruction to return to the top of the execution block. The test for completion will of course consist of an IF(. . .)THEN test. The Fortran statement that accomplishes a return to the top of the execution block is called an *unconditional GO TO*.

The Unconditional GO TO   The form of the unconditional GO TO statement is

$$\text{GO TO} \quad \langle statement\ label \rangle$$

where ⟨statement label⟩ is the statement number of an executable statement. The effect of this statement is to transfer control of the program to the statement bearing the indicated statement number, which may be any executable statement in the program, coming either before or after the GO TO. For example,

```
      NPASS = 0
   1  READ *,SCORE
          IF(SCORE .GT. 100.)THEN
              PRINT *,'ERROR ON INPUT - SCORE TOO LARGE'
              GO TO 99
          END IF
          IF(SCORE .GE. 60.)THEN
              NPASS = NPASS + 1
          END IF

          IF(SCORE .GE. 0.0)THEN
              GO TO 1
          END IF
*
      PRINT *,'NUMBER WHO PASSED = ',NPASS
  99  STOP
      END
```

This program has an abnormal exit if a score larger than 100 is entered and will continue to cycle back to the READ statement until a negative score is entered by the user.

> *Caution:* The unconditional GO TO is the statement that is primarily responsible for Fortran code that resembles spaghetti: numerous unrestrained branchings up and down in a program that result in code that is unreadable and unalterable. You should be extremely conservative in the use of this statement.

The CONTINUE Statement   The CONTINUE statement in Fortran is an executable statement that performs no operation and is used primarily as a marker for the end of a loop or as the target of a GO TO. The form of the CONTINUE statement is

⟨*statement label*⟩   CONTINUE

The CONTINUE statement should have a statement number label. Since Fortran 77 does not have an EndDo statement in its vocabulary, this statement is used in its place. An example of its use in a DoWhile structure is given below.

```
    NPASS = 0
 1  READ *,SCORE
    IF(SCORE .LT. 0.0)THEN        i.e., DoWhile
                                  score is ≥ 0
        GO TO 12
    END IF
        ...           ...
        ...           ...
    GO TO 1
12  CONTINUE                      This plays the
                                  role of EndDo.
*
    PRINT *,'NUMBER PASSED  =  ',NPASS
        ...       ...
99  STOP
    END
```

3.4.2   Examples of Loop Structures—The Repetitive Program

Very frequently we will want to design a program that executes a calculation for a given set of parameter values and then simply repeats the calculation with a slightly modified set of values, and so on. The loop control is often accomplished by means of a counter that is incremented after each calculation,

```
    IRUN = 1
 1  READ *, ⟨set of input data⟩

        perform calculation
        PRINT *, ⟨results⟩ ,IRUN
```

```
      IRUN = IRUN + 1
      IF(IRUN .LE. 10)THEN
           GO TO 1
      ELSE
           STOP
      END IF
```

This code will execute ten independent calculations.

An alternative, when you are uncertain of how many calculations are required, is to use a so-called trailer data line,

```
 1 READ *,ITEST, <list of the rest of the input data>

      IF(ITEST .LT. 0)THEN
           STOP
      ELSE
               perform calculations
               PRINT *, <results>
           GO TO 1
      END IF
```

This code will continue to loop until a negative integer is entered on the last line of data.

As a final example of a loop structure, consider the situation where it is desired to examine the results of a calculation for a variety of values of a single parameter over a limited range. For example, the temperature dependence of the speed of sound in air is given approximately by the expression

$$v = 331\left(1 + \frac{T}{273}\right)^{1/2} \quad \text{(m/sec)}$$

where T is expressed in °C. If we are interested in obtaining values for the sound velocity for temperatures in the range $20°\ \text{C} \le T \le 35°\ \text{C}$ in steps of $1°$ C, the Fortran code would be

```
      T = 20.
      DT = 1.
 10 V = 331. * SQRT(1. + T/273.)
      PRINT *,'FOR T = ',T,' THE SOUND SPEED IS ',V
      T = T + DT
      IF(T .LT. 36.0)THEN
           GO TO 10
      ELSE
           PRINT *,'CALCULATION COMPLETE -- JOB TERMINATED'
           STOP
      END IF
      END
```

Example Program—Auto Loans When buying a new car some of us must borrow money. If the amount of money borrowed is P, at a yearly interest rate of R for Y years, and the payments are monthly, then the amount of the monthly payment, PAYMNT, is given by

$$\text{PAYMNT} = rP \frac{1}{1 - (1 + r)^{-n}}$$

where r is the monthly interest rate ($r = R/12$) and n is the total number of installments ($n = 12Y$).

The problem is to write a program to calculate PAYMNT and the total cost of the loan for all the possible combinations of

$$Y = 2, 2\tfrac{1}{2}, 3, 3\tfrac{1}{2}, 4 \text{ years}$$

$$R = 10, 11, \ldots, 18\%$$

with $P = \$9500$. The flowchart for this program is given in Figure 3-7. This program requires that two independent loops, one for the values of R and one for the values of Y, be nested, one completely inside the other. For each value of Y, the term of the loan, a complete cycle of interest rates is executed. The R loop is called an *inner loop* and the Y loop is the *outer loop*. As with nested block IF structures, nested loops must never overlap. The complete Fortran program is given in Figure 3-8. Notice that the output from the example program is not particularly neat or orderly. It is difficult to produce tables using the PRINT * statement. However, vertical spacing can be achieved by inserting blank lines in the output by printing a blank, i.e., PRINT *, '' .

3.4.3 Examples of Loop Structures—The Iterative Program

You may be familiar with a method, resembling long division, of calculating the square root of a number. It is very awkward and tedious. However, if you have a calculator available to handle ordinary division, there is an alternative procedure devised by Isaac Newton that is an excellent example of an iterative calculational method.

Newton's algorithm for the calculation of square roots is: If x_0 is a guess for the value of the square root of a number C, then an improved value is x_1, where

$$x_1 = \frac{1}{2}\left(x_0 + \frac{C}{x_0}\right)$$

The idea is to read a number C whose square root is to be computed, guess the square root x_0, and improve the guess with Newton's algorithm. If the improved guess x_1 is still not accurate enough (i.e., $x_1^2 - C$ is not small), the value is again improved by a further application of the algorithm. This

Figure 3-7 A flowchart for calculation of auto loan payments. This program has two nested loops.

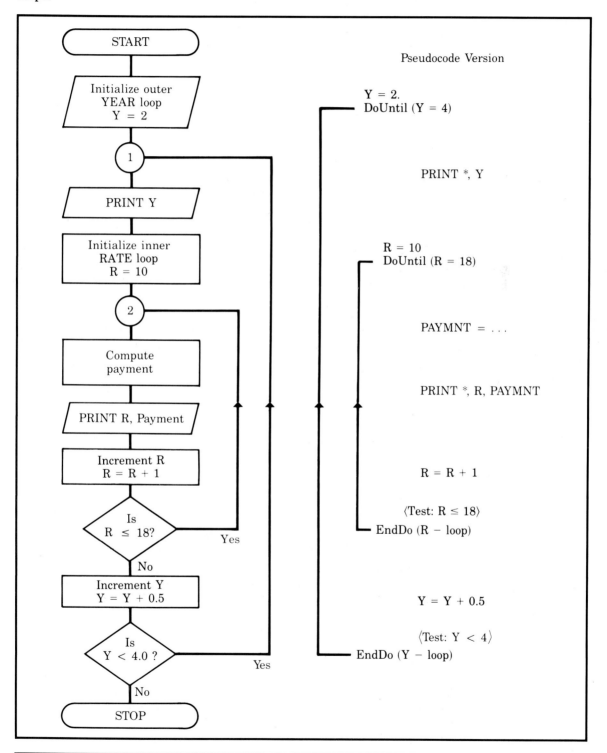

START

Initialize outer
YEAR loop
Y = 2

Pseudocode Version

Y = 2.
DoUntil (Y = 4)

1

PRINT Y

PRINT *, Y

Initialize inner
RATE loop
R = 10

R = 10
DoUntil (R = 18)

2

Compute
payment

PAYMNT = ...

PRINT R, Payment

PRINT *, R, PAYMNT

Increment R
R = R + 1

R = R + 1

Is
R ≤ 18?

Yes

⟨Test: R ≤ 18⟩
EndDo (R − loop)

No

Increment Y
Y = Y + 0.5

Y = Y + 0.5

Is
Y < 4.0 ?

Yes

⟨Test: Y < 4⟩
EndDo (Y − loop)

No

STOP

Figure 3-8 The Fortran code and results for the auto loan program.

```
ENTER PRINCIPAL 9500.00
ENTER THE MAXIMUM LENGTH OF THE LOAN 4.0
ENTER MINIMUM, MAXIMUM INTEREST RATES(PERCENT) 10,18

THE PRINCIPAL OF THE LOAN IS 9500. DOLLARS

  THE CALCULATION LIMITS ARE
     YEARS - FROM 2. TO 4. YEARS
     INTEREST - FROM 10 TO 18 PERCENT

THE RESULTS FOR A LOAN OF 2. YEARS ARE
```

INTEREST RATE	NUMBER OF PAYMENTS	PAYMENT (DOLLARS)	TOTAL COST OF LOAN
10	24	438.3768002065	1021.043204956
11	24	442.774462867	1126.587108808
12	24	447.1979861211	1232.751666905
13	24	451.6473145904	1339.535550171
14	24	456.1223910498	1446.937385195
15	24	460.6231564459	1554.955754702
16	24	465.1495499172	1663.589198013
17	24	469.7015088122	1772.836211492
18	24	474.2789687103	1882.695249047

```
THE RESULTS FOR A LOAN OF 2.5 YEARS ARE
```

INTEREST RATE	NUMBER OF PAYMENTS	PAYMENT (DOLLARS)	TOTAL COST OF LOAN
10	30	359.2083966108	1276.251898323
11	30	363.6416066668	1409.248200004
12	30	368.1070755506	1543.212266518
13	30	372.6047178087	1678.141534262
14	30	377.1344443752	1814.033331256
15	30	381.6961626159	1950.884878478
16	30	386.2897763744	2088.693291232
17	30	390.9151860176	2227.455580529
18	30	395.5722884845	2367.168654536

```
THE RESULTS FOR A LOAN OF 3. YEARS ARE
```

INTEREST RATE	NUMBER OF PAYMENTS	PAYMENT (DOLLARS)	TOTAL COST OF LOAN
10	36	306.5382783415	1535.378020295
11	36	311.0178126115	1696.641254013
12	36	315.5359432221	1859.293955996
13	36	320.0925440313	2023.331585128
14	36	324.6874827013	2188.749377248
15	36	329.3206207897	2355.542348431
16	36	333.9918138453	2523.705529843
17	36	338.700911504	2693.232814145
18	36	343.4477575912	2864.119273283

```fortran
      PROGRAM CARS
*  ---
*  ---   THIS PROGRAM WILL COMPUTE THE AMOUNT OF THE MONTHLY
*  ---   PAYMENT FOR AN AUTO LOAN FOR LOANS OF 2 THROUGH 4
*  ---   YEARS IN HALF-YEAR STEPS AND FOR INTEREST RATES FROM
*  ---   10 TO 18 PERCENT. THE PROGRAM ILLUSTRATES NESTED
*  ---   LOOPS.
*  ---
* VARIABLES
*  ---
      REAL P,YMAX,R,XY
      INTEGER N,RMIN,RMAX,RATE
*  ---
*  ---   P      -  PRINCIPAL, THE AMOUNT OF THE LOAN - INPUT
*  ---   YMAX   -  MAXIMUM LENGTH OF THE LOAN IN YEARS - INPUT
*  ---   RMIN   -  STARTING VALUE OF THE INTEREST RATE
*  ---             IN PERCENT - INPUT
*  ---   RMAX   -  MAXIMUM INTEREST RATE - INPUT
*  ---   Y      -  CURRENT VALUE OF LOAN LENGTH IN YEARS
*  ---   RATE   -  CURRENT VALUE OF THE INTEREST RATE
*  ---   R      -  CURRENT VALUE OF THE INTEREST RATE
*  ---             PER MONTH EXPRESSED AS A DECIMAL
*  ---   X      -  THE COMPUTED VALUE OF THE MONTHLY PAYMENT
*  ---   N      -  THE NUMBER OF PAYMENTS (12 * YEARS)
*  ---
* INITIALIZATION
*  ---
*  ---   READ IN THE PRINCIPAL AND THE LIMITS ON THE INTEREST
*  ---   RATES AND THE LENGTH OF THE CALCULATION. ECHO PRINT.
*  ---
      PRINT *,' '
      PRINT *,'ENTER PRINCIPLE'
      READ *,P
      PRINT *,'ENTER THE MAXIMUM LENGTH OF THE LOAN'
      READ *,YMAX
      PRINT *,'ENTER MINIMUM, MAXIMUM INTEREST RATES(PERCENT)'
      READ *,RMIN,RMAX
      PRINT *,' '
*  ---
*  ---   PRINT THE OVERALL HEADINGS FOR EACH SET OF
*  ---   CALCULATIONS (I.E., ONE FOR EACH VALUE OF Y)
*  ---
      PRINT *,'THE PRINCIPAL OF THE LOAN IS ',P,' DOLLARS'
      PRINT *,' '
      PRINT *,'  THE CALCULATION LIMITS ARE'
      PRINT *,'     YEARS - FROM 2 TO ',YMAX,' YEARS'
      PRINT *,'     INTEREST - FROM ',RMIN,' TO ',RMAX,' PERCENT
      PRINT *,' '
```

```
* ---
* ---      NEXT PRINT THE TABLE HEADINGS
* ---
      Y = 2.0
      PRINT *,' '
      PRINT *,' '
    1 PRINT *,'THE RESULTS FOR A LOAN OF ',Y,' YEARS ARE'
      PRINT *,' '
      PRINT *,'   INTEREST      NUMBER OF      PAYMENT      TOTAL COST'
      PRINT *,'     RATE        PAYMENTS      (DOLLARS)       OF LOAN'
*---------------------------------------------------
* COMPUTATION
* ---
* ---      THE CURRENT MONTHLY INTEREST RATE IS COMPUTED
* ---
      RATE = RMIN
      N = 12. * Y
* ---
* ---      THE TOP OF THE INNER LOOP
* ---
    2 R = RATE/100./12.
      TOP = R * P
      BOT = 1. - (1. + R)**(-N)
      X = TOP/BOT
* ---
* ---      PRINT THE CURRENT TABLE ENTRY
* ---
      PRINT *,'     ',RATE,'     ',N,'     ',X,'     ',
     +           N*X - P
* ---
* ---      INCREMENT RATE AND IF STILL WITHIN LIMITS LOOP
* ---      BACK TO TOP OF THE INNER LOOP, ELSE PROCEED TO
* ---      NEXT CALCULATION
* ---
      RATE = RATE + 1.
      IF(RATE .LE. RMAX)THEN
          GO TO 2
* ---
* ---      AT THE END OF THE RATE LOOP, INCREMENT THE YEAR
* ---      VARIABLE AND TEST FOR COMPLETION OF THE CALCULATION.
* ---
      ELSE
          Y = Y + 0.5
          IF(Y .GT. YMAX)THEN
              PRINT *,' '
              PRINT *,'   CALCULATION COMPLETED'
              STOP
          ELSE
              GO TO 1
          END IF
      END IF
* ---
      END
```

INTEREST RATE	NUMBER OF PAYMENTS	PAYMENT (DOLLARS)	TOTAL COST OF LOAN
10	42	269.0097837627	1798.410918035
11	42	273.5416838825	1988.750723064
12	42	278.1184473382	2180.974788204
13	42	282.7398947456	2375.075579313
14	42	287.4058370092	2571.045154387
15	42	292.11607496	2768.875170833
16	42	296.8704022155	2968.55689305
17	42	301.6686000047	3170.081200199
18	42	306.5104427214	3373.438594298

THE RESULTS FOR A LOAN OF 4. YEARS ARE

INTEREST RATE	NUMBER OF PAYMENTS	PAYMENT (DOLLARS)	TOTAL COST OF LOAN
10	48	240.9445426301	2065.338046247
11	48	245.5324648088	2285.558310823
12	48	250.1714366033	2508.22895696
13	48	254.8612109908	2733.33812756
14	48	259.601526703	2960.873281744
15	48	264.3921085315	3190.82120951
16	48	269.2326678455	3423.168046983
17	48	274.1229019203	3657.899292174
18	48	279.0624962775	3894.999821321

CALCULATION COMPLETED

process is continued until the computed value for x_1 is sufficiently accurate, or until the program exceeds its maximum cycle limit which is supplied by the programmer.

All iterative programs should have built-in checks for potential problems, such as

1. The algorithm is diverging. That is, successive answers are getting worse, not better.
2. The convergence is too slow. After a prescribed maximum number of iterations, satisfactory accuracy has not been attained.

The pseudocode outline of a program to implement Newton's method is given below,

```
READ in number whose square root is wanted (C)
Echo PRINT
Determine initial guess for square root (X)
      ⟨initial guess is (C + 1)/2⟩
Initialize iterations counter (ITER)
   Using Newton's algorithm, improve the guess by the replacement
      [X ← 0.5 * (X + C/X)]
   Increment the iterations counter by one
      Check for excessive number of iterations
         ⟨IF(ITER > 100) quit and print
            failure statement⟩
      Check for convergence: is |C − X * X| small?
         IF yes → success, quit and PRINT results
         IF no →  has not yet converged, return to improvement statement
      Success path: PRINT results
      Failure path: PRINT diagnostic
```

A flowchart version of the program is given in Figure 3-9 and the complete Fortran code is illustrated in Figure 3-10. In preliminary versions of this program the prudent programmer would have the program print the latest value of X in each iteration to see how the calculation is progressing. A failure to converge to an answer could be caused by an incorrect algorithm or by using too small a value for the convergence criterion. For example, if your machine only retains eight digits in a computer word, a test for success of the form

```
IF(ABS(C - X * X)) .LT. 1.E - 12)THEN
```

would be inappropriate and would always fail if C is larger than 0.001. If the number entered is 200., the output from this program is

```
ENTER NUMBER WHOSE SQUARE ROOT IS WANTED
200.
THE SQUARE ROOT OF 200.0000 IS 14.14214
THE NUMBER OF ITERATIONS WAS      6
```

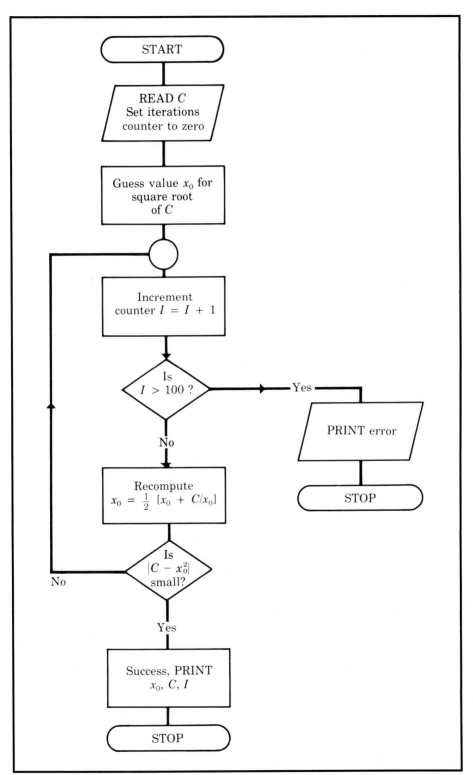

Figure 3-9
Flowchart for
square roots via
Newton's algo-
rithm.

START

READ C
Set iterations
counter to zero

Guess value x_0 for
square root
of C

Increment
counter $I = I + 1$

Is
$I > 100$?

Yes

No

Recompute
$x_0 = \frac{1}{2} [x_0 + C/x_0]$

PRINT error

STOP

Is
$|C - x_0^2|$
small?

No

Yes

Success, PRINT
x_0, C, I

STOP

Figure 3-10 The Fortran code for calculating square roots via Newton's algorithm.

```
        PROGRAM SQROOT
C --DEMONSTRATION PROGRAM ILLUSTRATING NEWTON'S ITERATIVE
C --ALGORITHM FOR COMPUTING SQUARE ROOTS
C --            XNEW = 0.5 * (XOLD + C/XOLD)
C -------------------------------------------------------
C VARIABLES
        REAL X, C
        INTEGER I
C                         C -- THE NUMBER WHOSE SQUARE ROOT
C                              IS DESIRED (INPUT)
C                         X -- THE SUCCESSIVE VALUES OF THE
C                              SQUARE ROOT OF C
C                         I -- ITERATIONS COUNTER
C -------------------------------------------------------
C   INITIALIZATION
C
        PRINT *,'ENTER NUMBER WHOSE SQUARE ROOT IS WANTED'
        READ *,C
        I = 0
C                         FOR AN INITIAL GUESS FOR SQRT(C)
C                         WE TRY SIMPLY (C + 1)/2
        X = (C+1.)/2.
C -------------------------------------------------------
C   COMPUTATION ALGORITHM
C
C                         IMPROVE THE GUESS AND INCREMENT I
C
    1   X = .5 * (X + C/X)
        I = I + 1
C                         TEST FOR EXCESSIVE ITERATIONS
C
        IF(I .GT. 100)THEN
            PRINT *,'NOT CONVERGING IN 100 STEPS'
            STOP
C                         TEST FOR CONVERGENCE BASED ON ABSOLUTE
C                         DIFFERENCE BETWEEN X * X AND C
C
        ELSE IF(ABS(X * X - C) .LT. 1.E-6)THEN
C
C                         SUCCESS - PRINT RESULTS
C
            PRINT *,'THE SQUARE ROOT OF ',C,' IS ',X
            PRINT *,'THE NUMBER OF ITERATIONS WAS ',I
            STOP
C                         CURRENT VALUE OF X IS NOT ACCURATE
```

```
C                           ENOUGH, REPEAT THE ALGORITHM
        ELSE
             GO TO 1
        END IF
C       -------------------------------------------------------
        END
```

3.5   A NOTE ON GOOD PROGRAMMING STYLE

Every so often, one reads in the newspapers accounts of individuals who
have been recorded as dead by the Social Security Administration and who
show up one day, in person, requesting the error be corrected and are told
that the mistake is irreparable, the computer code has no entry for
"resurrected." These stories are examples of poorly documented computer
programs, assembled piecemeal over many years by numerous individuals.
These codes are often a mysterious black box to those who use them and are
almost impossible to modify. This is a serious problem everywhere computers are used. It cannot be stressed too strongly that the programs you write
must be understandable, now and anytime in the future, to any potential
user.

   I have already suggested inserting comment lines throughout your code
to explain the operation of your program. There are several additional steps
you can take to improve a program's readability.

1. Segment the program into a sequence of blocks or modules. For example,

Variable dictionary:	A list of all the variables in the program and an explanation of their meaning (perhaps including units).
Initialization block:	A segment of the program where the constants are assigned values and the data are read in.
Computation block:	Contains the body of algorithms used in obtaining the numerical results.
Success path:	After a successful completion of the program it branches to here and the results are printed.
Failure path:	If problems are encountered during execution, the program branches here and some form of diagnostic message is printed.

2. Explicitly declare each variable in the program to be either INTEGER,
REAL, or CHARACTER.
3. Echo print all numbers read in to verify accurate assignments. This
usually means that a corresponding PRINT statement should immedi-

ately follow each READ statement. This is often relaxed somewhat provided that *every* value read is eventually printed somewhere later in the program.

4. Before you begin writing the program, outline it by means of a simplified flowchart or pseudocode description. After the program has been tested and successfully executed, prepare a detailed flowchart as part of the documentation for the program.

5. The fundamental commandment of structured programming is: Spend more time on the design of the program *before* you begin to write the code. You may think you can sit down at a terminal and compose a program, and often you can. But 90 percent of the time you will end up frustrated. Write out the program first, in block form, using either a flowchart or pseudocode. Include ample comments in the program and liberally add error diagnostics.

3.6 ADDITIONAL CONTROL STATEMENTS

As promised, several of the alternative forms of the Fortran control statements are described in this section.

3.6.1 The Logical IF Test

Frequently the block of statements that is to be executed as a result of a single option IF(. . .)THEN—END IF structure will consist of a single executable Fortran statement. For example,

```
IF(ITER .GT. 100)THEN
         GO TO 99
END IF
```

Fortran permits simple block IFs of this type to be written as a single line with the executable Fortran statement following the IF test, as

```
IF(<logical expression>)   <an executable Fortran statement>
```

If the logical expression is evaluated as ⟨true⟩, the Fortran statement is executed.

If the logical expression is evaluated as ⟨false⟩, the Fortran statement is ignored and the program proceeds to the next line of code.

The Fortran statement can be almost any of the executable statements that we have encountered, such as READ, PRINT, assignment, GO TO,

STOP. It cannot, however, be another IF-type statement (or a DO statement, which will be discussed in Chapter 6).

It is usually acceptable to replace single option–single statement block IF structures by a one-line logical IF statement. However, if you find your program contains numerous statements of the form

```
IF(...)GO TO
```

you are probably creating a code that contains excessive branchings. Structured programming means that you should, whenever possible, be constructing all computational algorithms out of the basic decision and loop structures discussed in this chapter. If the logical IF statement is part of one of these larger structures, it is acceptable; if not, it is likely that it violates the normal style considerations of structured programming.

3.6.2 The Computed GO TO Statement

The form of the computed GO TO statement is

$$\text{GO TO } (k_1, k_2, k_3, \ldots, k_n), \text{Iexp}$$

where $k_1, k_2, \ldots, k_n$ are statement numbers of existing executable statements and Iexp is an integer, integer variable, or integer arithmetic expression. This statement transfers control to one of the statements in the parentheses depending on the value of Iexp. If

Iexp = 1	*the program branches to statement k_1*
Iexp = 2	*the program branches to statement k_2*
...	...
Iexp = n	*the program branches to statement k_n*
Iexp > n	*no transfer is made, the next statement is executed*
Iexp < 1	*no transfer is made, the next statement is executed*

For example,

```
K = 15
I = 6
GO TO (4,83,15,1,9,15,6),K/I + 1
```

has the same effect as GO TO 15. The comma after the parenthesis is optional.

3.6.3   The Arithmetic IF Statement

The form of the arithmetic IF statement is

$$IF(exp)k_-, k_0, k_+$$

where k_-, k_0, k_+ are statement numbers of executable statements existing somewhere in the program, and exp is an integer or real constant, variable, or expression. The arithmetic IF transfers control to one of the three statement labels depending upon the value of exp.

If exp < 0.0   *the program branches to statement k_-*

If exp = 0.0   *the program branches to statement k_0*

If exp > 0.0   *the program branches to statement k_+*

For example,

```
X = 2.
IF(X**2 - 3.)17,35,108     ⟨transfer is made to statement 108⟩
```

But beware of the following improper code,

```
X = 2.
Y = SQRT(X)
Z = Y**2 - X
IF(Z)7,8,9
```

will probably cause a transfer to statement 7, not statement 8. This is of course the recurring problem of approximate arithmetic when using real numbers.

In the earliest versions of Fortran the arithmetic IF statement was the only comparison statement available, and as a consequence many of the "old-timers" use it almost exclusively. In modern usage it has been replaced by IF(. . .)THEN structures or the logical IF test.

PROBLEMS

1. Identify the compilation errors in the following. If none write OK.

```
a. IF(A .EQ. 5)
      THEN STOP
   END IF
```

```
b. IF(A .LT. 5.)THEN
      A = 5.
   ELSE
      A = 5.
   END IF


c. IF(I .GT. 10 .AND. X .LT. 1.)THEN
      X = X/2.
      I = I + 1
   ELSE IF(I .LE. 10 .OR. X .GE. 1.)THEN
      X = 2. * X
      I = I - 1
   ELSE
      PRINT *,'HOW DID WE GET HERE'
   END IF
```

2. The following code will print how many of the values A, B, C match. By simply indenting portions, rewrite the code in a more readable form.

```
IF(A .EQ. B)THEN
IF(B .EQ. C)THEN
PRINT *,'ALL THREE MATCH'
ELSE
PRINT *,'TWO MATCH'
END IF
ELSE IF (B .EQ. C .OR. C .EQ. A)THEN
PRINT *,'TWO MATCH'
ELSE
PRINT *,'NONE MATCH'
END IF
```

3. Identify the errors in the following. If none, write OK.

```
a. GO TO END
b. GO TO (5,4,3,2,1),INDEX
c. GO TO (5,5,5,5,2),INDEX
d. IF(A = 0.)STOP
e. IF(X .AND. Y .EQ. 1.5)GO TO 5
f. IF(C .LT. 5. .AND. C .GT. 5.)C = 5.
g. IF(A .EQ. 0.)A .EQ. 0.
h. IF(I .LT. 10)IF(I .EQ. 6)GO TO 9
i. IF(GE .GE. LE)PRINT *,GE,LE
j. IF(EQ)1,5,3
k. IF(SQRT(B**2 - 4. * A * C))2,2,3
```

4. Write a program that will read a person's age and print the word TEEN-AGER if $13 \le AGE \le 19$ or the word RETIRED if $AGE \ge 65$.

5. Determine whether the following logical expressions are ⟨true⟩ or ⟨false⟩. Use the following values for the variables.

$$I = 2$$
$$K = 4$$
$$X = -2.$$
$$Y = 1.0$$

a. (I**4 ,EQ, 2 * K ,AND, K/I ,EQ, I * I ,OR, Y ,GT, X)
b. (I ,GT, K ,AND, Y ,GT, X ,OR, K - I * I ,EQ, 0)
c. ((I ,GT, K) ,AND, (Y ,GT, X ,OR, K - I * I ,EQ, 0))

6. Write a pseudocode description of the program in Figure 3-2.
7. Write a program that will determine whether an integer is even or odd.
8. Write a program that will print the squares of all odd integers less than 50.
9. Write a program that will determine whether
   a. Four sides that are read in could form a polygon. *Note:* if the sides are labeled a, b, c, d, the conditions that must be satisfied are

$$a \leq b + c + d$$
$$b \leq a + b + c$$
$$c \leq a + b + d$$
$$d \leq a + b + c$$

   b. If a polygon can be formed, whether the polygon could be a rectangle or a square.

10. An algorithm to compute the inverse of a number C without using any division is

$$x_{new} = x_{old}(2 - Cx_{old})$$

provided that the initial guess for the inverse (x_{old}) is chosen such that $(2 - Cx_{old})$ is not negative.
   a. Write a complete flowchart to

   - Read a positive number C
   - Specify an initial guess for the inverse of C.
   - Check that $(2 - Cx)$ is not negative. If it is negative, reduce x and try again.
   - Set an iteration counter to zero.
   - Improve the guess for the inverse and increment the counter.
   - Test for success by determining if $|1. - Cx| < 1 \times 10^{-8}$.
   - Test for excessive iterations.
   - Print C and x if successful.

   b. Write and execute a Fortran program corresponding to your flowchart. Also carry out the calculation on your pocket calculator for a variety of numbers.

11. A nonlinear equation can sometimes be solved by an iterative procedure called *the method of successive substitutions,* provided the equation can be written in the form

$$x = f(x)$$

The procedure is to guess a value for the solution, say x_{old}, and compute a new value for the solution via

$$x_{new} = f(x_{old})$$

This is continued until the difference between successive values of x are smaller than some prescribed small number or until it is clear that the procedure is diverging. (That is successive x's differ by more than some prescribed large number.) Write a complete program to solve for a root of the following functions (i.e. find an x such that $f(x) = 0.0$). The program should have safeguards to handle a diverging solution.

a. $F(x) = \dfrac{x}{3} - e^{-x^2}$

First write the equation, $F(x) = 0$ as $x = [\ln (3/x)]^{1/2}$ and then try writing it as $x = 3e^{-x^2}$. The root is near $x = 1.0$

b. $g(x) = x^{10} + 5x^3 - 7$

First write the equation $g(x) = 0$ as $x = (7 - 5x^3)^{1/10}$ and then try writing it as $x = ((7 - x^{10})/5)^{1/3}$. The root is near $x = 1.0$

c. $h(x) = a_0 + a_1 x + a_2 x^2$

The values a_0, a_1, a_2 should be read in.

12. The commission earned by a used-car salesman is determined by the following rules:

If the amount of the sale is less than $200 there is no commission.
If the amount of the sale is between $200 and $2500, the commission
   is 10 percent of the sale.
If the amount of the sale is greater than $2500, then the commission
   is $250 plus 12 percent of the amount above $2500.

The amount of the sale is the price of the car sold less the value of any trade-in. Write a complete program which reads in the value of the sale, the value of the trade-in, and calculates the commission.

13. Write a program to determine whether an integer N, where $100 < N < 10,000$, is a prime number. The number N is to be read in and your program should be as efficient as you can make it. *Note:* An integer N is divisible by an integer I if

$$(N/I * I) .EQ. N$$

Do you understand why? Also, you only need to test whether N is divisible by numbers I from $I = 2$ to $I = N^{1/2}$.

PROGRAMMING ASSIGNMENT I

I.1 INTRODUCTION

To learn Fortran, you must write programs. The short programming exercises in the problem sections after each chapter are meant to illustrate specific elements of Fortran; but to really develop skill in Fortran programming, you must construct more complicated programs. A moderately long program is usually much more challenging than several short programs. There are eight major programming assignments in this text. Each is designed so that it can be completed in a week or so and is sufficiently challenging to be interesting.

In addition, these programming assignments will be used to familiarize you with some of the methods of engineering analysis associated with the subfields of engineering. Each branch of engineering uses the computer to solve a variety of problems, many of which can be understood by a novice and can provide a good illustration of the ideas used in that area of engineering. Each of the programming assignments is constructed in such a way that the understanding of the background material concerning engineering concepts is not essential to the problem's solution. Also, each major programming assignment will begin with a sample problem similar to the assignment. The sample problem is completely solved and can be used as a model for constructing your programs.

I.2 SAMPLE PROGRAM

Civil Engineering: Pressure Drop in a Fluid Flowing Through a Pipe

Background: When an incompressible fluid is pumped at a steady rate through a pipe from point 1 to point 2, the pressure drop is given by

94

$$dP = P_1 - P_2$$
$$= \rho(gh + W)$$

where W is the energy lost per kilogram due to internal friction in the fluid and with the pipe walls. The fluid density is ρ and the gravitational acceleration is g. All units are SI (i.e., mks). The expression for the energy loss is

$$W = \frac{4fv^2L}{D}$$

where L = pipe length (m)

D = pipe diameter (m)

v = velocity of fluid flow (m/sec)

f = friction factor

Q = fluid flow rate (m^3/sec)

For smooth pipes the friction factor f depends only on the Reynold's number R.

$$R = \frac{\rho v D}{\mu}$$

where μ is viscosity of the fluid.

If ($R < 2000$), the flow is laminar and $f = 8/R$

If ($R > 2000$), the flow is turbulent and $f = .0395R^{-1/4}$

Problem: Write a program to do the following:

1. Initialize a counter for each run as IRUN = 1
2. Read D, h, ρ, μ, and echo print with labels along with the run number (IRUN).
3. Read Q_{start} and L_{start} and print.
   a. Let $Q = Q_{start}$
   b. Let $L = L_{start}$
      (1) From the values of L, D, and Q determine v (and print).
      (2) Compute R, determine f (print)
      (3) Determine W and dP (print)
      (4) Increment $L = L + L_{start}$ and if L is less than three times L_{start}, return to (1) and repeat the calculation, otherwise
   c. Increment $Q = Q + 0.5Q_{start}$ and if Q is less than two times Q_{start}, return to (b) and repeat the calculation, otherwise
4. Increment IRUN = IRUN + 1 and if IRUN < 3 return to the READ statements and repeat the entire calculation for a different set of parameter values, otherwise STOP.

The output of the program should be neat. Suggested data follows:

Run No.	$\rho(kg/m^3)$	$\mu(kg/m\text{-}sec)$	h	D	Q_{start}	L_{start}
1	1500.	0.03	3.0	0.02	0.01	30.0
2	1500.	0.03	3.0	0.60	0.25	100.0
3	1500.	0.001	25.0	0.60	2.50	100.0

Sample Program Solution The Fortran program for this problem is given below.

```
      PROGRAM FLUID
*--
*--   This program computes the pressure drop for a fluid flowing
*--   through a pipe. The flow is either laminar or turbulent,
*--   depending on the Reynold's number.
*--------------------------------------------------------------------
* Variables
*
      REAL DENSITY,VISCOS,HEIGHT,DIAM,FRICT,L,LSTRT,Q,QSTRT,R,W,VEL,
     +     DP
      INTEGER IRUN
*--
*--            IRUN     -  Run Number
*--            DENSITY  -  Fluid density (kg/m**3)
*--            VISCOS   -  Fluid viscosity (kg/m-sec)
*--            DIAM     -  Pipe Diameter (m)
*--            HEIGHT   -  Distance fluid is pumped above original
*--                        position
*--            LSTRT    -  Length of pipe for first calculation (m)
*--            QSTRT    -  Flow rate in first calculation
*--            L        -  Current pipe length
*--            Q        -  Current flow rate
*--            VEL      -  Velocity of flowing fluid
*--            FRICT    -  Pipe friction factor (dimensionless)
*--            W        -  Energy loss/kg due to friction
*--            R        -  Reynold's number
*--            DP       -  Pressure drop
*--------------------------------------------------------------------
* Initialization
*--
      PI = ACOS(-1.)
      IRUN = 1
    1 PRINT *,'RUN NUMBER = ',IRUN
         PRINT *,'  '
         PRINT *,'ENTER DIAMETER,HEIGHT,DENSITY,VISCOSITY'
         READ *,DIAM,HEIGHT,DENSTY,VISCOS
         PRINT *,'DENSITY = ',DENSTY,'  VISCOSITY = ',VISCOS,
```

```
        +                    '     DIAMETER = ',DIAM,'  HEIGHT = ',HEIGHT
                 PRINT *,'ENTER QSTART AND LSTART'
                 READ *,QSTRT,LSTRT
                 PRINT *,'Q-START = ',QSTRT,' AND L-START = ',LSTRT
                 PRINT *,'  '
*------------------------------------------------------------------------
*------------------------------------------------------------------------
* Calculation
                 Q = QSTRT
   2             PRINT *,'  '
                 PRINT *,'  '
                 PRINT *,'  THIS CALCULATION IS FOR A FLOW RATE = ',Q
                 L = LSTART
*--
*--              The velocity of flow is determined from
*--              Q and DIAM
*--
                 VEL = Q/(PI * .25 * DIAM**2)
*--
*--                  Next compute the Reynold's number
*--                  and the friction factor
*--
                 R = DENSTY * VEL * DIAM/VISCOS
                 IF(R .LT. 2000.)THEN
                         FRICT = 8./R
                 ELSE
                         FRICT = 0.395/R**.25
                 END IF
                 PRINT *,'       VELOCITY = ',VEL
                 PRINT *,'  FRICTION FAC. = ',FRICT
                 PRINT *,'   REYNOLDS NO. = ',R
                 PRINT *,'  '
*--
*--                  Compute the energy loss (W) and
*--                  the pressure drop (DP). Also print
*--                  table heading for L loop.
*--
                 PRINT *,'  '
                 PRINT *,'   PIPE       ENERGY       PRESSURE'
                 PRINT *,'  LENGTH      LOSS/KG        DROP'
   3             W = 4. * FRICT * VEL**2 * L/DIAM
                         DP = DENSTY * (9.8 * HEIGHT + W)
*--
*--                  Print intermediate results
*--
*--                  PRINT *,'  ',L,'      ',W,'      ',DP
```

Continued

```
*--
*--                                             Increment the pipe length
*--                                             and loop back
*--                                             to stmt. 3
*--

                           L = L + LSTRT
                   IF(L .LT. 4. * LSTRT)GO TO 3

*--
*--                                             We have completed the calculations for
*--                                             all the desired lengths. Next increment
*--                                             the flow rate and loop back to statement
*--                                             2 to repeat the calculation.
*--

                           Q = Q + .5 * QSTRT
                   IF(Q .LT. 2.*QSTRT)GO TO 2
*--
*--                                             This calculation is complete. Next loop
*--                                             back to stmt 1 for the next data set.
*--

                           IRUN = IRUN + 1
                   IF(IRUN .LE. 3)THEN
                           GO TO 1
                   ELSE
                           PRINT *,'CALCULATION COMPLETED -- JOB TERMINATED'
                           STOP
                   END IF
                   END
```

The output from this program is

```
RUN NUMBER = 1


ENTER DIAMETER,HEIGHT,DENSITY,VISCOSITY
0.02,  3.0,  1500.0,  0.03
DENSITY = 1500.000  VISCOSITY = 3.00000E-02 DIAMETER = 2.00000E-02   HEIGHT = 3.0000
ENTER QSTRT AND LSTRT
0.01,  30.0
Q-START = 1.00000E-02   AND L-START = 30.00000


   THIS CALCULATION IS FOR A FLOW RATE = 1.00000E-02
        VELOCITY  = 31.83099
    FRICTION FAC. = 2.9572263E-02
      REYNOLDS NO. = 31830.99
```

```
   PIPE          ENERGY         PRESSURE
  LENGTH         LOSS/KG          DROP
 30.00000       179777.8      2.6971080E+08
 60.00000       359555.6      5.3937750E+08
 90.00000       539333.4      8.0904421E+08


 THIS CALCULATION IS FOR A FLOW RATE = 1.5000000E-02
        VELOCITY = 47.74648
  FRICTION FAC. = 2.6721556E-02
   REYNOLDS NO. = 47746.48


   PIPE          ENERGY         PRESSURE
  LENGTH         LOSS/KG          DROP
 30.00000       365507.1      5.4830470E+08
 60.00000       731014.1      1.0965653E+09
 90.00000       1096521.      1.6448259E+09
RUN NUMBER = 2


ENTER DIAMETER,HEIGHT,DENSITY,VISCOSITY

          <etc.  Output continues for two more runs>


CALCULATION COMPLETED -- JOB TERMINATED
```

You should note that there are three nested loops in this program.

```
        IRUN = 1
   1    PRINT ... ─────────────────────────────┐
                                                │
        Q = QSTRT                               │
                                                │
   2        PRINT ... ──────────────────┐       │
            L = LSTRT                    │       │
                                         │       │
   3            VEL = ... ───────┐       │       │
                                 │       │       │
                L = L + LSTRT    ↑  ↑  ↑ │       │
                IF(...) GO TO 3 ─┘       │       │
                                         │       │
            Q = Q + .5 * QSTRT           │       │
            IF(...) GO TO 2 ─────────────┘       │
                                                 │
        IRUN = IRUN + 1                          │
        IF(...) GO TO 1 ─────────────────────────┘
```

The Fortran code is made somewhat more readable by indenting each layer of loop.

The output from this program is not particularly readable or attractive. This will be corrected in Chapter 6 when we discuss formatted output.

I.3 PROGRAMMING PROBLEMS

Mechanical Engineering

Generally speaking, mechanical engineers are concerned with machines or systems that produce energy or its application. The spectrum of technological activities that carries the banner of mechanical engineering is probably broader than any other engineering field. The field can be roughly subdivided into:

Power: The design of power-generating machines and systems such as boiler-turbine engines for generating electricity, solar power, heating systems, and heat exchangers.

Design: Innovative design of machine parts or components from the most intricate and small to the gigantic. For example, mechanical engineers work alongside electrical engineers to design automatic control systems such as robots.

Automotive: Design and testing of transportation vehicles and the machines to manufacture them.

Heating, Ventilation, Air Conditioning, and Refrigeration: Design of systems to control our environment both indoors and out, pollution control.

The mechanical engineer usually has a thorough background in subjects like thermodynamics, heat transfer, statics and dynamics, and fluid mechanics.

Programming Problem A: Most Cost-Efficient Steam Pipe Insulation

When deciding on the amount of insulation to be installed on a long steam supply line (see Figure I-1), the amount of money to be saved from lower fuel bills must be compared with the initial insulation purchase and installation costs; that is, excessive insulation can be just as wasteful as too little. This problem will allow you to determine the thickness of the insulation that will give the greatest savings for the least cost. The heat flow through the insulation is given by

(I.1)
$$Q_1 = 2\pi kL \frac{(T_a - T_b)}{\ln(b/a)} \quad \text{(watt)}$$

While the heat transfer from the insulation to the air is given approximately by

(I.2)
$$Q_2 = 2\pi bF(T_b - T_{air})L \quad \text{(watt)}$$

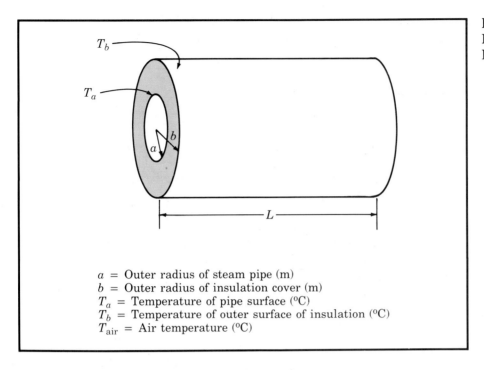

Figure I-1
Insulated Steam
Pipe

a = Outer radius of steam pipe (m)
b = Outer radius of insulation cover (m)
T_a = Temperature of pipe surface (°C)
T_b = Temperature of outer surface of insulation (°C)
T_{air} = Air temperature (°C)

where k = Thermal conductivity of the insulation
    = 0.1 watt/(m °C)

F = Convection coefficient for the air-insulation interface
    = 3.0 watt/(m² °C)

In a steady-state situation $Q_1 = Q_2$, so that T_b can be eliminated from Equations (I.1) and (I.2), and by combining the two equations we obtain

$$Q = \left[\frac{2\pi bkFL}{k + bF \ln(b/a)} \right] (T_a - T_{air}) \tag{I.3}$$

(Verify that this is true.)

Now for the costs. The pipe insulation costs $325.00 per cubic meter ($C_{vol}$ = 325.0) and the installation costs amount to $1.50 per meter of pipe ($C_L$ = 1.50), independent of thickness. The cost of heat is 0.4 cents per kilowatt-hour or $1.11 × 10⁻⁹ per watt-sec (C_{heat} = 1.11 × 10⁻⁹). Assuming a pipe of length L, the volume of the insulation used is $\pi(b^2 - a^2)L$ and the total insulation cost is

$$C_{insul} = \pi(b^2 - a^2)LC_{vol} + LC_L \tag{I.4}$$

To obtain the amount of fuel savings we need the difference between the heat loss with no insulation, i.e.,

$$Q_3 = 2\pi aF(T_a - T_{air})L \tag{I.5}$$

and the heat loss with insulation [Equation (I.3)], i.e., $dQ = Q_3 - Q$, or

(I.6)
$$dQ = Q_3 \left[1 - \frac{b/a}{1 + (bF/k)\ln(b/a)} \right]$$

(Verify this equation.)

The fuel savings then over a 5-year period $(1.578 \times 10^8 \text{ sec})$ is then

(I.7)
$$C_2 = dQ(1.578 \times 10^8)C_{\text{heat}}$$

The outer radius of the pipe is 5 cm $(a = 0.05)$ and insulation is available in thicknesses t ranging from 1 to 10 cm in 1-cm steps (i.e., $b = a + t$, $t = 1$, $2, \ldots, 10$ cm). For air temperatures of $T_{\text{air}} = -10°$ C and $+10°$ C, determine the most cost-effective thicknesses of insulation.

Details

Write a program that will

1. Read the REAL variables

A	- pipe radius	= 0.05 m
L	- pipe length	= 500.0 m
TP	- pipe temperature	= 150.0° C
TAIR	- air temperature	= −10.0° C
RK	- conductivity constant	= 0.10 watt/(m °C)
RF	- convection constant	= 3.0 watt/(m² °C)
CVOL	- cost per volume of insulation	= $325.0
CL	- cost per meter for installation	= $1.50
CH	- cost per kilowatt-hour for heat	= $0.004

and echo print them all with labels.

2. Read the air temperature and print table headings of the form

```
FOR RUN NUMBER xxx
THE AIR TEMPERATURE IS xxx (OC)

AND THE RESULTS OF THE COST-EFFECTIVENESS
COMPUTATIONS ARE

THICKNESS        INSULATION        SAVINGS OVER
 (METERS)           COST            FIVE YEARS
```

3. Start with the insulation thickness, T = 0.01 and calculate the total cost

of insulation [Equation (I.4)], and the dollar savings over 5 years [Equation (I.7)]. Print these quantities as one line in the table. (Be sure to convert CH to cost per watt-second.)

4. Increment the thickness by 0.01 (T = T + 0.01), and if T is less than or equal to 0.10 GO TO step 3 and repeat the calculation. [IF(T .LE. 0.10)GO TO. . . .]

5. When T > 0.10 the above IF test fails and the program should then GO TO step 2 and read a second value for the air temperature (+10.00° C). The remaining parameters stay the same.

6. Add a run counter (IRUN) to the program that is incremented after each complete set of calculations. The program should STOP if IRUN > 2.

7. By inspecting the printed output from your program, determine the most cost-effective insulation thickness for each of the two air temperatures. Indicate this optimum thickness in pencil on your output. Include a flowchart with your program.

Civil Engineering

The field of civil engineering is primarily concerned with large-scale structures and systems used by a community. A civil engineer designs, constructs, and operates bridges, dams, tunnels, buildings, airports, roads, and other large-scale public works. Civil engineers are also responsible for the effects on society and the environment of these large-scale systems. Thus, civil engineers are involved in water resources, flood control, waste disposal, and overall urban planning. The field can be subdivided into

Structures: Design, construction, and operation of large-scale edifices such as dams, buildings, and roads. The properties of materials, geology, soil mechanics, and statics and dynamics are important elements of the background training. For example, how tall a building can be constructed before it will buckle under its own weight is a question involving all of these subjects.

Urban Planning: Planning, design, and construction of transportation systems (roads, railroads, river development, airports) and general land use. Surveying and mapmaking are necessary skills.

Sanitation: Waste treatment, water supply, and sewage systems. Fluid mechanics, hydrology, pollution control, irrigation, and economics are important considerations.

Programming Problem B: Oxygen Deficiency of a Polluted Stream

The determination of the variation with time of the dissolved oxygen in a polluted stream is important in water resources engineering. Organic matter in sewage decomposes through chemical and bacterial action. In this process, free oxygen is consumed and the sewage is deoxygenated. A standard procedure for determining the rate of deoxygenation of sewage involves diluting a sewage sample with water containing a known amount of dissolved oxygen and measuring the loss in oxygen after the mixture has been maintained at a temperature of 20° C for a period of 20 days. This loss is called the first-stage biochemical oxygen demand (B_{20}). The subscript refers to the temperature. For any temperature T, the first-stage demand B_T can be computed from

(I.8)
$$B_T = B_{20}(0.02T + 0.6)$$

As mentioned, when sewage is discharged into a stream, oxygen is consumed in the decomposition of organic matter. At the same time, oxygen is absorbed from the air. However, deoxygenation and reoxygenation take place, in general, at different rates. Usually, reoxygenation lags behind deoxygenation, the dissolved oxygen decreases with time, reaches a minimum, and then increases. As the dissolved oxygen is at a minimum, the oxygen deficit is at a maximum.

The oxygen deficit of the polluted stream may be computed from

(I.9)
$$D(t) = \frac{K_d B_T}{K_r - K_d}(e^{-K_d t} - e^{-K_r t}) + D_0 e^{-K_r t}$$

where $D(t)$ = Oxygen deficit of the stream at time t (mg/L)

K_d = Coefficient of deoxygenation

K_r = Coefficient of reoxygenation

D_0 = Initial oxygen deficit (mg/L)

B_T = First-stage biochemical oxygen demand of steam at temperature T (mg/L)

t = Elapsed time in days (when $t = 0$, $D(t) = D_0$)

The constants K_r and D_0 are known and tabulated for this stream; however, K_d depends on the amount and type of sewage dumped into the stream. If $K_d(20)$ is the measured coefficient of deoxygenation at 20° C, then the value at a temperature T is given by

(I.10)
$$K_d(T) = K_d(20)(1.047)^{T-20}$$

Furthermore, The first-stage biochemical oxygen demand of the mixture of stream plus pollutants is given by

(I.11)
$$B_{20} = [(B_{20,\text{upstream}})Q_s + (B_{20,\text{sewage}})Q_{\text{sewage}}]/Q$$

where $B_{20,\text{upstream}}$ = First-stage biochemical oxygen demand of stream above point at which the sewage is discharged

$B_{20,\text{sewage}}$ = First-stage biochemical demand of sewage

Q_s = Flow rate of the stream

Q_{sewage} = flow rate of the sewage

Q = Net flow rate of sewage plus stream

$ = Q_s + Q_{\text{sewage}}$

Thus using Equations (I.11) and (I.8) we can calculate B_T for the stream at the temperature T.

Details

Write a Fortran program that will

1. For each case, read in and echo print the parameters listed in Table I-1.
2. For each case compute the oxygen deficit $D(t)$, starting at $t = 0$ days and continuing in time steps of 0.1 days until the oxygen deficit reaches a maximum and then decreases. That is, stop if the current value is less than the previous one.
3. Print out the results for the three separate cases. In your printout each case is to be identified as CASE1, CASE2, etc. Where appropriate, include units in your labels. Avoid excessive recalculations.

Case no.	Q_{stream} (L/sec)	Q_{sewage} (L/sec)	D_0 (mg/L)	$K_d(20)$	K_r	$B_{20,\text{stream}}$ (mg/L)	$B_{20,\text{sewage}}$ (mg/L)	T (°C)
1	1500	150.0	1.3	0.23	0.50	0.8	145.0	17.6
2	2000	150.0	1.3	0.23	0.50	0.8	145.0	17.6
3	1500	150.0	1.3	0.23	0.50	0.8	145.0	21.1

Table I-1 Input parameters for pollution problem.

Programming Problem C: Cost Comparisons for Purchase of a Fleet of Cars (Industrial Engineering)

The increasing cost of automobile fuel has caused a surge in the demand for smaller cars with greater fuel economy. A counter argument being used for larger cars with greater fuel use is based on the economics of a lower long-term maintenance cost for the larger automobile.

The executive officer of a large company is faced with the decision of selecting between two automobiles that will constitute a fleet purchase of over 8000 cars to be used by the company over the next four years. He turns to the engineering department for guidance. To provide a recommendation, the group must first restate the problem in mathematical terms amenable to a computation. Since different user groups in the company will drive different distances, it is necessary to determine an estimate of the annual cost for a variety of cars as a function of the distance driven.

The annual cost of a car is composed of three parts: a) Loan repayment (LOAN), b) Operating costs (OPERC), and c) Maintenance costs (MAINT).

$$ACOST = LOAN + OPERC + MAINT$$

Loan Repayment: The loan payment depends on the amount borrowed (the principal (P), the yearly interest rate (r), and the number of years of the term of the loan $(n = 4)$. The equation for the amount of the loan repayment is

(I.12)
$$Loan = P\frac{r}{1 - (1 + r)^{-n}}$$

Operating Costs: The operating cost of a car depends on the distance driven, d (in kilometers), the fuel cost (dollars/liter, labeled FUELC), and the fuel use rate (km/l). The fuel use rate for city driving will be labeled QCITY and that for highway driving, QHGHW. From reviewing company records and from surveys, it is known that the first 10,000 kilometers (labeled d_0) of annual driving is about 70% city and 30% highway. Any additional distance $(d - d_0)$ is about 20% city and 80% highway. The fuel operating costs are then expressed as

(I.13)
$$Operc = \left[\frac{d_0}{0.7Q_{city} + 0.3Q_{hghw}} + \frac{d - d_0}{0.2Q_{city} + 0.8Q_{hghw}}\right]Fuelc$$

Maintenance Costs: The average maintenance costs for a large fleet of automobiles generally depend on the distance driven (d). The maintenance costs increase dramatically with distance driven and are essentially zero while the car is under warrantee $(d < 20,000$ km$)$. For distances beyond the warrantee limit the following expression is found to roughly summarize the maintenance costs:

(I.14)
$$Maint = M_0(d - 20,000)^\beta \quad \langle\text{for } d > 20,000 \text{ km}\rangle$$

The two parameters M_0, β are known for each type of car.

Annual distance driven by automobiles in the company fleet range from 10,000 km to 25,000 km. The problem is to determine the annual cost for each of two cars for the first-through-fourth years and the total four-year cost for each car. This is to be done for a range of annual yearly driving distances. The parameters to use in this problem are given in Table I-2.

Variables		Automobile VW-Rabbit	Type Gran Prix
P	—Original cost	7250.00	9625.00
R	—Annual interest rate	0.12	0.12
N	—Length of loan	4	4
QCITY	—Fuel use rate (city)(km/l)	10.6	7.1
QHGHW	—Fuel use rate (highway)	16.0	10.2
FUELC	—Fuel cost (dollars/liter)	0.55	0.55
M_0	—Maintenance cost coefficient	0.95	15.84
β	—Maintenance cost exponent	0.70	0.333

Table I-2
Values of parameters to use in the annual car cost problem.

Details

Your program should:

1. Type-declare all the variables used in the problem.
2. Read the parameters common to both calculations—for example, R, N, FUELC.
3. Initialize a counter called CASE that will distinguish the two calculations.
4. Read the name and the parameters associated with the first car. Print these quantities with appropriate labels.
5. Compute the loan payments for this car.
6. Start with $d = d_0 = 10,000$ km/yr. Compute the first-through-fourth-year cost and the total cost for this car. Print these along with the operating and maintenance costs.
7. Increment $d \rightarrow d + 1000$.

```
        IF(d ≤ 25,000.)THEN
                repeat the computation and print
        ELSE
                increment the CASE counter and
                    IF(Case < 2)then
                            return to the READ for the next
                            car parameters
                    ELSE
                            PRINT a termination statement
                            and STOP
                    END IF
        END IF
```

8. After a successful computer run, inspect the printed output and determine the range, if any, in the annual driving distance where the car with the greater fuel economy also has the lower total four-year cost.

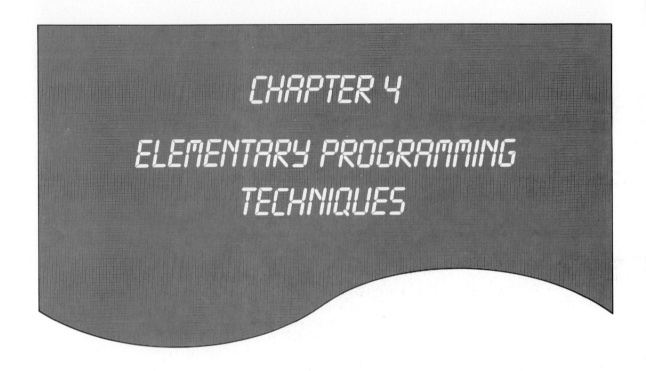

CHAPTER 4

ELEMENTARY PROGRAMMING
TECHNIQUES

4.1 INTRODUCTION

With the introduction of the flow control structures to our Fortran vocabulary, a vast array of problems become amenable to our analysis. It is the purpose of this chapter to concentrate on a few elementary techniques that form the basis for a great many computational algorithms. In Section 4.3 we will discuss methods for scanning a list or for monitoring a function to find the maximum or minimum. In Section 4.4 the computer algorithm for summations is described. Each of these procedures will frequently be used in subsequent, more elaborate programs.

4.2 THE STATEMENT FUNCTION

As your computer programs become more complicated, the need for clear and concise Fortran code becomes more critical. In algebraic problems this is achieved by symbolically replacing a collection of frequently occuring operations by a function, $f(x)$, representing the operations. One manner of accomplishing this in Fortran is by means of the statement function, which is discussed in this section.

4.2.1   Statement Function Definition

The form of the Fortran statement defining a statement function is

FNC(A,B,X) = Any algebraic combination, including intrinsic functions, of the variables A, B, X, and constants

The name of the function (here FNC) may be any valid Fortran name. It is strongly suggested that any and all variables appearing on the right side of the replacement operator should also appear as one of the variables in the argument list (here, A,B,X), although this is not required.

Two examples follow:

1. $f(x) = 35x^2 - \sin(\pi x) - \sqrt{x}$ would be written as a statement function

```
FNC(X) = 35.0 * X * X - SIN(3.14159 * X) - SQRT(X)
```

2. The function for $x^3 + ax^2 + bx + c$ could be written

```
CUBE(A,B,C,X) = X**3 + A * X**2 + B * X + C
```

The type of the value returned by the function is determined by the type of the name of the function. Thus with default typing, F(X) would be REAL and N(X) would be INTEGER. The function name can also be explicitly typed by including it in a prior type statement.

> *Important:* **A statement function definition is not considered an executable statement and it must appear before the first executable statement and after the type declarations (i.e., INTEGER, REAL, and CHARACTER).**

4.2.2   Use of the Statement Function

The definition of the statement function given above allows us almost limitless possibilities for constructing our own set of intrinsic-like functions in addition to those listed in Table 2-3. Statement functions are often used to reduce long and complicated program statements and make the code easier to read. This in turn makes the program much less prone to error. A statement function is used, or *referenced,* in precisely the same manner that intrinsic functions are referenced. After the function has been introduced in the defining line, the name of the function, along with its argument list, can be used in any arithmetic expression or executable Fortran statement.

For example, when the need arises in a program to calculate the square root of a number, we could insert the line defining a statement function called SQROOT that employs one iteration of Newton's algorithm to compute $\sqrt{C}$ using an initial guess X0 (see Section 3.4.2).

```
SQROOT(C,XO) = 0.5 * (XO + C/XO)        (statement function
                                         definition)
```

The code using this function to obtain a rough value for $\sqrt{7}$. then might be

```
X = SQROOT(7.,2.)
```

or a better value would be

```
X = SQROOT(7.,2.)
X = SQROOT(7.,X)
```

Of course a much simpler and more accurate approach would be to use the intrinsic function SQRT(C), which, incidentally, also uses Newton's algorithm.

When the compiler encounters a Fortran name followed immediately by a parenthesis, as in

```
SQROOT(C,XO)
```

a search is initiated for the definition of the name SQROOT. The compiler has been written to recognize that a "name–left parenthesis" structure like FA(C + D) *does not* imply multiplication but instead that the name FA must have a special meaning. To determine precisely what is represented by FA, the compiler looks at the beginning of the program to see if this name has been defined as a statement function. If so, a value is computed for FA as defined by the code in the statement function and is returned to the line in the program that contained the reference to FA. The program then continues from that point. If FA is not found at the top of the program, a search of the library is made, and if a function named FA is found, it is linked with the main program and that function is used whenever the name is referenced. If the function is not found among the intrinsic functions an execution time error results.

Notice, this implies that if we had named our square root function SQRT[1], the program would use that function and not the intrinsic function, since it finds ours first. Obviously, to minimize confusion you should try to avoid defining functions with names identical to intrinsic functions.

---

[1]A word of caution: In some installations of Fortran 77 it is forbidden to define statement functions that have the same name as existing intrinsic functions; and in all circumstances it is flagrantly poor programming style.

4.2.3 The Argument List

The variable names that appear in the definition of a statement function are called *dummy* arguments, meaning that they are not to be thought of as representing numerical quantities in the same sense as ordinary Fortran variable names. That is, instead of designating a location in memory for a numerical value, a dummy argument name merely reserves a position in the arithmetic procedure defined in the statement function. Only later, when the statement function is referenced in the program, will an actual numerical value be inserted in the function expression. This is analogous to the ordinary symbolic manipulation of variable names in algebra.

When the function is referenced, the names, numbers, or arithmetic expressions that appear in the arguments list are called *actual* arguments, implying that they are then expected to have numerical values at that point in the program. Consider the following code:

```
F(A,B) = (A**2 - B**2) * PI
PI = ACOS(-1.)
X = F(3.,2.)                        Output
Y = F(X + 1.,X)
PRINT *,'X= ',X,' Y= ',Y            X=15.7080  Y=101.8376
                                    (i.e., x = 5π, y = 10π² + π)
PRINT *,'A= ',A,' B= ',B            Output is unpredictable
```

The variables A and B were never *assigned* values in this program. The statement function definition merely defines a procedure for symbolically manipulating the variables in its argument list. Thus, in the first reference, the value of F will be determined after replacing the symbols A by 3.0 and B by 2.0. Notice that at this point in the program PI has been assigned a value and since it will not be changed it need not appear in the argument list. The value used for PI will be whatever value currently resides in that memory location. In the second reference to F, the arithmetic expressions in the argument list are first evaluated before the function expression is actually used.

In summary, actual arguments represent numbers that are transferred to the dummy arguments in the statement function definition. The transfer is determined by the *position* in the argument list, not by the name of the variable. That is, in the example above, the value associated with B is whatever number appears in the second position when the function F is referenced. In addition, the argument list in any reference to a statement function must agree in *number* and *type* with the argument list in the function's definition. In our example this means that any reference to function F must have two numbers, variable names, or expressions representing numbers in the argument list and both must be of type REAL. We will return to considerations regarding dummy and actual arguments in Section 9.3.2.

4.3 FINDING THE MINIMUM AND MAXIMUM OF A SET OF NUMBERS

A very common and useful programming procedure is a code that will scan a list of numbers and determine the minimum and/or the maximum number in the list. The basic algorithm simply mimics the steps you or I would follow in scanning a list of numbers. Starting at the top of the list we define the current maximum to be the first number. We then check the next element in the list. If this number is larger than the current maximum, it becomes the current maximum and we discard the earlier value. If not, the current maximum is still valid and we proceed to the next number in the list. The Fortran code to accomplish this would be

```
        READ *,X
        XMAX = X
   1    READ *,X
   *--
   *--                      COMPARE CURRENT X AND XMAX
   *--                      IF NEED BE REDEFINE XMAX
   *--
        IF(X .GT. XMAX)XMAX = X
        IF(X .LT. 0.)THEN
   *--
   *--                      END OF LIST MARKED WITH
   *--                      NEGATIVE VALUE OF X
   *--
            PRINT *, 'MAXIMUM VALUE = ',XMAX
            STOP
        ELSE
            GO TO 1
        END IF
   END
```

The most important statement in the program is the IF test where the current maximum (XMAX) is compared with the most recently read value of X. If X is greater than the current maximum, the current maximum is redefined.

This method can also be used to find the maximum or minimum of a function of a single variable x over some interval, $a < x < b$. A program to accomplish this is illustrated in the following example.

4.3.1 Example—Minimizing the Costs

In an automobile manufacturing plant, door handles are made in large lots, placed in storage, and used as required in assembling the cars. A common problem is the determination of the most economical lot size x. The following parameters enter into the considerations.

w = work days per year   = 242   days/yr

s = setup cost to produce one
     lot of door handles   = 947.0   dollars/lot

u = usage rate (door handles/day) = 190.0   no./day

m = material + labor costs
     per door handle   =   2.05   dollars

a = annual storage costs   =   9.65   dollars/yr

Thus the number of lots used per year is

$$\frac{w\,(\text{day/yr})u\,(\text{no./day})}{x\,(\text{no./lot})} = \frac{wu}{x}\,(\text{lots/yr})$$

and so the yearly equipment setup cost is

$$\frac{swu}{x} \quad (\text{dollars/yr})$$

The storage costs increase with the size of a lot and the annual cost is ax. Finally, the annual production cost is given by

$$umw \quad (\text{dollars/yr})$$

Adding all the costs we obtain the following expression

$$c(x) = \frac{swu}{x} + umw + ax$$

We next turn to the computer to find the minimum of this expression as a function of the lot size x. The procedure will be built around the algorithm given earlier for finding the minimum and/or maximum of a list of numbers.

The Fortran program to determine the lot size that results in a minimum overall cost would be constructed along the following lines:

Pseudocode description of optimum lot size code.
_____

Type all variables as they are encountered in writing the code
    REAL X, W, S, U, M, A, COST, CC, CZERO
    INTEGER STEP

Statement function for costs
    COST(X) = . . .

_____

Initialize:
    X—starting x value
    STEP—step counter
    CZERO—current minimum
        [start with COST(X)]

DoWhile (costs are still decreasing)

1   Increment X → X + 10.
        STEP → STEP + 1

      Compare COST (new X) with COST (previous X)
        If latest COST is less, THEN
            proceed to next STEP
            GO TO 1
      ELSE
            We have found the minimum, print the results for X,
            COST (X) and STOP
      END IF

END

The results of a Fortran program constructed from this pseudocode outline are shown in Figure 4-1.

Figure 4-1 The output of a program that computes the optimum lot size.

```
INPUT WORK DAYS/YR AND USAGE RATE 242., 190.
INPUT COSTS FOR SETUP, MATERIAL, AND STORAGE 947.0, 2.05, 9.65


    A COMPUTATION OF THE OPTIMUM LOT SIZE

            INPUT PARAMETERS
                    WORK DAYS/YEAR        --    242.
                    USAGE RATE PER DAY  --    190.
                    SETUP                 -- 947.00 DOLLARS
                    MATERIAL-LABOR        --    2.05 DOLLARS/ITEM
                    ANNUAL STORAGE        --    9.65 DOLLARS/ITEM



    THE OPTIMUM LOT SIZE IS   2120.0

    ANNUAL COSTS FOR THIS LOT SIZE ARE 135256.10 DOLLARS
```

4.3.2 A Reconsideration of the Optimum Lot Size Problem

The previous program is an example of one of the most common errors in programming: Computing without first carefully thinking the problem through. Even though the program finds a valid estimate of the optimum lot size, and does so with relative ease, the calculation can be greatly improved in both speed and accuracy with the introduction of more intelligence into the algorithm. The expression obtained for the cost was a simple function of the lot size and is graphed in Figure 4-2.

At the lot size corresponding to the minimum cost, the curve has a tangent line that has zero slope; i.e., it is horizontal. Since the slope of the tangent line at a point is equal to the derivative, we have as the requirement for the minimum point on the curve

$$\frac{dc(x)}{dx} = 0 = swu \frac{d}{dx}\left(\frac{1}{x}\right) + 0 + a \frac{d(x)}{dx}$$

$$= a - \frac{swu}{x^2} \qquad \left(\frac{d}{dx}x^{-1} = -x^{-2}\right)$$

Figure 4-2 Annual production costs (c) as a function of lot size (x).

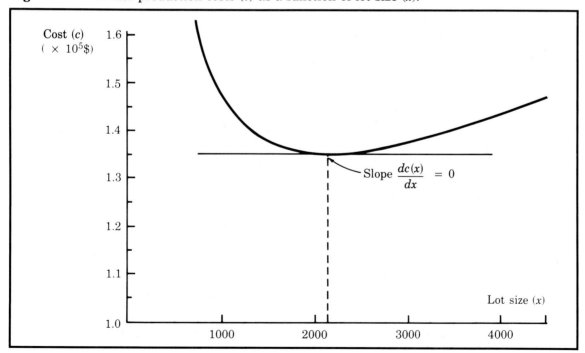

Solving this equation for x yields the optimum lot size x_{best}

$$x_{\text{best}} = \left(\frac{swu}{a}\right)^{1/2} = 2124.202 \to 2124 \qquad \langle \textit{truncated to an integer} \rangle$$

with no need for a computer at all. This example illustrates the fundamental tenet of all of numerical analysis on a computer: Intelligence is more important than computing power. It is not at all rare to replace a computer program that costs thousands of dollars per run with an improved version that obtains more accurate results for only pennies. This is done by programmers experienced in the analysis of the problem as well as in writing Fortran code.

4.4   PERFORMING A SUMMATION

Perhaps the single most useful task a computer is called upon to handle is to quickly perform a summation. The ability to transcribe the algebraic equations for summations in Fortran code is an absolutely essential element in developing programming skills.

4.4.1   The Structure of the Summation Algorithm

If we represent the current value of the sum as SUM and the current value of an individual term in the sum as TERM, then the Fortran code for executing a summation consists of four parts.

1. *Initialize*[2] the variables SUM, TERM, and a counter, say K.

```
SUM = 0.0
K = 1
TERM = ...
```

2. *Summation* by the replacement

```
SUM = SUM + TERM
```

---

[2]Some installations of Fortran preset all memory locations to zero before the program begins, but most do not. If a variable that has not been assigned a value within a program is used in an arithmetic expression, the results will be unpredictable. Thus a fundamental commandment of Fortran is: *All* variables that are used in a program *must* have been assigned values earlier within that same program.

---

3. *Increment* the counter and *redefine* TERM

```
K = K + 1
TERM = ...
```

4. *Test* for completion and loop

```
        IF(K .LT. KMAX)GO TO ...
or
        IF(ABS(TERM) .LT. EPS)GO TO
```

4.4.2 Examples of Summations

In Figure 4-3 the Fortran code is illustrated that will compute the sum of the squares of the integers from 1 to 100 (called ISUM) and the sum of the square roots of the integers from 1 to 100 (called XSUM). The algebraic notation for these quantities is

$$i_{sum} = \sum_{i=1}^{100} i^2 = 1^2 + 2^2 + 3^2 + \cdots + 100^2$$

$$x_{sum} = \sum_{x=1}^{100} \sqrt{x} = \sqrt{1} + \sqrt{2} + \sqrt{3} + \cdots + \sqrt{100}$$

Figure 4-3   Fortran program for summations.

```
      PROGRAM SUMS
* ---
* --- THIS PROGRAM PERFORMS TWO SUMMATIONS, ONE USING REAL ARITH-
* --- METIC, THE OTHER INTEGER ARITHMETIC. THERE ARE 100 TERMS IN
* --- EACH SUM.
*-----------------------------------------------------------------
* INITIALIZATION
*
      REAL X,XSUM,XTERM
      INTEGER I,ISUM,ITERM,COUNT

      COUNT = 0
      X = 0.0
      I = 0
      XSUM = 0.0
      ISUM = 0
*
    1 XTERM = SQRT(X)
      ITERM = I * I
*-----------------------------------------------------------------
*            SUMMATION CONSISTS IN THE REPLACEMENT
*                    SUM => SUM + TERM
```

Continued

```
*
         XSUM = XSUM + XTERM
         ISUM = ISUM + ITERM
*
*           INCREMENT COUNTER, X AND I
*
         COUNT = COUNT + 1
         X = X + 1.0
         I = I + 1
*
*           TEST FOR COMPLETION
*
         IF(COUNT .LE. 100)THEN
            GO TO 1
         ELSE
*
*           THE LOOP HAS BEEN EXECUTED 100 TIMES,
*           PRINT THE RESULTS


            PRINT *,' '
            PRINT *,'THE SUM OF THE SQUARE ROOTS OF THE INTEGERS'
            PRINT *,'FROM ONE TO ONE HUNDRED IS'
            PRINT *,'                      ',XSUM
            PRINT *,' '
            PRINT *,'THE SUM OF THE SQUARES OF THE INTEGERS'
            PRINT *,'FROM ONE TO ONE HUNDRED IS'
            PRINT *,'                      ',ISUM
            PRINT *,' '

         END IF

         STOP
         END


         THE SUM OF THE SQUARE ROOTS OF THE INTEGERS
         FROM ONE TO ONE HUNDRED IS
                           671.4629471032

         THE SUM OF THE SQUARES OF THE INTEGERS
         FROM ONE TO ONE HUNDRED IS
                           338350
```

4.4.3 Infinite Summations

A great many functions in mathematics are represented in a form involving the summation of an infinite number of terms. For example, the series expansion for the exponential function can be found in any book of mathematical tables.

$$e^x = 1 + x + \frac{x^2}{2!} + \frac{x^3}{3!} + \cdots + \frac{x^n}{n!} + \cdots = \sum_{n=0}^{\infty} \frac{x^n}{n!} \qquad \textbf{(4.1)}$$

where $n!$ (n factorial) means

$$n! = n(n-1)(n-2) \ldots 2 \times 1$$

That is $2! = 2$, $3! = 6$, etc. Also, $0!$ is defined to be unity. Equation (4.1) is exact only if an infinite number of terms are included in the summation. However, you will notice that eventually the terms in the sum will become extremely small (i.e., $n!$ will be very large) and we may be justified in terminating the summation when this is the case.[3]

The relation between successive terms in this sum can be expressed in terms of their ratio,

$$R = \frac{(\text{term})_{n+1}}{(\text{term})_n} = \frac{x^{n+1}}{(n+1)!} \frac{n!}{x^n} = x \frac{n!}{(n+1)!} \qquad \textbf{(4.2)}$$

Since $(n+1)! = (n+1)n!$, this can be simplified to

$$R = \frac{x}{n+1} \qquad \textbf{(4.3)}$$

This ratio approaches zero as $n \rightarrow \infty$ for *any* finite value of x. The key step in the algorithm to compute the sum is then the line that calculates the *next* term in the series by using Equations (4.2) and (4.3).

$$\texttt{TERM => TERM * RATIO = TERM * (X/(N + 1.))} \qquad \textbf{(4.4)}$$

The Fortran code to calculate e^x for a specific x by summing the series expansion until the absolute value of a term is less than 10^{-6} is given in Figure 4-4. However, I should warn you that this program will have difficulty obtaining accurate results if x is large and negative. (See Problem 4-4.)

As a final example, consider the problem of finding both the maxima and minima and performing a summation in the same program. Specifically, the problem is:

---

[3] The determination of whether or not dropping an infinite number of small quantities is justified comes under the heading of "convergence of series" and is covered in a calculus course. For the present, you can test the convergence of a series numerically by evaluating the summation twice: first using a convergence criterion on the magnitude of individual terms (e.g., $|(\text{term})_n| < \text{EPS}$), and then repeating the summation simply using twice the number of terms in the first calculation. If the two results are roughly the same, you are very likely justified in assuming that the series converges to the computed result plus or minus the difference in the two calculations. (See Problem 4.6.)

Figure 4-4   The Fortran code to evaluate e^x.

```
      PROGRAM ETOX
* ---
* --- THIS CODE EVALUATES THE SERIES EXPANSION FOR EXP(X). THE
* --- SUMMATION IS TERMINATED WHEN ABS(TERM) .LT. 1.E-6. EACH
* --- TERM IS RELATED TO THE PREVIOUS TERM IN THE SUMMATION BY
* ---
* ---               TERM-(N + 1) = TERM-(N) * RATIO
* ---
* --- WHERE RATIO IS AN ALGEBRAIC EXPRESSION FOR THE RATIO OF
* --- TERMS.
*-------------------------------------------------------------
* VARIABLES
* ---
      REAL X,TERM,SUM,RATIO
      INTEGER I
*-------------------------------------------------------------
* INITIALIZATION
* ---
      PRINT *,' '
      PRINT *,'ENTER VALUE OF X'
      READ *,X
* ---
      I = 0
      SUM = 0.0
      TERM = 1.0
* ---
* ---       NOTE: THE TERMS ARE LABELED BY THE POWER OF X. THUS
* ---             THE ZERO-TH TERM (X**0 => 1) IS 1 AND THE I = 1
* ---             TERM IS X.
*-------------------------------------------------------------
* SUMMATION
* ---
    1 SUM = SUM + TERM
* ---
*-------------------------------------------------------------
* REDEFINE TERM
* ---
      RATIO = X/(I+1.0)
      TERM = TERM * RATIO
*-------------------------------------------------------------
* INCREMENT I AND TEST FOR TERMINATION OF SUMMATION
* ---
      I = I + 1
      IF(I .GT. 100)THEN
* ---
* ---          FAILURE PATH--EXCESSIVE NUMBER OF TERMS
* ---
      PRINT *,'PROGRAM FAILED--TERMS NOT SMALL AFTER 100 TERMS'
      STOP
* ---
      ELSE IF(ABS(TERM) .GT. 1.E-6)THEN
* ---
* ---          SERIES NOT YET CONVERGED, PROCEED TO NEXT TERM
      GO TO 1
```

```
*  ---
      ELSE
*  ---
*  ---                    THIS IS THE SUCCESS PATH
*  ---
         PRINT *,'THE COMPUTED VALUE OF EXP(',X,') IS ',SUM
         PRINT *,'THE NUMBER OF TERMS INCLUDED WAS N = ',I
*  ---
      END IF
*  ---
      STOP
      END

ENTER VALUE OF X 1.0
THE COMPUTED VALUE OF EXP(1.) IS 2.718281525573
THE NUMBER OF TERMS INCLUDED WAS N = 10
```

Given a list containing the following information on each line: (a) a student ID number (e.g., Social Security number) and (b) a final exam score, the task is to write a program to compute the average exam score, the minimum score, the maximum score, and the ID numbers of the corresponding students. The program is given in Figure 4-5.

Figure 4-5 A program to compute the best, worst, and average exam scores.

```
      PROGRAM EXAMS
*--
*-- This program will read a set of student ID's and exam scores
*-- entered at the terminal. The list is scanned for the minimum
*-- and maximum scores. The ID of the students with these scores
*-- are then printed. Also the scores are summed and an average
*-- computed.
*------------------------------------------------------------------
* Variables
*--
      REAL AVG
      INTEGER EXAM,BEST,WORST,SUM,IDBEST,IDWRST,ID,N
*--
*--          ID      - Current student ID
*--          EXAM    - Current exam score
*--          BEST    - The maximum exam score
*--          WORST   - The lowest exam score
*--          IDBEST  - The ID of student with best exam
*--          IDWRST  - The ID of student with worst exam
*--          SUM     - Sum of all exam scores
*--          N       - The total number of exams
*--          AVG     - The average of all exams
*------------------------------------------------------------------
* Initialization
*--
      SUM = 0.0
      N = 0
```

Continued

```
          BEST = 0
          WORST = 100
*--
*--            Note, we start with impossible values for best/worst
*--
   1      READ *,ID,EXAM
*--
*--            The End-of-data is marked with a negative ID
*--
          IF(ID .LT. 0)GO TO 99
*------------------------------------------------------------------
* Summation
*--
          SUM = SUM + EXAM
          N = N + 1
*------------------------------------------------------------------
* Determine the maximum and minimum exam score
*--
          IF(EXAM .LT. WORST)THEN
                WORST = EXAM
                IDWRST = ID
          END IF
          IF(EXAM .GT. BEST)THEN
                BEST = EXAM
                IDBEST = ID
          END IF
*--
*--            Return to read next data line
*--
          GO TO 1
*------------------------------------------------------------------
* Output section
*--
  99      AVG = SUM/N
*--
*--            (Note the use of intentional mixed-mode arithmetic)
*--
          PRINT *,'THE AVERAGE OF THE ',N,' EXAMS IS ',AVG
          PRINT *,' '
          PRINT *,'STUDENT ID = ',IDBEST,' HAD THE HIGHEST SCORE OF ',BEST
          PRINT *,'STUDENT ID = ',IDWRST,' HAD THE LOWEST SCORE OF ',WORST
*--
          STOP
          END
```

This program will not handle tie scores. You should attempt to modify the program to cover the situation of a few students tying for the best or worst scores. You will note that the program will very quickly get very complicated.

An important use of the computer's skill in summation is in the numerical evaluation of integrals. The integration of a function of a single variable can be thought of as either the opposite of differentiation—that is, the anti-derivative—or as the area under a curve (Figure 4-6). The former is ordinarily discussed in depth in a calculus course. We shall concentrate here on the second approach.

The interpretation of the integral

$$I = \int_a^b f(x)\, dx$$

as the area under the curve of f vs. x from $x = a$ to $x = b$ lends itself so naturally to numerical computation that the most effective way to understand the process of integration is to learn the numerical approach first and later have these ideas reinforced by the more formal concepts of the anti-derivative.[4]

Most algorithms for numerically evaluating an integral, that is, for estimating the area under a function $f(x)$ from $x = a$ to $x = b$, proceed by dividing the original interval into many smaller subintervals, approximating the area over the subintervals, and then summing these areas.

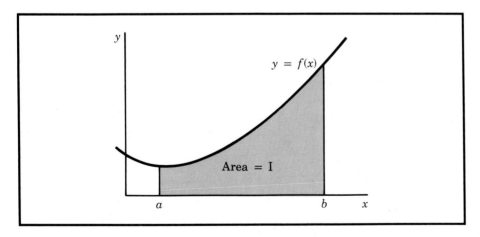

Figure 4-6 The integral as an area under a curve.

[4] This interpretation defines regions where the function falls below the x-axis to have *negative* area. Thus the area "under" the sine function from $\theta = 0$ to $\theta = 2\pi$ is zero.

4.5.1 The Trapezoidal Rule

An approximation to the area under a complicated curve can be obtained by assuming that the function can be replaced by simpler functions over a limited range. A straight line, the simplest approximation to a function, is the first to be considered and leads to what is called the Trapezoidal Rule.

The area under the curve $f(x)$ from $x = a$ to $x = b$ is approximated by the area beneath a straight line drawn between the points (x_a, f_a) and (x_b, f_b). (See Figure 4-7.) The shaded area is then the approximation to the integral and is the area of a trapezoid, which is:

$$I = \langle\text{average value of } f \text{ over interval}\rangle * \langle\text{width of interval}\rangle$$

or

(4.5)
$$I = \tfrac{1}{2}(f_a + f_b)(b - a) \equiv T_0$$

This is the trapezoidal rule for one panel, identified as T_0.

The Formula for the Trapezoidal Rule for _n_ Panels   To improve the accuracy of the approximation to the area under a curve, the interval is next divided in half and the function approximated by straight line segments over each half (Figure 4-8). The area in this case is approximated by the area of two trapezoids.

$$I \sim T_1 = [\tfrac{1}{2}(f_a + f_1)\,\Delta x_1] + [\tfrac{1}{2}(f_1 + f_b)\,\Delta x_1]$$

or

Figure 4-7
Approximating the area under a curve by a single trapezoid.

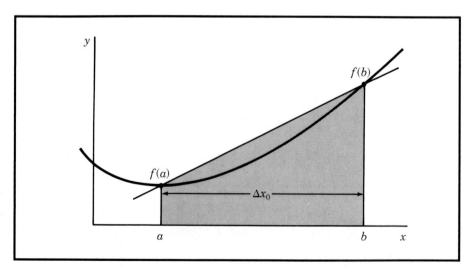

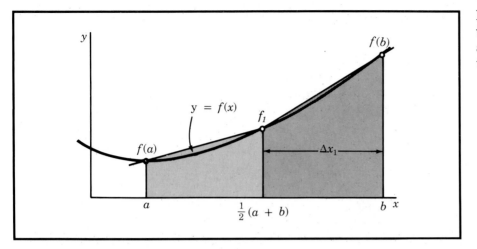

Figure 4-8 A two-panel approximation to the area.

$$T_1 = \frac{\Delta x_1}{2}[f_a + 2f_1 + f_b] \tag{4.6}$$

where

$$\Delta x_1 = (b - a)/2$$

$$f_1 = f(x = a + \Delta x_1)$$

Note that, when the areas of the trapezoids are added, the sides at f_a and f_b are sides of only the first and last trapezoid, while the side at f_1 is a side of two trapezoids and thus "counts twice," explaining the factor of two in Equation (4.6).

Furthermore, the two-panel approximation T_1 can be related to the earlier one-panel result T_0 as

$$T_1 = \frac{T_0}{2} + \Delta x_1 f_1 \tag{4.7}$$

To increase the accuracy further, the interval is simply subdivided into a larger number of panels. The result for n panels is clearly

$$I \sim T_n = \frac{1}{2} \Delta x_n \left[f_a + 2 \sum_{i=1}^{n-1} f_i + f_b \right] \tag{4.8}$$

where $\Delta x_n = (b - a)/n$, and f_i is the function evaluated at all the interior points,

$$f_i = f(x = a + i\,\Delta x_n) \tag{4.9}$$

The reason for the extra factor of two in Equation (4.8) is the same as in the two-panel example.

An Alternate Form of the Trapezoidal Rule   Equation (4.8) was derived assuming that the widths of all the panels are the same and equal to Δx_n. This is not required in the derivation, and the equation can easily be generalized to a partition of the interval into unequal panels of width Δx_i, $i = 1, \ldots, n - 1$. However, for reasons to be explained a bit later I will not only restrict the panel widths to be equal but the number of panels to be a power of two—that is,

$$n = 2^k$$

The number of panels is n, the order of the calculation will be called k, and the corresponding trapezoidal rule approximation will be labeled as T_k. Thus T_0 is the result for $n = 2^0 = 1$ panel. The situation for $k = 2$ or $2^2 = 4$ panels is illustrated in Figure 4-9. In Figure 4-9 the width of a panel is $\Delta x_2 = (b - a)/2^2$ and the value of the $k = 2$ trapezoidal rule approximation is

(4.10) $$T_2 = \frac{\Delta x_2}{2}[f_a + 2f(a + \Delta x_2) + 2f(a + 2\,\Delta x_2) + 2f(a + 3\,\Delta x_2) + f_b]$$

However, since $2\,\Delta x_2 = \Delta x_1$, we see that

(4.11) $$f(a + 2\,\Delta x_2) = f(a + \Delta x_1)$$

and $f(a + \Delta x_1)$ was already determined in the previous calculation of T_1, (Equation (4.7)). The point is, once a particular order trapezoidal rule approximation has been computed, to proceed to the next order trapezoidal rule approximation, the only new information that is required is the evaluation of the function at the *midpoints* of the current intervals. This is, of course, true only if the number of panels is successively doubled in each stage.

Figure 4-9 The four-panel trapezoidal approximation T_2.

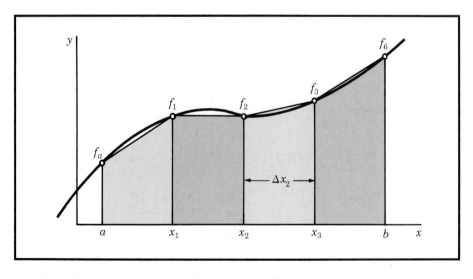

To exploit this fact further, Equations (4.6) and (4.11) can be used to rewrite Equation (4.10) in the form

$$T_2 = \frac{\Delta x_1}{4}[f_a + 2f(a + \Delta x_1) + f_b] + \Delta x_2[f(a + \Delta x_2) + f(a + 3\,\Delta x_2)]$$

$$= \frac{T_1}{2} + \Delta x_2[f(a + \Delta x_2) + f(a + 3\,\Delta x_2)]$$

(4.12)

This can easily be generalized to yield

$$T_k = \frac{1}{2}T_{k-1} + \Delta x_k \sum_{\substack{i=1 \\ i=\text{odd only}}}^{n-1} f(a + i\,\Delta x_k)$$

(4.13)

where

$$\Delta x_k = \frac{b - a}{2^k}$$

The procedure for using Equation (4.13) to approximate an integral by the trapezoidal rule is:

1. Compute T_0 by using Equation (4.5).
2. Repeatedly apply Equation (4.13) for $n = 1, 2, \ldots$ until sufficient accuracy is obtained.

An Example of a Trapezoidal Rule Calculation To illustrate the ideas of this section, consider the integral

$$I = \int_1^2 \left(\frac{1}{x}\right) dx$$

The function $f(x) = 1/x$ can, of course, be integrated analytically to give $\ln(x)$ and since $\ln(1) = 0$, the value of the integral is $\ln(2) = 0.69314718$. The trapezoidal rule approximation to the integral with $b = 2$ and $a = 1$ begins with Equation (4.5) to obtain T_0.

$$T_0 = \frac{1}{2}\left[\frac{1}{1} + \frac{1}{2}\right](2 - 1) = 0.75$$

Repeated use of Equation (4.13) then yields

$$k = 1, \Delta x_1 = 1/2$$

$$T_1 = \frac{T_0}{2} + \frac{1}{2}\left[f\left(1 + \frac{1}{2}\right)\right] = \frac{0.75}{2} + \frac{1}{2}\left(\frac{1}{1.5}\right)$$

$$= 0.708333$$

$$k = 2, \Delta x_2 = 1/4$$

$$T_2 = \frac{T_1}{2} + \frac{1}{4}\left(\frac{1}{1.25} + \frac{1}{1.75}\right)$$

$$= 0.6970238$$

$$k = 3, \Delta x_3 = 1/8$$

$$T_3 = \frac{T_2}{2} + \frac{1}{8}\left(\frac{1}{1.125} + \frac{1}{1.375} + \frac{1}{1.625} + \frac{1}{1.875}\right)$$

$$= 0.69412185$$

Continuing the calculation through $k = 5$ yields

k	T_k
0	0.75
1	0.70833
2	0.69702
3	0.69412
4	0.69339
5	0.693208
. . .	. . .
exact	0.693147 . . .

The convergence of the computed values of the trapezoidal rule is not particularly fast, but the method is quite simple.

The Structure of the Fortran Code Implementing the Trapezoidal Rule A pseudocode outline of the program to obtain a trapezoidal rule approximation for an integral using Equation (4.8) follows. The actual Fortran code is easily constructed from the outline and is not presented here.

Statement function for $f(x)$
F(X) = . . .

READ interval limits (a, b) and
        the number of points (n)

Initialize panel width $[\Delta x = (b - a)/(n - 1)]$

Perform summation $\sum\limits_{i=1}^{n-1} f(a + i\,\Delta x)$

```
    SUM = 0
    I = 1
99 TERM = F(A + I*DX)
        SUM = SUM + TERM
        I = I + 1
        IF (I .LT. N) THEN
            GO TO 99
```

```
      ELSE
            <answer = ½ Δx[f(a) + 2 * SUM + f(b)]>
            PRINT<answer>
            STOP
      END IF
```

The Fortran code for the iterative form of the trapezoidal rule given in Equation (4.13) is more interesting and the pseudocode version of this program follows. You should construct a Fortran program from this outline and use it to evaluate the integral of a variety of test functions.

Statement function for integrand function
$F(X) = \ldots$

READ integration limits (A, B) and the
      maximum order (Imax) of the calculation

Compute the zeroth order (one panel) approximation
      I = initial order = 0
      n = number of panels = 2^I
      $T = \frac{1}{2}(b - a)[f(a) + f(b)]$

1 Increment the order: $I = I + 1$

IF(I .LE. IMAX)THEN

      Number of panels now is $n = 2^i$
      Compute new panel width $\Delta x = (b - a)/n$
      Sum the function over midpoints of old panels $\displaystyle\sum_{\substack{i=1 \\ i=\text{odd only}}}^{n-1} f(a + i * \Delta x_k)$
        $K = 1$
        Sum = 0.
    2 Sum = Sum + $F(a + k \Delta x)$
          $K = K + 2$
          IF(K .LE. N − 1)GO TO 2

      Compute the next order trapezoid approx.
        T = $\frac{1}{2}$T + $\Delta x \times$ Sum

      Print result for this value of I
      Return for next order
        GO TO 1

ELSE

      Print final results and STOP

END IF

4.5.2 Simpson's Rule

There are numerous computational algorithms that are refinements of the trapezoidal rule. One of the most popular is called Simpson's rule. Simpson's rule proceeds in the same manner as the trapezoidal approximation, except that, instead of the function being replaced by straight lines, the function is replaced by parabolas through three adjacent points. For the derivation of the Simpson's rule formula see Problem 4.8. The result is quoted as follows:

(4.14)
$$S_n = \frac{1}{3} \Delta x_n \left[f_a + 4 \sum_{\substack{i=1 \\ \text{odd only}}}^{n-1} f(a + i\,\Delta x_n) + 2 \sum_{\substack{i=2 \\ \text{even only}}}^{n-2} f(a + i\,\Delta x_n) + f_b \right]$$

where n = number of panels {must be even}

$\Delta x_n = (b - a)/n$

$f_a = f(a)$

$f_b = f(b)$

Using the basic summation algorithm, computer codes can easily be written to numerically evaluate an integral using Equation (4.14). To illustrate the improved accuracy of Simpson's rule as compared with the trapezoidal rule, we will again consider the integral:

$$I = \int_1^2 \left(\frac{1}{x} \right) dx$$

Using Equation (4.14) first for two panels yields

$$n = 2^1 = 2 \qquad \Delta x_2 = \frac{b - a}{2} = \frac{1}{2}$$

$$S_2 = \frac{1}{3} \left(\frac{1}{2} \right) \left[1 + 4 \left(\frac{1}{1.5} \right) + \frac{1}{2} \right]$$

$$= 0.6944444$$

Repeating for $n = 4, 8, 16, \ldots$ panels (that is, $n = 2^k$)

$$k = 2 \qquad n = 2^2 = 4 \qquad \Delta x_4 = \frac{1}{4}$$

$$S_4 = \frac{1}{3} \left(\frac{1}{4} \right) \left[1 + 4 \left(\frac{1}{1.25} + \frac{1}{1.75} \right) + 2 \left(\frac{1}{1.5} \right) + \frac{1}{2} \right]$$

$$= 0.69325397$$

Continuing the calculation we obtain the values listed in Table 4-1. For comparison, I have also included the results for the same integral obtained in the previous section by the trapezoidal rule. Clearly, Simpson's rule converges much faster than the trapezoidal rule, at least for this example.

Table 4-1 The trapezoidal and Simpson's rule results for the integral $I = \int_1^2 dx/x$.

k order	n No. of panels	T_n	S_n
0	1	0.75	---
1	2	0.7083	0.6944
2	4	0.69702	0.69325
3	8	0.69412	0.69315
4	16	0.69339	0.6931466
5	32	0.693208	0.6931473
6	64	0.693162	0.6931472

exact result = ln(2) = 0.6931471806 . . .

4.6   THE BISECTION TECHNIQUE FOR FINDING ROOTS OF EQUATIONS

A very common feature found in almost every program concerned with numerical analysis is to have the computer repeatedly monitor some property of a function and to take some action when a particular condition is satisfied. A rather nice example of this is found in the determination of the roots of an equation by the bisection method.

4.6.1   Roots of Equations

Engineering and scientific problems very often require the calculation of the roots of a function or equation for their solution. That is, it is required to find a value of x, such that $f(x) = 0$. The function may be a polynomial like

$$p(x) = x^7 - 5x^5 + 6x^4 - 2x^2 + 1 = 0$$

or a more complicated expression involving transcendental functions,

$$t(x) = e^{-x} - \sin\left(\tfrac{1}{2}\pi x\right) = 0$$

The problem could be

Since the polynomial is of degree 7, find all seven roots of $p(x)$.
If the polynomial $p(x)$ has less than seven real roots, find only the real roots or only the positive real roots.
The second equation, $t(x) = 0$, has an infinite number of positive roots. So perhaps we need only the first five, or only the first.
Given an approximate value for a root, find a more precise value.

For the moment, we will concentrate on the last, least ambitious project.

As with any program, we begin by gathering as much information as we can before we make any attempt at constructing a Fortran code. In functional analysis this almost always means you should attempt a rough sketch of the function under consideration. The second equation above can be written as

$$e^{-x} = \sin\left(\tfrac{1}{2}\pi x\right)$$

A root of the equation then corresponds to any value of x such that the left side and the right side are equal. If the left and right sides are plotted independently, the roots of the original equation then are given by the points of intersection of the two curves (see Figure 4-10). From the sketch we see that the roots are approximately

$$\text{Roots} = 0.4,\ 1.9,\ \ldots$$

and since the sine oscillates there will be an infinite number of positive roots. We will concentrate first on improving the estimate of the first root near 0.4. We begin by establishing a procedure, or algorithm, that is based on the most obvious method of attack when using a pocket calculator. That is, we begin at some value of x just before the root (say 0.3) and step along the x axis, carefully watching the magnitude and particularly the sign of the function.

Step	x	e^{-x}	$\sin\left(\tfrac{1}{2}\pi x\right)$	$f(x) = e^{-x} - \sin\left(\tfrac{1}{2}\pi x\right)$
0	0.3	0.741	0.454	0.287
1	0.4	0.670	0.588	+0.082
2	0.5	0.606	0.707	−0.101

The function has changed sign between 0.4 and 0.5, indicating a root and that we have stepped too far. We therefore

Figure 4-10 The intersection of e^{-x} and $\sin(\tfrac{1}{2}\pi x)$.

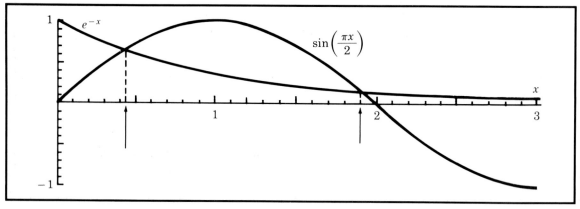

1. backup one step
2. reduce the step size by half, and continue

Step	x	e^{-x}	$\sin(\frac{1}{2}\pi x)$	$f(x) = e^{-x} - \sin(\frac{1}{2}\pi x)$
3	0.45	0.638	0.649	-0.012
4	0.425	0.654	0.619	$+0.0347$
5	0.4375	0.6456	0.6344	$+0.01126$
6	0.44365	0.6417	0.6418	-0.00014

.
.
.

The key element in the procedure is the monitoring of the *sign* of the function. When the sign changes, we take specific action to refine the estimate of the root. This will also form the key element of the computer code.

You should try this rather unsophisticated root-solving method on some simple equation. For example,

$$f(\alpha) = \sin(\alpha) - \frac{\alpha}{3} = 0 \qquad (\alpha \text{ is in radians})$$

This equation may be written as

$$\sin(\alpha) = \frac{\alpha}{3}$$

Again plotting left and right sides independently (see Figure 4-11), we see that the first positive root is near 2.2. The answer you should get for the root is

$$\alpha = 2.2788626602$$

Figure 4-11

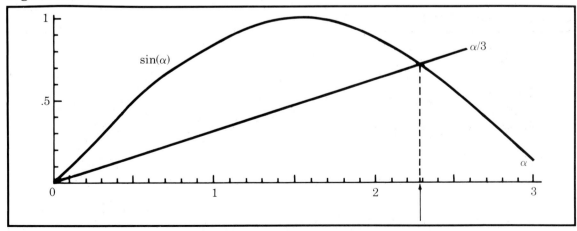

4.6.2 The Bisection Method

The root-solving procedure illustrated in the previous section is quite suitable for hand calculations; however, a slight modification will make it somewhat more "systematic" and easier to adapt to computer coding.

Suppose we already know that there is a root between $x = a$ and $x = b$. That is, the function changes sign in this interval. For simplicity I will assume that there is only one root between $x = a$ and $x = b$, and that the function $f(x)$ is continuous in this interval. The function might then resemble the sketch in Figure 4-12. If I next define $x_1 = a$ and $x_3 = b$ as the left and right ends of the interval, and $x_2 = 0.5(x_1 + x_3)$ as the midpoint, consider the question: In which half-interval does the function cross the axis? In the drawing, the crossing is on the right, so I *replace* the full interval by the right half-interval,

$$x_1 \rightarrow x_2$$

$$x_3 \rightarrow x_3$$

$$x_2 = 0.5(x_1 + x_3)$$

and ask the question again. After determining a second time whether the left half or the right half contains the root, the interval is once more replaced by either the left or right half-interval. This is continued until we narrow in on the root to within some previously assigned accuracy. Each step halves the interval, and so after n iterations, the size of the interval containing the

Figure 4-12

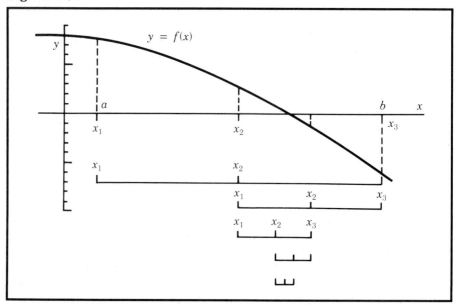

root will be $(b - a)/2^n$. If we are required to find a root to within a tolerance δ; i.e., $|x - \text{root}| < \delta$, the number of iterations n required can be determined from

$$\frac{(b - a)}{2^n} < \delta$$

For example, the initial search interval in the example of Section 4.5 was $(b - a) = 0.1$. If the root was required to an accuracy of $\delta = 10^{-5}$, then

$$\frac{0.1}{2^n} < 10^{-5} \qquad \text{or} \qquad 2^n > 10^4$$

So

$$n > \frac{\log(10^4)}{\log(2)} > 13$$

The only element of the bisection method that has been omitted is how the computer is to determine which half of the interval contains the axis crossing. To that end, consider the product of the function evaluated at the left end, $f_1 = f(x_1)$, and the function evaluated at the midpoint, $f_2 = f(x_2)$.

IF   THEN

1. $f_1 f_2 > 0$. f_1 and f_2 are *both* positive
or *both* negative. In either
case there is no crossing between
x_1 and x_2.

2. $f_1 f_2 < 0$. $f(x)$ has changed sign between
x_1 and x_2 and thus the
root is in the left half.

The program to compute the root of an equation by this procedure is given in Figure 4-13. This program illustrates most of the ideas of this chapter and should be carefully studied. You will notice the following features:

1. In each iteration after the first, there is only *one* function evaluation. It would be highly inefficient to reevaluate $f(x_1)$, $f(x_2)$, and $f(x_3)$ for each iteration since two of them are already known. If the function were extremely complicated, this could be a serious problem. A great deal of computer time is wasted every day by careless programs that contain unnecessary function evaluations.
2. The program contains several checks for potential problems along with diagnostic messages (e.g., excessive iterations, no root in interval, etc.) even though the programmer may think these possibilities are remote.

Generally, the more of these checks a program contains, the better. They only take a few minutes to add to a code and they can save you hours of debugging time.

3. The criterion for success is based on the size of the interval. Thus, even if the function were not close to zero at this point, x is changing very little and continuing would not substantially improve the accuracy of the root.

One final comment: This method is an example of a so-called brute force method, that is it possesses a minimum of finesse. It is an excellent example of Fortran techniques, but much more powerful and clever procedures are usually used to find the roots of functions. One of these is described in the next section.

Figure 4-13 The Fortran program for the bisection algorithm.

```
        PROGRAM BISEC
*--
*-- The interval a < x < b is known to contain a root of f(x). The
*-- estimate of the root is successively improved by finding in
*-- which half of the interval the root lies and then replacing
*-- the original interval by that half.
*-------------------------------------------------------------------
* Variables
*--
        REAL X1,X2,X3,F1,F2,F3,A,B,EPS,D,D0
        INTEGER I,IMAX
*--
*--       X1,X3,X2 - The left, right, and midpoint of the
*--                  current interval.
*--       F1,F3,F2 - The function evaluated at these points
*--       A,B,D0   - The left and right ends of the original
*--                  interval and its width (b-a).
*--       EPS      - Convergence criterion based on the size
*--                  of the current interval
*--       D        - The width of the current interval (x3-x1)
*--       IMAX     - Maximum number of iterations
*--       I        - Current iteration counter
*-------------------------------------------------------------------
* Statement function for the function f(x)
*--
        F(X) = EXP(-X) - SIN(PI * X/2.)
*--
*--                  (Or any other function)
*-------------------------------------------------------------------
* Initialization
*--
        PI = ACOS(-1.)
        PRINT *,'ENTER LIMITS OF ORIGINAL SEARCH INTERVAL-A,B'
        READ *,A,B
        PRINT *,'ENTER CONVERGENCE CRITERION AND MAX ITERATIONS'
        READ *,EPS,IMAX
        PRINT *,' THE ORIGINAL SEARCH INTERVAL IS FROM ',A,' TO ',B
        PRINT *,' THE CONVERGENCE CRITERION IS FOR INTERVAL < ',EPS
        PRINT *,' THE MAXIMUM NUMBER OF ITERATIONS ALLOWED IS ',IMAX
*--
```

```
        X1 = A
        X3 = B
        F1 = F(X1)
        F3 = F(X3)
        I  = 0
        D0 = (B - A)
        D  = 1.0
*--
*--     First verify that there is indeed a root in the interval
*--
        IF(F1 * F3 .GT. 0.0)THEN
                PRINT *,'NO ROOT IN ORIGINAL INTERVAL'
                STOP
        END IF
*--
   1    X2 = (X1 + X3)/2.
        F2 = F(X2)
*-----------------------------------------------------------------
*                       Convergence test
*--
        IF(D .LT. EPS)THEN
                PRINT *,' A ROOT AT X = ',X2' WAS FOUND'
                PRINT *,' IN ',I,' ITERATIONS'
                PRINT *,' THE VALUE OF THE FUNCTION IS ' F2
                STOP
        ELSE IF(I .GT. IMAX)THEN
*--
*--             Failure path - excessive iterations
*--
                PRINT *,'FAILURE--NO CONVERGENCE IN ',I,' STEPS'
                STOP
        END IF
*--
*--             Check for crossing in left half
*--
        IF(F1 * F2 .LT. 0.)THEN
                D = (X2 - X1)/D0
                F3 = F2
                X3 = X2
*--
*--                     Or in the right half
*--
        ELSE IF(F2 * F3 .LT. 0.)THEN
                D = (X3 - X2)/D0
                F1 = F2
                X1 = X2
*--
*--             If no crossing in either half,
*--             either f(x2) = 0.0 or an error
*--
        ELSE IF(F2 .EQ. 0.0)THEN
                PRINT *,'F(',X2,') IS IDENTICALLY ZERO'
                PRINT *,' FOUND IN ',I,'STEPS'
                STOP
```

Continued

```
          ELSE
                    PRINT *,'THE CURRENT INTERVAL ',X1,' TO ',X3
                    PRINT *,'DOES NOT CONTAIN A ROOT, THE'
                    PRINT *,'CODE FOR F(X) IS PROBABLY WRONG'
                    STOP
          END IF
*--
*--                 Increment interations counter and repeat
*--
          I = I + 1
        GO TO 1
*---------------------------------------------------------------
      END
```

4.7 IMPROVED ROOT SOLVING TECHNIQUES— NEWTON'S METHOD

There are a number of alterations one can make to the bisection algorithm to improve its rate of convergence and, if you are interested, you will find a discussion in Conte and deBoor or in Hornbeck (see References in the appendix). A quite different approach is due to Isaac Newton and may be familiar to you from your calculus class. This procedure starts with an *initial guess* for the root, say x_0, then, by approximating the function near x_0 by a straight line, computes an improved estimate of the root x_1. The process is repeated until sufficient accuracy is obtained.

The basic algorithm is most easily derived by considering a graphical statement of the problem (Figure 4-14). Near the initial estimate of the root x_0, the function is replaced by a straight line that goes through the point $(x_0, f((x_0))$ and that has the same slope as the tangent to the curve at x_0. Since the slope of the tangent at x_0 is equal to the derivative of $f(x)$ evaluated at x_0, this straight line approximation to $f(x)$ is clearly

(4.15)
$$f(x) \simeq f(x_0) + \frac{df}{dx}\bigg|_{x=x_0} (x - x_0) \qquad \langle\text{for } x \text{ near } x_0\rangle$$

[*Note:* When $x = x_0$, $f(x) = f(x_0)$.]

The fundamental element in Newton's algorithm is then to find the root not of $f(x)$ but of the straight line approximation to $f(x)$. That is,

(4.16)
$$f(x_0) + \frac{df}{dx}\bigg|_{x_0} (x - x_0) = 0 \rightarrow x = x_0 - f(x_0)/f'(x_0)$$

where

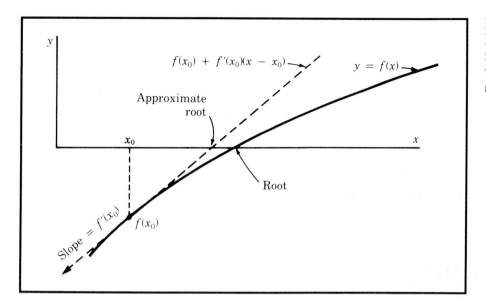

Figure 4-14
Replacing a function by two terms of a Taylor series.

In the figure:
$f(x_0) + f'(x_0)(x - x_0)$
$y = f(x)$
Approximate root
x_0
x
Root
Slope $= f'(x_0)$
$f(x_0)$
y

$$f'(x_0) = \frac{df}{dx}\Bigg|_{x=x_0}$$

If the straight line approximation to $f(x)$ is a good one, this should then be an improved estimate of the root of $f(x)$.

The procedure is next repeated, now using the improved value of the root as the guess.

$$x_1 = x_0 - f(x_0)/f'(x_0)$$
$$x_2 = x_1 - f(x_1)/f'(x_1)$$
$$\text{etc.}$$

Notice that Newton's algorithm uses much more information about the function in each step than does the bisection algorithm, and so it is reasonable to expect the convergence to be significantly improved, as is indicated in the example calculation that follows.

To compare the two root-solving techniques, consider the same function that was used in the analysis of the bisection method.

$$f(x) = e^{-x} - \sin\left(\frac{1}{2}\pi x\right)$$

The derivative of this function is

$$f'(x) = -e^{-x} - \frac{\pi}{2}\cos\left(\frac{1}{2}\pi x\right)$$

Starting with an initial guess for the root of $x_0 = 0.4$, we obtain

$$x_0 = 0.4$$

$$f(x_0) = 0.8253479$$

$$f'(x_0) = -1.9411212$$

$$\Delta x = -f(x_0)/f'(x_0) = +0.04251914$$

$$x_1 = x_0 + \Delta x = 0.44251914$$

Continuing the calculation, we obtain the results below:

i	xi	$f(x_i)$	$f'(x_i)$	$\Delta x = -f/f'$
0	0.400	0.0825348	-1.941121	$+0.04251914$
1	0.44251914	0.0019481	-1.848762	$+0.00105372$
2	0.44357287	0.0000128	-1.846420	$+0.00000067$
3	0.44357353	5.0E-11	-1.846420	$+2.7E-11$

Obviously the convergence rate is dramatically improved in comparison with the bisection method. But there is a price to pay.

4.7.1 Comparison of the Bisection and Newton's Algorithms

The principal advantage of Newton's method over the bisection algorithm is in the accelerated rate of convergence. A second advantage is the feature that the method does not require a starting interval in which the function changes sign. Thus, the bisection method could not be used to find a root of a function such as $f(x) = (x - 1)^2$, which never changes sign. Newton's method, however, will successfully find the root of functions of this type.[5]

The disadvantages of Newton's algorithm are all related to the potential dangers of replacing the function $f(x)$ by straight lines. For example, if the initial guess x_0 is near a point where the function has a horizontal slope, the next computed value will likely be thrown far out of the region of interest. (See Figure 4-15.)

The success of the method is very much dependent on the quality of the initial guess. Whereas the bisection method is guaranteed to eventually converge to a root of any desired accuracy (within the limits of round-off error), Newton's method may or may not converge depending on the details of the function near the initial guess. This means that the programmer must supply numerous diagnostic checks within the program to anticipate a variety of potential problems. These would include:

---

[5] A function such as this is said to have a multiple root of multiplicity 2 at $x = 1$. That is, since the function is a quadratic, it must have two roots, both of which in this case are equal to one. Newton's method will handle functions with multiple roots; however the convergence rate will be significantly slower.

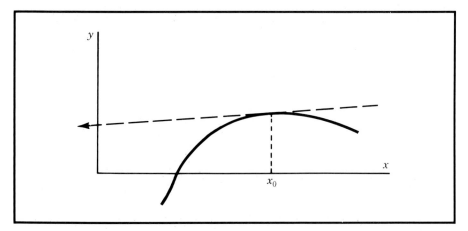

Figure 4-15
Newton's
algorithm will
fail near a point
of horizontal
slope.

1. A limit on the total number of iterations. (This is required in all programs that make use of an iterative procedure.)
2. A check for horizontal slope near the current guess. Since the improvement is $\Delta x = -f(x)/f'(x)$, the method will certainly fail if the derivative is zero. The program should STOP and print a message. If the derivative is small but not zero, the program should probably stop, or at least print a warning that the calculation appears to be diverging.
3. The ultimate convergence of the method should be based on the size of Δx. Even if the function is not zero, successive iterations have not altered the estimate of the root significantly and continuing further will not help.

4.7.2 The Fortran Code To Implement Newton's Algorithm

A program to find the root of a function using Newton's Algorithm must begin with *TWO* statement functions: one for $f(x)$ and a second for the derivative of this function $f'(x)$. If the function is quite complicated, this can be a major inconvenience of the method.

In addition the programmer will have to supply the initial guess x_0, the maximum number of iterations permitted IMAX, and the convergence tolerance EPS, (that is, if $(|\Delta x| < EPS) \rightarrow$ success). The pseudocode outline of the program follows.

```
      Statement functions for f(x) and f'(x)
      F(X) = ...
      DF(X) = ...

      READ initial guess, IMAX, EPS

      Set iteration counter to zero and compute Δx.
         I = 0
1        TOP = F(X)
         BOT = DF(X)
            IF( |Bot| = 0.)THEN
```

Continued

```
                method fails; print current values and a failure statement, and STOP,
        ELSE IF( |Bot| < 1.E-5)THEN
                {1.E-5 chosen arbitrarily}
                method is likely diverging; Print warning and perhaps STOP,
        END IF
    DX = -TOP/BOT
Compute improved estimate and test for completion,
    X = X + DX
    I = I + 1
        IF(I .GT. IMAX)THEN
                method did not converge within iterations limit; print failure notice and
                current values of I, DX, X, F(X), and STOP,
        ELSE IF( |Δx| < EPS)THEN
                method has converged; print results,
                and STOP,
        ELSE
                method has not yet converged;
                return to statement 1 for next iteration,
                    GO TO 1
        END IF
```

PROBLEMS

1. Write statement functions for the following:
   a. $f(x) = 3x^2 + x - 1$
   b. $g(z) = ax^2 + bx + c$
   c. $h(x) = e^{-ax} + \ln(\sin(\pi x))$
   d. Index $= 3i + 2j$

2. Write a complete program to be run at a terminal that will
   a. Read a series of exam scores (integers ≤ 100), entered one at a time and preceded by a prompt to the user.
   b. After each score is entered the computer will respond with:

```
              AVERAGE SO FAR = ...
              MAX, SCORE SO FAR = ...
              MIN, SCORE SO FAR = ...
```

   c. The read loop should terminate by reading a score of -1.
   d. The program should contain validity checks on the data entered and repeat the request for an exam score if bad data is entered (for example, scores > 100 or < -1).

3. Write a program to verify the following identities:

a. $\displaystyle\sum_{i}^{N} i = \frac{N(N+1)}{2}$

b. $\displaystyle\sum_{i=1}^{N} i^2 = \frac{N(N+1)(2N+1)}{6}$

c. $\displaystyle\sum_{i=1}^{N} i^3 = \frac{N^2(N+1)^2}{4}$

d. $\displaystyle\sum_{i=1}^{N} i^4 = \frac{N(N+1)(2N+1)(3N^2+3N-1)}{30}$

Test the program for $N = 2, 10, 25, 100$.

4. Write programs to evaluate the following, terminating the sums when the absolute value of a term is less than 1.E-6. (Note that Problem 4(f) is an infinite product and must be handled somewhat differently.)

a. $\displaystyle\frac{1}{1^2} + \frac{1}{2^2} + \frac{1}{3^2} + \cdots + \frac{1}{n^2} + \cdots \qquad \left(= \frac{\pi^2}{6}\right)$

b. $\displaystyle 1 - \frac{1}{3} + \frac{1}{5} - \frac{1}{7} + \cdots \qquad \left(= \frac{\pi}{4}\right)$

c. $\displaystyle 1 + \frac{1}{2} + \frac{1}{4} + \frac{1}{8} + \cdots + \frac{1}{2^k} + \cdots \qquad (= 2)$

d. $\displaystyle\frac{1}{4} + \frac{1}{16} + \frac{1}{64} + \cdots + \frac{1}{4^k} + \cdots \qquad \left(= \frac{1}{3}\right)$

e. $\displaystyle 1 + \frac{1}{2!} + \frac{1}{3!} + \frac{1}{4!} + \cdots + \frac{1}{n!} + \cdots \qquad (= e)$

f. $\displaystyle\frac{(2)(2)}{(1)(3)} \times \frac{(4)(4)}{(3)(5)} \times \frac{(6)(6)}{(5)(9)} \times \cdots \times \frac{(2n)(2n)}{(2n-1)(2n+1)} \times \cdots \qquad \left(= \frac{\pi}{2}\right)$

5. The Fortran program to sum the series for e^x given in Figure 4-4 is not an acceptable method to evaluate e^x when the magnitude of x is large. If x is large and positive, successive terms in the series will continue to grow until n, the order of the term, is greater than x, and a great many terms will have to be included to obtain an accurate result. For x large and negative, the situation is much worse. Not only will a great many large initial terms have to be accumulated, but since the result is nearly zero (for example, $e^{-10} \sim 0.000045$) the near cancellation of large terms will introduce considerable round off error. Rewrite the program to read a value of X, and if X is larger than 10 in magnitude, first scale to a value between 0 and 1. Once the series is evaluated, e^x is then scaled back up. Test the program for $x = +50.$ and $x = -50.$

6. Write a program to evaluate the series

$$\sum_{n=1}^{\infty} \frac{1}{n^2}$$

terminating the series when the magnitude of an individual term is less than 1.E-6. Let NTERMS be the total number of terms included. Next repeat the summation, this time including twice as many terms. Compare the two results. What is your conclusion regarding the convergence of this sum? Finally, repeat the problem for the sum

$$\sum_{n=1}^{\infty} \frac{1}{n}$$

7. The Golden Mean of antiquity is defined to be a solution of the equation

$$r = \frac{1}{1 + r}$$

Solve this equation iteratively—that is,

$$r_{new} = \frac{1}{1 + r_{old}}$$

Start with $r = 1.0$. Also, show analytically that the limit of the iterative solution is $r = (\sqrt{5} - 1)/2$.

8. A derivation of Simpson's rule:
   a. Show that

$$\int_a^b f(x) \, dx = \left(\frac{b - a}{6}\right) \left[f(a) + 4f\left(x = \frac{b + a}{2}\right) + f(b) \right]$$

if $f(x) = 1, x$, or x^2. Use this result to prove that the area under a general parabola,

$$y_2(x) = c_0 + c_1 x + c_2 x^2$$

from $x = a$ to $x = b$ is

$$\text{Area} = \frac{1}{3} \Delta x_2 [y_2(a) + 4y_2\left(\frac{a + b}{2}\right) + y_2(b)]$$

where Δx_2 is the width of one of the two panels—that is, $(b - a)/2$. If a function $F(x)$ is evaluated at the points $x = a, x = b$, and at the midpoint $x = (a + b)/2$, the area under the function $F(x)$ may be approximated by the area under a *parabola* drawn through these points. This is the lowest order Simpson's rule result. Show that

$$\int_a^b F(x) \, dx \simeq S_2 = \frac{\Delta x}{3} \left[F(a) + 4F\left(\frac{a + b}{2}\right) + F(b) \right]$$

Next, subdivide the interval into four panels and approximate the integral by *two* parabolas: one through the first three points and another through the last three points and show that now

$$\int_a^b F(x)\,dx \simeq S_4 =$$

$$\frac{\Delta x_4}{3}[F(a) + 4(F(a + \Delta x_4) + F(a + 3\,\Delta x_4)) + 2F(a + 2\,\Delta x_4) + F(b)]$$

Generalize this result to an arbitrary even number of panels.

9. Using a pocket calculator and the bisection technique, solve for the smallest positive roots of the following functions. Find the root accurate to three significant figures.

   a. $f(x) = x^2 + 2x - 15$ ⟨initial interval $2.8 < x < 3.1$
                                          answer = 3.0⟩

   b. $g(x) = \sin(x)\sinh(x) + 1$ ⟨elliptic gear equation,
                                           $\sinh(x) = (e^x - e^{-x})/2$.
                                         initial interval $1.0 < x < 4.0$
                                         x is in radians⟩

   c. Predict the number of steps needed to obtain the answer to the specified accuracy in both problems.

10. Write a program using the bisection algorithm of Section 4.6 to determine the root of the following functions accurate to six figures.

   a. $f(x) = xe^{-x^2} - \cos(x)$ $a = 0,\ b = 2$
                                           exact root = 1.351491185 . .

   b. $g(x) = x^2 - 2x - 3$ $a = 0,\ b = 4$
                                           exact root = 3.0

   c. $h(x) = e^x - (1 + x + x^2/2)$ $a = -1,\ b = 1$
                                           exact root = 0.0

   d. $F(x) = x^3 - 2x - 5$ $a = 1,\ b = 3$
                                           exact root = 2.0945514815 . .

   e. $G(x) = 10\ln(x) - x$ $a = 1,\ b = 2$
                                           exact root = 1.1183255916 . .

11. Construct a program to find a maximum of a function $f(x)$. Do this by starting at x_0 with a step size Δx. Evaluate $f_1 = f(x_0)$, $f_2 = f(x_0 + \Delta x)$. IF $f_1 < f_2$, continue, ELSE reduce Δx by half and repeat the comparison.

12. Newton's method to compute square roots was illustrated in Figure 3-10. The procedure was to guess a value x_0 for the square root of number C. This estimate was then improved by using the algorithm

$$x_1 = \frac{1}{2}\left[x_0 + \frac{C}{x_0}\right]$$

This equation is derived by expressing the problem in the form of a root-solving problem. Thus the search for the square root of C becomes a search for the root of the function

$$f(x) = x^2 - C$$

     i. With this as a start, derive the equation for square roots via Newton's method.

     ii. Generalize the result to obtain the nth root of C—that is, solve the equation $f(x) = x^n - C$.

13. In the not too distant past, mechanical calculators were available that could multiply but not divide. Use Newton's method to devise a scheme that does not employ division to iteratively calculate the inverse of a number C, beginning with an initial guess x_0. [The condition on the initial guess is that $(2 - x_0 C)$ be positive.]

14. Write a program implementing the pseudocode outline of Newton's method given in Section 4.7 and use this program to find the root of the elliptic gear equation given in Problem 9. Start with an initial guess of $x_0 = 2.0$ and limit the calculation to 20 iterations. Use EPS = 1.E-6.

PROGRAMMING ASSIGNMENT II

II.1 SAMPLE PROGRAM

Chemical Engineering

Chemical engineering is the application of the knowledge or techniques of science, particularly chemistry, to industry. The chemical engineer is responsible for the design and operation of large-scale manufacturing plants for all those materials that undergo chemical changes in their production. These include all the new and improved products that have so profoundly affected society, such as petrochemicals, rubbers and polymers, new metal alloys, industrial and fine chemicals, foods, paints, detergents, cements, pesticides, industrial gases, and medicines. In addition, chemical engineers play important roles in pollution abatement and the management of existing energy resources.

The field has grown to be so broad, it is difficult to classify the activities of a chemical engineer. A rough subdivision is into engineers concerned primarily with large-scale production systems or chemical processing or with smaller-scale or molecular systems.

Chemical Processing Concerns all aspects of the design and operation of large chemical processing plants. Example are

Petrochemicals: The distillation and refinement of fuels such as gasoline, synthetic natural gas, coal liquefaction and gasification, and the production of the infinite variety of products from petroleum from cosmetics to pharmaceuticals.

Synthetic Materials: The process of polymerization, a joining of simpler molecules into large complex molecules, is responsible for a great many modern materials such as nylon, silicon, synthetic rubbers, polystyrene, and a great variety of plastics and synthetic fibers.

Food and Biochemical Engineering: The manufacture of packaged food, improved food additives, sterilization, and the utilization of

147

industrial bacteria, fungi, and yeasts in processes like fermentation.

Unit Operations: The analysis of the transport of heat or fluid such as the pumping of chemicals through a pipeline or the transfer of heat between substances. Also, the effect of heat transfer on chemical reactions such as oxidation, chlorination, etc.

Cryogenic Engineering: The design of plants operating at temperatures near absolute zero.

Electrochemical Engineering: The use of electricity to alter chemical reactions such as electroplating or the design of batteries or energy cells.

Pollution Control: A rapidly growing field which seeks to monitor and reduce the harmful effects of chemical processing on the environment. Topics of concern are waste water control, air pollution abatement, and the economics of pollution control.

Molecular Systems   Application of laboratory techniques to large-scale processes. Examples are

Biochemical Engineering: Application of enzymes, bacteria, etc. to improve large-scale chemical processes.

Polymer Synthesis: Molecular basis for polymer properties and the chemical synthesis of new polymers adapted for large-scale production.

Research and development in all areas of chemical processing.

Preparation for a career in chemical engineering requires a thorough background in physics, chemistry, and mathematics, and a knowledge of thermodynamics, and physical, analytic, and organic chemistry. Though extensively trained in chemistry, chemical engineers differ from chemists in that their main concern is the adaptation of laboratory techniques to large-scale manufacturing plants.

Sample Program: Optimum Depth of a Fluidized-Bed Reactor

A fluidized-bed chemical reactor is a structure that is used extensively in chemical engineering to provide more uniform contact for chemical reactions such as catalytic cracking in petroleum processing or heat transfer in combustion operations. A fluidized reactor contains a bed of granular material through which a fluid is flowing at a rate sufficiently high to suspend the material (akin to "quicksand" conditions). An important design concern is to provide sufficient height of reactor to prevent the loss of the bed particles. In addition, an operating constraint is to keep the rate of fluid flow through the reactor sufficient to suspend the material but not enough to "flush" the material out of the reactor. The reactor is sketched in Figure II-1.

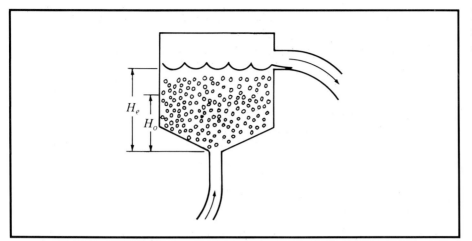

Figure II-1   A fluidized-bed chemical reactor.

Once the fluid is flowing in the reactor, the total height of the fluid plus material increases to the expanded height H_e, which is given by the expression

$$H_e = H_0(1 + f) \sum_{i=1}^{n} \left[\frac{p_i}{(1 - \varepsilon_i)} \right] \qquad \text{(II.1)}$$

where H_0 = static (unexpanded) bed height (m)
  f = void fraction of the unexpanded bed
  p_i = fraction of bed particles with diameter d_i (m)
  ε_i = porosity or void fraction of the expanded bed part that is made up of particles of size d_i

The individual void fractions ε_i for the expanded bed for particles of size d_i can be found from

$$\frac{\varepsilon_i^3}{(1 - \varepsilon_i)} = F(q, d_i) \qquad \text{(II.2)}$$

where $F(q, d_i)$ is a function that depends on the flow rate q and the particle size d_i, and is given by

$$F(q, d_i) = \frac{180}{g} \frac{\mu}{(\rho_s - \rho)} \frac{q}{d_i^2} \qquad \text{(II.3)}$$

where g = gravitational acceleration, 9.8 m/sec^2
  μ = fluid viscosity (N $-$ sec/m^2)
  ρ_s = particle density (kg/m^3)
  ρ = fluid density (kg/m^3)
  q = reactor bed flow rate (m/sec)

The cubic equation for the void fractions can be written as

$$\varepsilon = F^{1/3}(1 - \varepsilon)^{1/3}$$

(II.4)

This equation is solved for a particular value of q and d_i by guessing a value for ε_i, inserting it into the expression on the right, and computing a new value for ε_i. If the new value is sufficiently close to the guess, a solution has been found; if not, the new value is inserted into the right and the process continued until successive values differ less than some small quantity. This process is known as successive substitutions.[1] For a given flow rate q, the void fractions ε_i for each of the constituent particles in the bed can be obtained by this means. Once all the ε_i's have been computed, the sum in Equation (II.2) can be evaluated and the bed height H_e determined.

---

[1] The method of successive substitutions does not always work. If we are attempting to find the root of an equation written in the form

$$x = f(x)$$

in an iterative manner—i.e.,

$$x_{k+1} = f(x_k)$$

and if the unknown root is designated by α, [i.e., $\alpha = f(\alpha)$], we begin by assuming that x_k is near α

$$x_k = \alpha + \delta_k$$

and that each successive iteration is an improvement ($|\delta_{k+1}| < |\delta_k|$). But if δ_k is small, then the derivative of $f(x)$ is approximately

$$f'(\alpha) = \frac{f(\alpha + \delta_k) - f(\alpha)}{\delta_k} = \frac{f(\alpha + \delta_k)}{\delta_k} - \frac{\alpha}{\delta_k}$$

or

$$\alpha + \delta_{k+1} = f(\alpha + \delta_k) = \alpha + f'(\alpha)\,\delta_k$$

Thus for the procedure to converge ($|\delta_{k+1}| < |\delta_k|$), the magnitude of the derivative of $f(x)$ near the root *must* be less than 1. For this reason, successive substitutions applied to the equation,

$$x^3 - \frac{1 - x}{8} = 0$$

written in the form

$$x = \frac{1}{2}(1 - x)^{1/3}$$

will work; while

$$x = 1 - 8x^3$$

will not. (The root is near 0.418).

Problem Specifics

There are particles of identical material but of four different sizes in the static bed. The static-bed reactor characteristics are given in Table II-1.

Static height H_0	Static void fraction f	Particle density ρ_s	Size distribution $(i = 1, 4)$ p_i	Particle size $(i = 1, 4)$ $d_i(10^{-3}$ m$)$
4.5	0.45	2666.	0.256	2.0
			0.350	8.4
			0.295	4.0
			0.099	1.3

Table II-1
Static-bed reactor parameters.

The properties of the fluid flowing through the reactor are:

$$\mu = \text{viscosity} = 8.13 \times 10^{-3}$$
$$\rho = \text{density} = 1000$$

Your program should:

1. Declare the types of all variables in the program.
2. Have statement functions for Equations (II.3) and (II.4).
3. Read the values for all the parameters and neatly print with appropriate labels.
4. For flow rates $q = 0.004$ to 0.012 in steps of 0.004, evaluate the summation over the four particle types in Equation (II.1) in the following way:
   a. For each of the four particle types:
      i. Solve Equation (II.2) by successive substitution. Use $\varepsilon_{old} = 1/2$ as the initial guess and terminate the calculation when $|\varepsilon_{new} - \varepsilon_{old}| <$ EPS $= 1.\text{E-4}$. The number of iterations should be limited to thirty. Print p_i, d_i along with ε_i and the number of iterations required.
      ii. Using the computed value of ε, evaluate the current term in the summation of Equation (II.1).
   b. Add this term to the summation.
   c. Once the sum is completed, compute and print the value of H_e.
5. Increment q and if $q \leq 0.012$, repeat the calculation.

Your program should have ample diagnostics for potential problems. A detailed pseudo code outline of this program is given in Figure II-2 and the output of the program is given in Figure II-3. Your program should have ample error diagnostics.

Figure II-2
Outline of a
program to
compute the
height of a
fluidized-bed
chemical reactor.

Type and name variables associated with the parameters.
REAL q, H_0, H_e, f, ρ, ρ_s, p_1, p_2, p_3, p_4, d_1, d_2, d_3, d_4,
$\quad$ μ, g, ε_{new}, ε_{old}, EPS, SUM, size, fract, F, F_0, E
INTEGER ITER, I, IMAX
Define statement functions for Equations (II.3) and (II.4).

$$F(q, d_i) = \frac{180}{g} \frac{\mu}{(\rho_s - \rho)} \frac{q}{d_i^2}$$

$$\varepsilon_{new} = E(F_0, \varepsilon_{old}) = F_0^{1/3}[1 - \varepsilon_{old}]^{1/3}$$

READ μ, ρ, ρ_s, H_0, f, and PRINT with labels.
Initialize g = 9.8, IMAX = 30, EPS = 1.E-4.
READ the particle fractions and sizes: (p_1, d_1), (p_2, d_2), (p_3, d_3), (p_4, d_4)
Initialize the flow rate q = 0.004

1 PRINT q along with overall table headings.
$\quad$ Begin the summation, initialize SUM = 0., counter I = 0.
2 $\quad$ Based on the counter, specify the particle size and fraction.
$\quad$ IF(I = 1)THEN
$\quad\quad$ size = d_1
$\quad\quad$ fract = p_1
$\quad$ ELSE IF(i = 2)THEN
$\quad\quad\quad$ . . . $\quad$ etc. $\quad$ . . .
$\quad$ END IF
$\quad$ Solve Equation (II.2) by successive substitution.
$\quad$ initial guess ε_{old} = 1/2
$\quad$ iterations counter ITER = 0
3 $\quad$ $F_0 = F(q, \varepsilon_{old})$
$\quad\quad$ $\varepsilon_{new} = E(F_0, \varepsilon_{old})$
$\quad\quad$ IF($|\varepsilon_{new} - \varepsilon_{old}| >$ EPS)THEN
$\quad\quad\quad$ ITER = ITER + 1
$\quad\quad\quad$ IF(ITER > IMAX)Print diagnostic and STOP
$\quad\quad\quad$ $\varepsilon_{old} = \varepsilon_{new}$
$\quad\quad\quad$ GO TO 3 for next value of ε
$\quad\quad$ ELSE
$\quad\quad\quad$ ε_i has been successfully computed.
$\quad\quad\quad$ Evaluate the term in the summation.
$\quad\quad\quad$ term = fract/$(1 - \varepsilon_{new})$, and sum
$\quad\quad\quad$ SUM = SUM + term
$\quad\quad\quad$ PRINT I, p_i, d_i, ε_i, ITER as elements of a table
$\quad\quad$ END IF
$\quad$ Increment particle counter: I = I + 1
$\quad$ IF(I < 4)THEN
$\quad\quad$ GO TO 2 to compute next term in SUM.
$\quad$ ELSE
$\quad\quad$ Summation completed.
$\quad\quad$ Compute H_e and PRINT result.
$\quad$ END IF

Increment flow rate, $q = q + 0.004$ and
IF($q \leq 0.012$)THEN
      GO TO 1 to repeat calculation for next flow rate,
ELSE
      calculation completed,
        PRINT termination statement and STOP.
END IF

```
INPUT PARAMETERS FOR PROGRAM TWO

   THE STATIC HEIGHT OF THE REACTOR =    4.5 M
   THE CHARACTERISTICS OF THE REACTOR FLUID ARE
        STATIC VOID FRACTION    = 0.45
        FLUID DENSITY           = 1000. KG/M3
        FLUID VISCOSITY         = 0.00813


   THE CHARACTERISTICS OF THE BED PARTICLES
        PARTICLE DENSITY           = 2666. KG/M3
          SIZE              PARTICLE
    I     DISTR.            DIAMETER
    1     0.256            0.002
    2     0.35             0.0084
    3     0.295            0.004
    4     0.099            0.0013


   THE COMPUTED RESULTS FOR PROGRAM TWO
             FLOW RATE          REACTOR BED HEIGHT
    I        (M3/SEC)               (METERS)
   -----    ----------        -------------------
    4        0.004            10.54674716952
    5        0.005            10.97321167927
    6        0.006            11.36376698323
    7        0.007            11.72691801755
    8        0.008            12.06906460384
    9        0.009            12.39351873729
   10        0.010            12.70446212164
   11        0.011            13.00357543073
   12        0.012            13.29253922276
```

Figure II-3
Output from the sample program.

II.2   PROGRAMMING PROBLEMS

Electrical Engineering

Electrical engineering deals with the application of the principles of electricity and electromagnetism to the manufacture of all forms of machines and devices that either make use of electricity or produce electrical energy. The field, the largest of all engineering fields, has evolved from its beginning in the mid-1800s with a concern for the generation of electrical energy, to its present very broad boundaries encompassing solid-state devices such as transistors, communication, and computers, and robotics.

Power: The generation of electrical energy in large fossil-fuel, nuclear, solar, or hydroelectric plants, or the efficient utilization of electrical energy by means of motors or illumination devices. Also important are the transmission and distribution of electrical energy through overhead lines, microwaves, light pipes, and superconducting lines.

Solid-State Electronics: In conjunction with modern physics and material science, exotic semiconducting materials are being developed and used to construct microcircuitry which is used in monitoring and controlling the operations of all manner of present-day devices from video games to assembly-line robots. The improved reliability, rapidly shrinking size, and reduced power requirements of modern miniaturized electrical components have created limitless opportunities for applications.

Communications: Design and construction of equipment used in transmission of information via electricity or electromagnetic waves (radio, light, microwaves, etc.). The use of the laser for communication is a topic of modern concern, while antenna characteristics and radar are somewhat older.

Computers and Robotics: While electronics deals with the principles associated with the functions of miniaturized components, the computer engineer is concerned with designing the complex circuitry that interweaves the components into a computer. Microprocessors, or small computers, are designed to constantly monitor and control the operations of a particular piece of equipment such as a lathe or an autopilot.

Programming Problem A: Semiconductor Diode

In a simple resistor, the current i that flows through the resistor is to a good approximation proportional to the voltage V that is applied to the resistor. This is represented schematically as

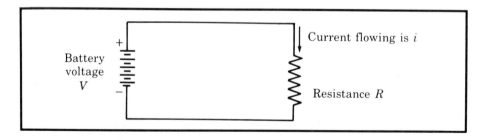

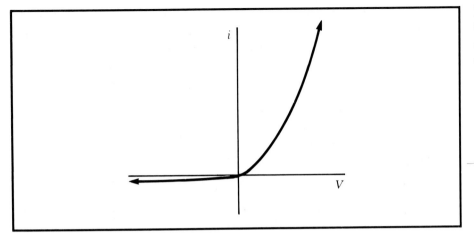

Figure II-4
Current vs. voltage for a typical diode.

$$i = \left(\frac{1}{R}\right)V \qquad \text{Ohm's law}$$

The proportionality constant is $(1/R)$, where R is the resistance (in ohms) of the resistor. The units of current are amperes and of voltage are volts. The current flows from the positive pole of the battery through the resistor and returns to the negative pole. Current is conserved, i.e., the current leaving the battery is equal to that returning.

The relation between current and voltage is not nearly so simple in solid-state devices. In a semiconductor diode the relationship is

$$i = I_s(e^{\lambda V} - 1) \tag{II.5}$$

where I_s = reverse saturation current (amperes)
        λ = diode characteristic (volts^{-1})

Both of these quantities vary from one diode to the next. Note that the diode has the property that current flows readily in only one direction as is shown in Figure II-4.

Problem Specifics

For the combination of diodes in Figure II-5 the following equations apply:

$$i = i_1 + i_2 \tag{II.6}$$
$$i_1 = I_{s1}(e^{\lambda_1 V} - 1) \tag{II.7}$$
$$i_2 = I_{s2}(e^{-\lambda_2 V} - 1) \tag{II.8}$$

Thus, in this circuit the relationship between current and voltage is

$$i = I_{s1}(e^{\lambda_1 V} - 1) + I_{s2}(e^{-\lambda_2 V} - 1) \tag{II.9}$$

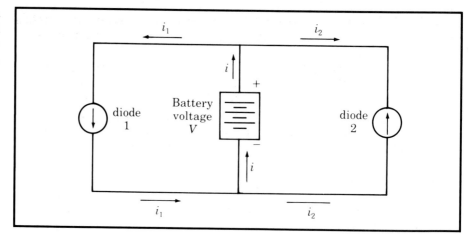

Figure II-5 A circuit diagram for a simple two-diode circuit.

1. Write a program to read in the two sets of diode parameters and a desired current I_0 and print with appropriate labels.

2. For each value of current $i = 0.1I_0, 0.2I_0, \ldots, I_0$ determine the necessary applied voltage V by using the bisection algorithm. For *each* value of the current i, you must first determine the initial search interval before you start the bisection code. Do this by evaluating the statement function based upon Equation (II.13) at $V = 0$ and then at steps of size $\Delta V = 0.01$ until the function changes sign. Use this last interval as the initial search interval.

3. Print the results in the form of a table for each set of parameters (see Table II-2) and plot i vs. V on a piece of graph paper.

Table II-2 Input parameters for diode problem.

Set	I_{s1}	I_{s2}	λ_1	λ_2	I_0
1	0.01	0.10	38.10	41.00	0.60
2	0.03	0.05	40.00	40.00	0.30

Metallurgical Engineering and Materials Science

Advances in many areas of engineering in the twentieth century have been made possible to a large extent by discoveries of new materials and a better understanding of properties of existing materials. Knowledge of the physical and chemical principles determining the electrical properties of exotic materials called semiconductors have resulted in the fantastic progress in the field of solid-state devices, from the transistor to the integrated circuit chip to large computers. Better understanding of the origins of metallic properties such as hardness, strength, ductility, corrosiveness, and others have led to improved design of automobiles, aircraft, spacecraft, and all types of machinery. The field is basically subdivided into metals and nonmetals, although there is often considerable overlap of interests and activities.

Materials Science: The behavior and properties of materials, both metals and nonmetals, from both microscopic and macroscopic perspectives.

Ceramics: Noncrystalline materials, such as glass, that are nonmetallic and which require high temperatures in their processing. Ceramics can be made brittle or flexible, hard or soft, stronger than steel, and to have a variety of chemical properties.

Polymers: Structural and physical properties of organic, inorganic, and natural polymers that are useful in engineering applications.

Materials Fabrication, Processing, and Treatment: All aspects of the manufacture of ceramics, metals, and polymer synthesis, from the growth of crystals and fibers to metal forming.

Corrosion: Reaction mechanism and thermodynamics of corrosion of metals in the atmosphere or submerged under water or chemicals whether standing or under stress.

Stress-Strain, Fatigue-Fracture of Engineering Materials: Physical properties governing the deformation and fracture of materials, their improvement and use in construction and design.

Metallurgical Engineering: The branch of engineering that is responsible for the production of metals and metal alloys, from the discovery of ore deposits to the fabrication of the refined metal into useful products. Metallurgical engineers are important in every step in the production of metal from metal ore.

Mining Engineering: Usually a separate branch of engineering; however, the concerns of the mining engineer and the metallurgist frequently overlap in the processes of extraction of metals from metal ores and the refinement into usable products. Extraction metallurgy makes use of physical and chemical reactions to optimize metal production.

Metals Fabrication: Metal forming into products such as cans, wires, and tubes, casting and joining of metals—for example, by welding.

Physical Metallurgy: Analysis of stress-strain, fatigue-fracture characteristics of metals and metal alloys to prevent engineering component failures.

Programming Problem B: Carburization

Introduction

To improve the hardness characteristics of steel, carbon is added to the steel in a controlled manner by a process called *carburizing*, which involves the gradual diffusing into the steel of atoms of carbon applied at the metal surface. For example, if a rod of pure iron is welded to a similar rod containing 1% carbon, the carbon content of the pure end is found to vary with time and position down the rod in a manner indicated in Figure II-6. At $t = 0$ the concentration of carbon in the right half is zero, while at some later time it is found that the concentration in the enriched half has been depleted near the boundary and carbon atoms have migrated into the pure end. After an infinite amount of time the distribution of carbon will be uniform throughout. The rate of the transport of the carbon atoms at a point x in the bar is found to be proportional to the negative of the slope of the concentration curve at that point, zero at the extreme ends, and large positive near the middle.

(II.10)
$$F = -D \frac{dC(x,t)}{dx}$$

where F is the volume concentration of atoms migrating per second, $C(x,t)$ is the concentration of carbon atoms at position x and time t, and the proportionality constant D is called the diffusion constant. It can be shown that combining this equation with the constraint that the total number of carbon atoms remain constant leads to the following expression for $C(x,t)$:

(II.11)
$$C(x,t) = \frac{C_0}{2}\left[1 - \frac{2}{\sqrt{\pi}} \int_0^{x/2Dt} e^{-y^2} dy\right]$$

where C_0 is the initial concentration at $x = 0$ and the integral is the area under the curve e^{-y^2} from zero to $y = x/(2Dt)$. This integral is called the error function and is defined as

(II.12)
$$\text{Erf}(\alpha) = \frac{2}{\sqrt{\pi}} \int_0^{\alpha} e^{-y^2} dy$$

Figure II-6   Diffusion of carbon between rods of differing composition.

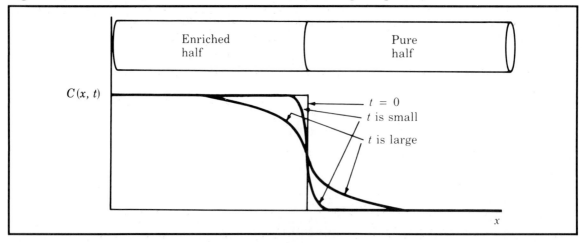

This integral cannot be evaluated in terms of elementary functions but occurs so frequently in engineering analysis that it has been extensively tabulated and even included as an intrinsic function in Fortran on some systems. If your local computer supports this function, the Fortran to compute Erf(3.6) would simply be

```
A=ERF(3.6)
```

The diffusion process is found to be extremely temperature dependent and this can be taken into account in the diffusion constant by expressing it as

$$D = D(T) = D_0 e^{-q/RT} \tag{II.13}$$

where D_0 is constant, q is called the activation energy and is constant, R is the ideal gas constant (8.31 J/molK). The units of D are m^2/sec. Thus once the constants D_0, R, q, C_0, and the temperature T have been specified, the concentration $C(x,t)$ can be determined for any value of x or t.

Problem Specifics

It is desired to allow the diffusion to take place until the average concentration in the right half reaches C_r, then disconnect the two halves and heat the right half until the concentration smooths out to a uniform concentration which would equal C_r.

The average concentration across the right half could be expressed as

$$C_{avg}(t) = [C(0,t) + C(\Delta x, t) + C(2\Delta x, t) + \cdots + C(L,t)]\frac{1}{n+1} \tag{II.14}$$

$$= \frac{1}{n+1} \sum_{i=0}^{n} C(i\Delta x, t)$$

where $\Delta x = L/n$.

Write a program with statement functions for Equations (II.11) and (II.13) that reads the parameters in Table II-3. The program should:

1. Have statement functions for Equations (II.11) and (II.14).

$$C(x, t, D) = \ldots$$

$$\text{DIF(Temp} = \ldots$$

2. Define the initial search interval in t to be $t_1 = 10$ days, $t_3 = 50$ days, and use the bisection method to find the root t of the equation

$$C_{avg}(t) - C_r = 0$$

where C_{avg} is computed using Equation (II.14) with $n = 4$.

Use IMAX = 20, EPS = 1.E−4 in the bisection algorithm. Print all parameters and the computed diffusion time with appropriate labels.

Table II-3   Input parameters for the carburizing problem.

C_0	= 0.25	(%)	= Initial end concentration
D_0	= 2.40×10^{-5}	m^3/sec	= Diffusion parameter
q	= 0.74×10^5	J	= Activation energy
T	= 1300	K	= Temperature
L	= 0.2	m	= Length of bar
C_r	= 0.030	(%)	= Desired average final concentration

Programming Problem C: Coexistence of Liquids and Gases

Introduction

The chemical engineer deals continually with chemical and physical interactions between gases and liquids, and it is essential that he or she have some form of approximate mathematical description of the properties of a substance as it undergoes a transition from gas to liquid phases and back. In elementary chemistry you are introduced to the ideal gas equation of state, which for 1 mole of gas may be written

(II.15)
$$P = \frac{RT}{V}$$

where P = pressure (n/m²)
V = volume of 1 mole (m³)
T = temperature (K)
R = ideal gas constant = 8.317 J/mole K

This equation is adequate for low pressures and high temperatures where the liquid state is not present. In fact the ideal gas law assumes that the substance remains a gas even down to a temperature of absolute zero. Over the years there have been hundreds of suggestions as to how to modify the ideal gas equation to incorporate the possibility of a gas condensing into a liquid. One of the earliest and still one of the best is the Van der Waal's equation of state for an imperfect gas, which may be written in a simplified form as

(II.16)
$$p = \frac{8t/3}{(v - \frac{1}{3})} - \frac{3}{v^2} = 0$$

where p, v, t are scaled pressure, volume, and temperature; i.e.,

$$p = \frac{P}{P_c} \qquad v = \frac{V}{V_c} \qquad t = \frac{T}{T_c}$$

and P_c, V_c, T_c are the values of the pressure, volume, and temperature at the critical point—i.e., the unique value of P, V, and T at which equal masses of the vapor phase and the liquid phase have the same density. These critical point values are extensively tabulated for most substances. The Van der Waal's equation of state is sketched for three temperatures in Figure II-7. As the temperature is reduced below the critical point temperature ($t = T/T_c < 1$), the gradual development of a hill and valley in the pv curve will be interpreted as a transition from gas to liquid. There is, however, a difficulty: the slope of the curve from point a to point b on the graph is positive,

$$\frac{\Delta p}{\Delta v} > 0 \qquad \text{between points } a \text{ and } b$$

Thus if the pressure is *increased*, $p \to p + \Delta p$, the volume is predicted to *increase*, $v \to v + \Delta v$. This is clearly unphysical and must be corrected. If p_a and p_b are the pressures at the points a and b, respectively, the procedure is to replace the unphysical part of the curve (where the slope is positive) with a horizontal straight line as shown in Figure II-8. Along this line the substance can change its volume with the pressure remaining constant by changing from gas to liquid or liquid to gas. The straight line is chosen in such a way that the area under the line is equal to the area under the curve

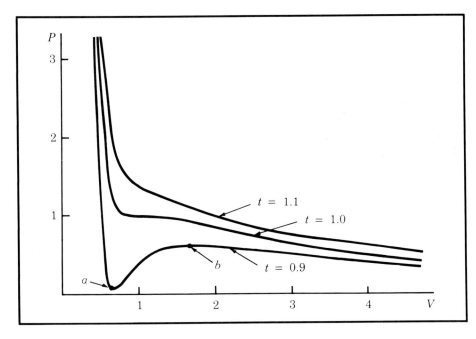

Figure II-7 Plot of Van der Waal's equation of state.

Figure II-8 Coexistence of liquid-gas phases on a Van der Waal's plot.

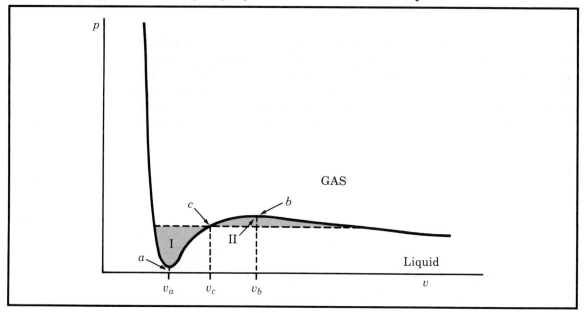

it replaces, namely $\int p\,dv$. That is, in Figure II-8 the area labeled I below the line is equal to the area above the line, II. This is a rather complicated condition to satisfy, so, for now, we will simply estimate the position of the line as follows: The points a and b on the Van der Waal's curve which correspond to a local minimum and maximum of the function are determined by a procedure explained below. The line is then drawn through the point c, which is specified by assuming that

$$v_c = \tfrac{1}{2}(v_a + v_b)$$

Once v_c is known, p_c can be obtained from the Van der Waal's equation (II.16). This value for p_c is a prediction for the pressure required to condense the gas into a liquid at the given temperature.

Note that the slope of the tangent line to the curve at points a and b is zero, that is, the tangent is horizontal. So

(II.17)
$$\frac{dp(v)}{dv} = 0$$

$$= -\frac{8t/3}{(v - \tfrac{1}{3})^2} + \frac{6}{v^3}$$

which may be rewritten as

(II.18)
$$f(v) = v^3 - C_2 v^2 + C_1 v - C_0$$

where

$$C_2 = \frac{9}{4t} \qquad C_1 = \frac{3}{2t} \qquad C_0 = \frac{1}{4t} \qquad\qquad \textbf{(II.19)}$$

Equation (II.18) has three positive real roots, $v = r_1, r_2, r_3$. The smallest, say r_1, will turn out to have a value less than $\frac{1}{3}$ and thus corresponds to a nonphysical region. [From Equation (II.16), we see that if $v < \frac{1}{3}$, the pressure is negative.] Discarding this root, the points a and b in Figure II-8 then correspond to the roots $v_a = r_2, v_b = r_2$. Once v_a and v_b are known, the predicted condensation pressure at this temperature is $p_c = p(v_c)$, where $v_c = (v_a + v_b)/2$.

Problem Specifics

Write a program to do the following:

1. Read in the critical point values for the first substance in Table II-4 and the value of the temperature to be used. Print with appropriate headings.
2. Use the bisection algorithm to find a root of Equation (II.18). Since $f(v = 0)$ is negative, you should step along the v axis in steps of 0.05 until $f(v)$ changes sign to find the initial bisection search interval. Then find this root to a tolerance $\delta = 10^{-8}$.
3. Next, on a separate piece of paper, verify that if r is a root of Equation (II.18), we may write

$$\begin{aligned} f(v) &= v^3 - C_2 v^2 + C_1 v - C_0 \\ &= (v - r)[v^2 - (C_2 - r)v + (C_1 - C_2 r + r^2)] \qquad \textbf{(II.20)} \\ &= 0 \end{aligned}$$

(Try multiplying this expression out.) Thus once a root of the cubic equation is obtained via the bisection algorithm, the remaining two roots can be found by applying the quadratic formula to the second term in Equation (II.20).
4. Use statement functions for Equations (II.16) and (II.18). Make sure your program has safeguards for potential problems like excessive iterations, complex roots, etc. Print out the limits of the liquid-gas coexistence region for this substance at the given temperature.
5. Repeat for the remaining substances in Table II-4.

Table II-4 Input parameters for liquid-gas coexistence problem.

Substance	T_c (K)	P_c (Pa)	V_c (m³/mole)	T (K) (given)
Carbon dioxide (CO_2)	304.26	7.40×10^6	2.02×10^{-5}	280.0
Benzene (C_6H_6)	561.66	4.83×10^6	2.37×10^{-5}	500.0
Nitrogen	126.06	3.39×10^6	4.36×10^{-6}	108.00
Water (H_2O)	647.56	22.0×10^6	7.21×10^{-6}	550.00

6. The variables in the problem are all scaled (i.e., divided by the critical point values). Your printed results should not be scaled, but rather values with the appropriate units. Note that the units of pressure in SI are called Pascals (Pa) (1 Pa = 1 N/m^2) These can be related to the more familiar units of atmospheres by

$$1 \text{ N/m}^2 = 1 \text{ Pa} = 9.87 \times 10^{-6} \text{ atmospheres}$$

CHAPTER 5
USE OF DATA FILES

5.1 INTRODUCTION

It is very likely that the programs you will be writing in the future will be designed to execute at a terminal and will require numerous variables to be assigned values by reading data. Every time the program is executed at the terminal, the data will have to be reentered and correctly positioned. During the debugging stage of a program's development, this can be an extremely annoying interruption to your work. It is possible in Fortran to set up a *data file* that contains all of the input lines that are to be read by the program and then instruct the program to read data from this file during execution. In addition, you could have the program write all or some of the results to a data file (distinct from output) and after execution of the program view some of the results on the terminal screen, and if it looks OK, send the results file to the line printer. Manipulation of data files is a rather advanced topic in Fortran and so I will only sketch some of the more elementary applications.

5.2 THE "LONG" FORM OF FORTRAN I/O STATEMENTS

The standard Fortran form of statements that read data into the machine or write numbers or text is

$$\text{READ}(i, j) \qquad \langle \textit{list of input variables}\rangle$$

$$\text{WRITE}(i, j) \qquad \langle \textit{list of output values, variables, or expressions}\rangle$$

The first symbol (i) in the parentheses is an integer constant that refers to a device or file. That is, $i = 3$, for example, may designate the card reader, or a paper tape reader, or the line printer, or any of several other I/O devices. The specific identification numbers of each of the I/O devices vary from one computing center to another. You must determine the particular number code for I/O devices at your computing center. An alternative procedure of assigning your own numbers to each will be discussed in Section 5.3.

The second item in the parentheses (j) is an integer constant that refers to an *existing format statement* in the Fortran code that bears that statement number. Format statements are used, for example, to specify the position of printed numbers on an output line and will be described in the next chapter. For now it is sufficient to know that if the format label (j) is replaced with an asterisk $(*)$, the READ or WRITE statement will be executed using a list-directed form of I/O; that is, the same form we are accustomed to when using the statements $\langle\text{READ} *,\rangle$ and $\langle\text{PRINT} *,\rangle$.

Following the parentheses is a list of those variables (separated by commas) to be read in or printed out. These statements would then be translated into English as:

Read (write), using device i, according to format number j, the variables . . . .

And the statements

```
READ(13,*)A,C,MAX
WRITE(13,*)A,C,MAX
```

would be interpreted as

Read(write), using I/O unit 13, by list-directed input (or output), the variables. . . .

The most common, though not universal, assignments are

$i = 5$ means card reader $\langle$using file named INPUT$\rangle$

$i = 6$ means line printer $\langle$using file named OUTPUT$\rangle$

I should point out that when a program is run interactively at a terminal, the normal I/O devices—the card reader and the line printer— are usually replaced by the terminal. That is, unless you specifically instruct the machine otherwise, any command to WRITE on the line printer will be diverted so that the output will appear on the terminal screen (and will *not* be printed). Likewise, instructions to READ data will be diverted to the terminal where you will be expected to enter the data.

The form of the READ and WRITE statements given above is valid in all versions of FORTRAN and is usually the preferred form. A somewhat shorter form is also available, namely

READ j,   ⟨list of input variables⟩

PRINT j,   ⟨list of output variables⟩

In this form it is assumed that printed output will go to the line printer and that data will be read from the card reader. (Again, all I/O to the line printer or from the card reader when working interactively at a terminal is diverted to the terminal.) These statements then read

Read (or print) according to format j, the variables . . . .

Replacing the format specification j by an asterisk in these statements results in the list-directed I/O statements that we have been using to this point. These forms are also available in all versions of Fortran but, as we shall see, are somewhat less flexible. Entering an extended data list at the terminal can be rather tedious and in these cases we will need a procedure whereby our programs can read data from a list already stored in memory. This can easily be done using the long form.

Finally, make careful note of the punctuation in the example I/O statements below

```
READ 7,A,B,C,M          No comma after READ
PRINT 52,IZ,DELTA
WRITE(6,33)X,DX         No comma after (6,33)
READ(5,111)TEMP
```

Of course, 7, 52, 33, 111 refer to existing format statements in the program. The list-directed versions of these same statements would be

```
READ *,A,B,C,M
PRINT *,IZ,DELTA
WRITE(6,*)X,DX
READ(5,*)TEMP
```

In summary, when using either the WRITE (. . .) or the long form of the READ(. . .) statement, the first number in the parentheses always refers to the associated I/O device or unit. The most common identifications are UNIT = 5 for input(card reader) and UNIT = 6 for output(line printer), and of course both input and output files are connected directly to the terminal if you are working interactively. Since these number identifications are not universal, you should have determined the unit numbers for files input and output used by your computing center.

Finally, if you are unaware of the correct unit numbers, Fortran provides an out. If the unit number is replaced by '*', all READ(*, . . .) commands are directed to file input and all WRITE(*, . . .) commands to file output. Thus the statements

```
READ(*,*)X,Y,Z
WRITE(*,*)X,Y,Z
```

read and write using files input and output respectively (the first *) by means of list-directed I/O (the second *).

5.3 THE OPEN STATEMENT—NAMING DATA FILES

The procedure in Fortran for linking with a program files other than the normal input and output files is via the OPEN statement. These files are normally written to or read from the secondary magnetic disk memory connected to the computer.

5.3.1 Disk Files

As mentioned in Section 1.2, there are two types of memory in a computer: main memory (or fast memory) and secondary memory (slower). Data files are always stored in secondary memory. In order to visualize the structure of data files it is useful to have some familiarity with the operation of two secondary memory devices, magnetic tape and magnetic disk.

One of the first devices used to store large amounts of data was magnetic tape. On the tape, each line of data or code is called a record and the information is copied onto the tape sequentially. The end of the data or program is then marked with an END OF FILE mark. These files are called SAM (*s*equential *a*ccess *m*ethod) files, and the information contained on these files can only be read in the same order in which it was written. Thus, it is not possible to skip to the middle of a large SAM file. This can be a significant disadvantage when huge data files are being used and manipulated. A second type of file is available in Fortran that makes use of the properties of a magnetic disk.

The magnetic disk, like magnetic tape, stores information compactly in terms of magnetized bits. The physical structure is quite similar to a phonograph record and the information is read from the disk by an access arm. The information recorded on the disk can be placed in a random order with each line (or record) assigned a sequencing record number. Such files are called DAM (*d*irect *a*ccess *m*ethod) files. Each record or line on a DAM file must be of the same length and the records on the file can be read in any order. This is especially useful when you are reading or updating large data files such as personnel records or inventory lists.

It is also possible to write SAM files on the disk, using the disk much like a magnetic tape. To do this you must inform the computer that the file you wish to create or use is a disk file and is of type sequential access. This is done in Fortran with the OPEN statement.

5.3.2   The Form of the OPEN Statement

The OPEN statement in Fortran is used to connect a disk file to a program and to define various attributes of the file. Some of the common disk file specifications that are prescribed in an OPEN statement are

1. To associate an existing disk file with a UNIT number. For example, the data for your program could be assigned to UNIT = 11. Then all READ(11, . . .) statements will read from this file. Or the output could be stored on UNIT = 12. Then WRITE(12, . . .) will place the results on a disk file which can be later printed at the line printer.
2. To associate a ⟨NAME⟩ of the file that is to be opened with the UNIT number. Thus the data file for your program could already reside on the disk and have a name like 'MYDATA'.
3. To define whether the file is type SAM or DAM.
4. To specify whether blanks in the file are to be interpreted as zeros or as blanks.
5. To specify whether the file already exists in the system (like a data file) or is to be a new file (e.g., results).

The form of the OPEN statement to accomplish all this is

```
OPEN(UNIT = u,ERR = sl,FILE = flname,
+      STATUS = stat,ACCESS = acc,BLANK = blnk)
```

where   u is an integer and specifies the unit of the file to be opened.

sl is a statement number of an executable statement to which the program will branch if there is an error encountered while opening the file. (Perhaps the data file is not present in the system.) The ERR = sl field is optional.

flname is a name (six or fewer characters, starts with a letter) that identifies the file. Since this is ordinarily a character expression it must be enclosed by apostrophes—for example, 'MYDATA'.

stat is a character expression that specifies whether the file already exists or is to be created. The valid values are

'OLD'   File ⟨flname⟩ already exists in the system
         Note, the apostrophes must be included.
'NEW'   The file does not yet exist in the system.

The STATUS = stat field is also optional and if it is omitted the value of stat is UNKNOWN.

acc is a character expression that specifies whether the file is of sequential or direct-access type. The valid values of acc are

'SEQUENTIAL'	file ⟨flname⟩ is SAM
'DIRECT'	file ⟨flname⟩ is DAM

The ACCESS = acc specification is optional and if it is omitted the file is assumed to be of SEQUENTIAL type.

blnk is a character expression used to specify whether blanks are to be interpreted as zeros or blanks. It may have the following values.

'ZERO'	All blanks, other than leading blanks are treated as zeros.
'NULL'	Blanks appearing in numerically formatted lines are ignored, except that a line of all blanks is treated as zero.

The default value is BLANK = 'NULL'.

The expression UNIT = 9 may be shortened to simply

```
OPEN(9,FILE = 'MYDATA')
```

which, because of the default assignments is the same as

```
OPEN(UNIT = 9,FILE = 'MYDATA',STATUS = 'UNKNOWN',
+        ACCESS = 'SEQUENTIAL', BLANK = 'NULL')
```

Notice the name of the file is a character expression and must be enclosed by apostrophes.

Most of the time we will be using a simplified form of the above statement like

```
OPEN(UNIT = 9,FILE = 'MYDATA')
```

The OPEN statement is an executable Fortran statement.

5.4   ADDITIONAL FILE MANIPULATION FORTRAN COMMANDS

In addition to connecting a data or results file to your program via the OPEN statement, Fortran provides capabilities for rewinding the entire file, for backspacing one line at a time on the file, for checking the attributes of a file (for example, whether it is SAM or DAM), and for disconnecting the file from your program when you are finished with it. These features are of great importance to nonengineering types whose programs are primarily concerned with input/output and the updating of files, not numerical computation. For this reason only a brief description of these file manipulation functions will be given here.

5.4.1 The REWIND Statement

When you use a disk data file it is always a good idea, after opening the file, to make sure that the computer is positioned at the beginning of the file. (It likely will be but it won't hurt to make sure.) This is done with the REWIND statement in Fortran, which is of the form

$$\text{REWIND} \quad \langle unit\ number \rangle$$

Another common application of this command is to read a large amount of data from a data file, execute the calculation, change a parameter or two, and reread the same data file after first REWINDing the data file. The REWIND statement is an executable Fortran statement.

5.4.2 The BACKSPACE Statement

A statement similar to the REWIND statement but that backspaces only one line at a time is the BACKSPACE command. The form of the BACKSPACE command is

$$\text{BACKSPACE (UNIT = i)}$$

or simply

$$\text{BACKSPACE i}$$

where i is the unit number of the file to be backspaced. Only sequentially accessed files (SAM) can be backspaced. If the file is already positioned at the beginning, a backspace command has no effect.

An obvious application of the BACKSPACE statement might be to read a data file down to a particular line, backspace, and then rewrite the data line with corrected information back into the same file. Sadly, this is *not* possible with sequentially accessed (SAM) files. You cannot both READ from and WRITE to the same file in a program.[1] Thus, SAM files cannot easily be updated using the BACKSPACE statement. In applications where the updating of large files is important, direct access (DAM) files should be used. (See Section 5.6.)

---

[1] Technically, writing to a SAM file is always possible. The difficulty is that the information on the file that follows the just-written line is destroyed. This is acceptable if we are always writing information sequentially but could be disastrous if an attempt is made to alter a line in the middle of the file.

The only real utility of the BACKSPACE statement is related to the diagnosis of data-reading errors. The idea is to include an "ERR = " option in the READ statement; then, if an error is detected during the read, branch to a section that backspaces and rereads the line with an A72 format. The line is then printed as is and perhaps the problem can be identified.

5.4.3 The CLOSE Statement

After you are finished with a file, the file may be disconnected from your program with the CLOSE statement. The specification options for the CLOSE statement are very similar to those for the OPEN statement. Thus, after reading data from the data file OPENed above, you could close the file with

```
CLOSE(9, FILE = 'MYDATA')
```

Closing a file is not required because all files are automatically closed when execution of the program is completed. Note: you should never attempt to OPEN, CLOSE, or REWIND the files INPUT and OUTPUT, for obvious reasons.

5.4.4 The ENDFILE Statement

The END-OF-FILE mark is useful when the 'END = ' option is employed in READ statements. An END-OF-FILE mark can be written on a SAM file with the Fortran statement

```
ENDFILE <unit number>
```

It is not required to specifically mark the end of your data files since an END-OF-FILE mark will automatically be placed on the file after it is written and disconnected from the program used to create it. The END-OF-FILE mark does not necessarily mean the end of information on a file and can be used to partition a large SAM file into subfiles. Reading past an END-OF-FILE mark is, however, somewhat tricky. First, the READ statement should contain the 'END = ' option. Then, if an END-OF-FILE mark is encountered, the data file is closed and reopened. The next READ statement will read the line (if any) after the END-OF-FILE mark.

```
      . . .        . . .       . . .
      . . .        . . .       . . .
      READ(24,*,END = 18)X,Y,Z
      . . .        . . .       . . .
18    CLOSE(24)
      OPEN(24,FILE = <FILE NAME>)
      READ(24,*)A,B,C
```
(The values for A, B, C follow the first END-OF-FILE mark on the data file.)

The steps involved in creating a data file at the terminal are similar to those you used in setting up a Fortran program. The specific computer instructions are quite system-dependent but basically they will involve

1. Log-on
2. CREATE a file (or some similar instruction). You will have to name the file and perhaps specify
   a. That the file is 'NEW'.
   b. The type of the file is "data" (as opposed to "Fortran")
3. After the file is created you may have to instruct the computer to save the file on the disk.

The following file was entered and saved as a data file with the name GRADES. The information on each line consists of a student's last name (enclosed in apostrophes), class (1 = freshman, 2 = sophomore, etc.), college (1 = Arts and Science, 2 = Business, 3 = Engineering, 4 = Education, 5 = Architecture), hour quiz grades(3), homework, and final exam.

```
column          1              2              3
        1234567890..........0..........0......
        'WILSON      ', 2,3,71,65,82,80,77
        'GREELEY     ', 2,1,95,92,91,30,85
        'NOVAK       ', 3,1,66,50,59,66,62
        'CHEN        ', 2,5,77,75,80,86,83
        'STRAUSS     ', 1,1,91,96,93,88,94
        'REEVES      ', 2,2,71,65,80,80,72
        'HUNSICHER   ', 2,3,82,89,91,75,84
        'LEVY        ', 2,3,61,68,60,42,57
        'BROWN       ', 2,4,71,80,77,65,73
        'CASSIDY     ', 2,3,82,71,88,56,71
        'TAYLOR      ', 3,2,66,75,77,67,82
        'STEPHENSON  ', 2,3,45,60,62,21,51
        'NELSON      ', 2,3,91,86,94,92,91
        'MCDERMITT   ', 2,1,71,65,66,61,69
        'END - DATA'
```

The program ASSIGN (Figure 5-1) reads the file GRADES (several times), computes a final score (the sum of the quizzes plus homework plus three times final exam) for each student. It rereads the file once for each college and lists the students in that college and his or her final grade. Note rereading the data file in this manner is extremely inefficient. We will discover more efficient methods for problems of this type in Chapter 7. A pseuodocode outline of the program follows.

Figure 5-1 Fortran program for computing student grades.

```
        PROGRAM ASSIGN
*
        CHARACTER GRADE*1, NAME*10
        INTEGER CLASS,COLLGE,Q1,Q2,Q3,HW,EXAM,TOTAL,K
*
        OPEN(2,FILE = 'GRADES',ERR = 99)
*
        K = 1
   1    REWIND(2)
          WRITE(*,*)'RESULTS FOR STUDENTS IN COLLGE = ',K
   2            READ(2,*)NAME,CLASS,COLLGE,Q1,Q2,Q3,HW,EXAM
               IF(NAME .EQ. 'END - DATA')THEN
                   K = K + 1
                   IF(K .LE. 5)THEN
                       GO TO 1
                   ELSE
                       PRINT *,'END OF ALL GRADE ASSIGNMENTS'
                       STOP
                   END IF
               END IF
* --
* --                COMPUTE STUDENT LETTER GRADE
* --
               TOTAL = Q1 + Q2 + Q3 + HW + 3*EXAM
               IF(TOTAL .GE. 600)THEN
                   GRADE = 'A'
               ELSE IF(TOTAL .GE. 540)THEN
                   GRADE = 'B'
               ELSE IF(TOTAL .GE. 475)THEN
                   GRADE = 'C'
               ELSE IF(TOTAL .GE. 410)THEN
                   GRADE = 'D'
               ELSE
                   GRADE = 'F'
               END IF
*
               IF(COLLGE .EQ. K)THEN
                   WRITE(*,*)'STUDENT-',ID,' GRADE = ',GRADE
               ELSE
                   GO TO 2
               END IF
*
  99 PRINT *,'ERROR IN OPENING FILE GRADES'
        STOP
        END
```

```
PROGRAM ASSIGN
OPEN data file
For class K = 1 (freshmen)
1       REWIND data file
        PRINT headings for college K
2       READ student data line
                End-of-data marked by name 'END - DATA'
                IF End-of-data THEN
                        Increment class, K = K + 1
                        IF class ≤ 5 THEN
                                GO TO rewind data, statement 1
                        ELSE
                                terminate program
                        END IF
                END IF
        Compute student letter grade
        IF student in class K THEN
                PRINT student grade
        ELSE
                return to READ next student line, statement 2
        END IF
```

Notice that, in addition to the inefficient rereading of the data file, this program recomputes each student's grade several times unnecessarily. How would you change the code to correct this?

5.6 THE USE OF DIRECT ACCESS (DAM) FILES

Sequentially accessed (SAM) files are the most common type of data file used in a typical engineering application. With a SAM file the information is read from or written to just as you commonly would with a list—that is, in order, one line after another. There are however two applications in particular for which the properties of SAM files are not well suited. The first is the reading of information directly from the middle or end of a very large data file. The second is the reading of a line of information and then correcting and rewriting the same line on the file—that is, updating the file. If these two types of tasks are or will be an important part of your work, you should be familiar with the manipulation of direct access (DAM) files.

5.6.1 The Concept of a "Record" on a DAM File

On a sequential file, a record is simply a line of information that may be of any length, limited only by the associated I/O devices. Ordinarily, the data lines or records will not have explicit line numbers and, to read the information on the 13th line, we have to first read through the preceding 12 lines.

On a direct access file, each record (or line) *must* have a corresponding record number (or line number). These records are then positioned on the file in random order. A particular record may then be accessed *directly* by specifying its record number. This is illustrated in Figure 5-2.

From the way in which information is stored on a DAM file we see that a command to REWIND the file would have no meaning. A DAM file does not have a conventional beginning nor end. Also, attempts to write an END-OF-FILE mark on a DAM file will cause an error.

Figure 5-2 A comparison of the structure of SAM and DAM files.

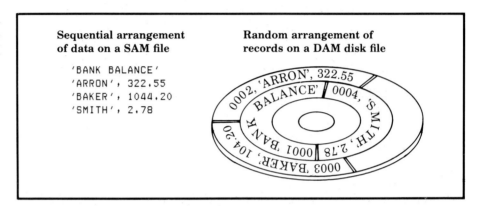

A very important constraint on the use of DAM files is the requirement that *all* records on the file have the same length. For this reason, list-directed I/O is *not possible* with direct access files. The record length is specified when the file is created and is ordinarily expressed as the total number of computer words (or on some computers, bytes) to appear in a record.[2] To determine the maximum record length you will need when creating a direct-access data file, simply count the maximum number of numerical values to be included in a record. Each numerical quantity of types REAL and INTEGER occupies one word (or *n*-bytes). Character variables are a bit more troublesome. Depending on the computer, from 4 to 10 characters are stored in a computer word. Thus if the variable NAME is of type CHARACTER*17, it would occupy from 2 to 5 computer words. If you don't know how many characters are in a computer word, assume it is 4 for safety. Obviously, unpredictable errors will be caused by writing beyond the specified length of a record. Also, each record need not be filled.

---

[2] A DAM file must be specified as either FORMATTED or UNFORMATTED. If the file is of type FORMATTED, the record length is given in terms of the maximum number of spaces or characters to be stored in each record. The DAM files described in this chapter will be of form UNFORMATTED. Reading from or writing to a FORMATTED direct access file is less efficient and requires the use of FORMAT statements that will be described in the next chapter. I should also point out that list-directed I/O (for example, PRINT *, . . .) is actually FORMATTED in the sense that the appropriate format specifications are supplied by the compiler rather than by the user.

5.6.2   Creating a DAM File

Whether you are creating a new file or reading from an existing data file, the OPEN statement must first be employed in the Fortran program to establish a link between the program and the file. The form of an OPEN statement used to access a DAM file is similar to that used in connection with a sequential file. The exceptions are, for a DAM file, that the argument of the OPEN statement must include the following additional information:

The ACCESS type of the file. Since the default is type sequential, the argument list for a direct-access file *must* contain the phrase

```
ACCESS = 'DIRECT'
```

The arrangement of the information must be specified as either FORMATTED or UNFORMATTED. So, for our applications, the OPEN statement should contain the phrase

```
FORM = 'UNFORMATTED'
```

If this specification is omitted, the default form is UN-FORMATTED.

The record length. As mentioned above, the required record length will depend on the problem at hand. Be sure that ample room is provided on each record. The record length is specified by including a phrase of the form

```
RECL = 100
```

This specification *must* be included when opening a direct-access file.

A typical OPEN statement to connect an existing data file of type DAM, named MYDATA and associated with unit number 43, would be

```
 OPEN(UNIT = 43,FILE = 'MYDATA',ACCESS = 'DIRECT',
+      FORM = 'UNFORMATTED',RECL = 100)
```

Once again, UNIT = 43 could be abbreviated as simply 43.

The procedure to write information to this file is now rather simple. In addition to the unit number, all READ/WRITE statements for DAM files must also include the *record number* of the associated line of information.

```
      READ(<unit>,REC = <integer>)...
      WRITE(<unit>,REC = <integer>)...
```

For example, to write names and bank balances to a DAM data file after they are entered at the terminal, we could employ the following code:

```
        CHARACTER NAME*20
        REAL       BAL
        INTEGER    LINE
*
        LINE = 0
        OPEN(16,FILE = 'BANKBL',ACCESS = 'DIRECT',FORM = 'UNFORMATTED',
     +       RECL = 50)
*
    1   PRINT *,'ENTER NAME (ENCLOSED IN SINGLE QUOTES)'
        PRINT *,'AND BANK BALANCE (SEPARATED BY A COMMA)'
        PRINT *,'ENTER ''END-OF-DATA'' TO TERMINATE'
*
        READ (*,*) NAME,BAL
        IF(NAME .NE. 'END-OF-DATA')THEN
            LINE = LINE + 1
            WRITE(16,REC = LINE)NAME,BAL
            GO TO 1
        END IF
        STOP
        END
```

5.6.3 Reading and Correcting Information on a DAM File

As indicated above, the only information that must be added to an ordinary READ statement when you are reading from a DAM file is the location (record number) of the line containing the data. Of course, the information must be read from the file in *exactly* the same manner in which it was written. The variables in the READ list *must* match the variables in the WRITE by type. The primary advantage of DAM files, at least for our purposes, is that the READ can be followed by a WRITE to the *same* file. (This is not permitted with sequential files.)

As an example, suppose you wish to correct the bank balance of John Smith. If we kept a list of the names and associated record numbers on file BANKBL, we would first scan the list and perhaps find that Smith's data line is on record 3017. After the bank balance file has been opened (unit = 88), the code to correct Smith's balance would simply be

```
READ *,BAL
WRITE(88,REC = 3017)'SMITH, JOHN            ',BAL
```
                                ⟨Note: this character constant
                                must be of length 20⟩

Direct access files are a relatively new addition to Fortran and engineers and scientists are just beginning to realize the advantages and convenience of DAM files for some applications.

1. A data file with the name of GRADES contains the results of an examination. The data on each line includes:
   1. Student ID
   2. Exam score

   Write a complete program to read the data file and:
   a. Write the ID's and exam scores of students who have failed, (score < 60) as they are encountered.
   b. Count the total number of exams and the number of each grade A–F assigned according to

$$F < 60 \leq D < 70 \leq C < 80 \leq B < 90 \leq A$$

   c. Determine the ID's of the students with the maximum and minimum score.
   d. Determine the average of the exam.
   e. Determine the class grade-point average (GPA) on the $A = 4.0$, $F = 0.0$ basis.

2. Write a program that
   a. Creates a data file of 101 lines. The first line contains values for dt, v_x, and v_y, and the numbers on successive lines are the values of t_i, x_i, y_i, where

$$x_i = v_x t_i$$
$$y_i = v_y t_i - (g/2)t_i^2$$
$$t_i = i\, dt \qquad \text{for } i = 1 \text{ to } 100$$

   and

$$v_x = 10.5 \qquad v_y = 51.0 \qquad dt = 0.1$$

   b. Write a program that reads and prints the data file.

3. Write a program that will read a file called *roster*. The first line of the file contains the total number of data lines that follow. On each of the subsequent lines is a student's ID number and the number of credits earned to date. If the number of credits is ≥ 130 the student has graduated, so rewrite the entire file deleting these students.

4. a. Write a program that will create a data file consisting of a line indicating a bridge element (integer) followed by lines containing a position on the element and the measured stress at that position. The end of the data for this bridge element is to be marked with an END-OF-FILE mark. Similar data values for 6 more bridge elements are to be included on the *same* file in a similar manner. The data is to be entered at the terminal.
   b. Write a program that will read an integer from the terminal and print the position-stress data for that element by reading the above data file.

5. a. Write a program that will read the SAM file used in program ASSIGN of Section 5.5 and write a more accessible DAM data file containing the same information. In addition the program should count and include in the DAM file the total number of students in each college. The information should be written to the DAM file so the record numbers on the file are of the form

$$\text{Rec. No.} = 1000(\text{COLLGE} - 1) + I$$

where COLLGE = college (1 → Arts + Science, etc.) and I is a counter from 1 to 999. Thus,

$$\text{Rec. No.} = 2006$$

includes the information for the sixth student in the engineering (3) college.

b. Rewrite the program ASSIGN to accomplish the same results using the newly established DAM data file.

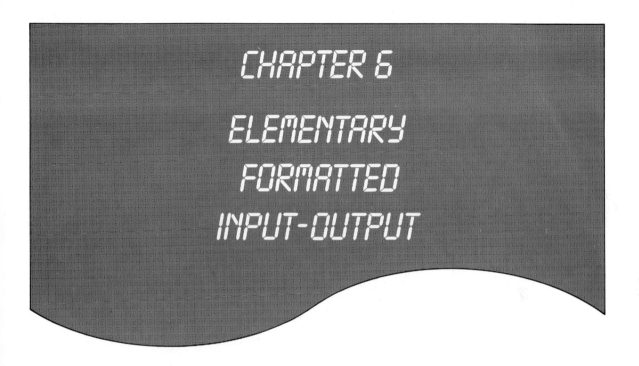

CHAPTER 6

ELEMENTARY FORMATTED INPUT-OUTPUT

6.1 INTRODUCTION

We have already seen a few simple examples of getting numbers in and out of the computer in Section 2.9. In almost every case, it is preferable to have the computer print the results in the form of tables with appropriate headings and some explanatory text. This necessitates a large number of editing decisions. Before the results are printed or the data file is read, the following questions must be addressed:

In what form are the numbers to be printed?
  Integers?
  Reals without exponents?
  Reals with exponents?
  How many significant figures are to be printed?
Where will the numbers appear on the page or where are the numbers to be found on an input data line?
What text appears with the numbers, table headings, etc.?

All of these decisions are made in the form of FORMAT statements, and clearly all of this arranging of output (and input) can take considerable time.

Formatting the input and output is the most tedious aspect of programming, but it is not terribly difficult to understand, and if you have spent a lot of time and effort in getting some intricate code to execute efficiently and correctly, it is certainly worth some additional time arranging the output in a neat, clear, and pleasing form.

6.2 FORMATTED I/O STATEMENTS

To this point, all of the I/O statements that we have used have been written in terms of the list-directed I/O option available in Fortran 77. This was effected by using an asterisk (*) in either the PRINT/WRITE or READ statements. The various forms of I/O statements employed thus far are

READ *,⟨input list⟩	*Simply separate numbers by commas. The values are read from file INPUT.*
READ(34,*)⟨input list⟩	*Same as above except the numbers are read from file opened with number 34.*
READ(*,*)⟨input list⟩	*The first asterisk means that the numbers are read from the file INPUT. This form is identical to READ *,⟨list⟩.*
PRINT *,⟨output list⟩	*List-directed output to file OUTPUT.*
WRITE(42,*)⟨output list⟩	*List-directed output to file opened with number 42.*
WRITE(*,*)⟨output list⟩	*The first asterisk means that the numbers will be written to the file OUTPUT. This form is identical to PRINT *,⟨list⟩.*

Each of the I/O statements listed above can be used to write or read information according to specific editing instructions that are included in a FORMAT statement. The asterisk that designates list-directed I/O is simply replaced by the statement number of the associated format statement. (Format statements are described in the next section.) Thus

	Means
READ 88,⟨input list⟩	*Read from file INPUT according to format number 88 the values for . . . .*
WRITE(2,2)⟨output list⟩	*Write to file number 2, according to format number 2 the values . . . .*

The form of a Format statement is:

$$j \quad \text{FORMAT}(\text{spec}_1, \text{spec}_2, \text{spec}_3, \ldots)$$

The integer j is a unique statement number and of course the word FORMAT begins in column 7 or later. Inside the parentheses is a list of formatting specifications separated by commas (or slashes, /, see Section 6.3.2) relating to the numbers, variables, or characters that are to be read in or printed out by the relevant READ or WRITE statement. The FORMAT statement is a *nonexecutable* statement and it may appear anywhere in the program, either before or after the associated READ/WRITE statement.

6.3.1   Format Specifications—Numerical Descriptors

The F Format(Floating-Point or Real Numbers Without Exponent)

The form of the F format specification is

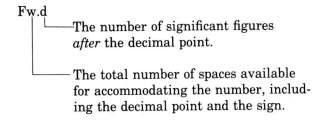

Fw.d
- The number of significant figures *after* the decimal point.
- The total number of spaces available for accommodating the number, including the decimal point and the sign.

For example,

```
        READ(5,66)X,Y
66      FORMAT(F10,3,F5,1)
```

If X = 271.736 and Y = 3.1, the data line or input line at the terminal would be

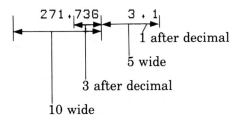

1 after decimal

5 wide

3 after decimal

10 wide

and the same numbers printed with

```
            WRITE(6,13)X,Y
      13    FORMAT(F7.2,F6.2)
```

would result in

```
                    1               2           3
      1234567890.........0.........0
       271.74  3.10
```

```
      |◄──────►|◄────►|
              6 wide
        |
      7 wide
```

Some Technical Points to Remember Regarding Formatted I/O

When printing either on the line printer or on the terminal screen, the first column of each line may not be displayed. Many computer systems use the first column of an output line for vertically positioning the output (carriage control). This will be described in Section 6.5.1. For the present, each line of output should skip over column 1.

When REAL numbers are printed, the values displayed are *rounded*, not truncated. In the above example, the value printed for X was 271.74. The number stored in memory for X is still 271.736, however.

Ordinarily when reading data the machine will read a blank space as a blank, i.e., it is ignored. However some installations have set up the read statement so that a blank is interpreted as a *zero*. Thus the above input line would actually be read as

      000271.7360003.1

and the same values would be assigned to X and Y. However, if the first number is accidentally misplaced so that it spills over into the field of the second number, as

```
                    1               2           3
      1234567890.........0.........0
        271.736  3.1
```

which is the same as

```
                    1               2           3
      1234567890.........0.........0
      0000271.73603.1
```

then X would be read as 271.73 and Y as 603.1, significantly different from what was intended. Most computers that are used for interactive computing read a blank as a blank, and in this case the assignments would be X = 271.73 and Y = 63.1, which is still wrong. For this reason it is essential that you verify that each value read in has resulted in the intended assignment. You should *always echo print* values read in. Moreover, whenever possible a list-directed READ is safer and easier to use than a formatted read.

If the number entered in a READ command fits in the prescribed width (w), and the number has a decimal point, then the placement of the decimal point overrides the d specification in the FORMAT statement. Thus the first number below

```
         1             2           3
1234567890.........0.........0
271.73600     3.1
```

is entered as an F10.5 and the original format was F10.3. This number would be read correctly and the assignment for X would again be 271.736.

If the numerical value being read does not contain a decimal point, the d specification in the F format will act as a negative power-of-10 scaling factor. Thus if the number 1234 were entered with an F5.2 format, the value stored would be

$$1234. \times 10^{-2} = 12.34 \qquad d = 2$$

It is suggested that you always include a decimal point when entering real numbers.

The I Format (Integers)

The form of the I format specification is

Iw
└── Total width of the field

For example,

```
        READ(5,21)N
21   FORMAT(I9)
```

If the value to be assigned to N is 23, then the data line or terminal input line would look like

```
           1               2               3
1234567890.........0.........0
           23
```

Of course, there can be no decimal point anywhere in an integer field. Notice, if the number were misplaced in the field as

```
           1               2               3
1234567890.........0.........0
           23
```

the number would still be read as 23 if blanks are ignored and as 230000 if a blank is read as a zero.

Attempts to read or print numbers by means of a format that conflicts with the type of the number will result in errors. For example, printing the integer M using an F-type format will cause an incorrect value to be printed. Other examples of I/O errors resulting from a mismatch of format and number type are listed below.

Variable type	If READ with format	Result
INTEGER	F	Value assigned will be incorrect
REAL	I	Execution time error— decimal point in I field
	If WRITE with format	
INTEGER	F	An output error may result. (The program will continue.)
REAL	I	An output error will result. (The program will continue.)

When printing either real or integer values some care must be exercised to ensure that the field width (w) is sufficiently large to accommodate the anticipated size of the numbers printed. If the field width is too small, the computer will instead fill the field with asterisks. For example, an attempt to print $X = 123.45$ and $K = 1234$ with formats F5.3 and I3, respectively, will result in output that resembles that below,

```
*****
***       * Indicates field-width overflow
```

The E Format (Floating-Point or Real Numbers with Exponent)

The form of the E format specification is

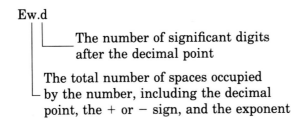

Ew.d

The number of significant digits after the decimal point

The total number of spaces occupied by the number, including the decimal point, the + or − sign, and the exponent

A typical real number with exponent is

$$-0.12345E-05$$

which occupies twelve spaces (w) and has five digits (d) following the decimal point. In addition to the mantissa digits (12345), seven spaces are required for the leading sign (1), leading zero (1), decimal point (1), E (1), exponent sign (1), exponent power (2). Thus when using E formats, $w \geq d + 7$. For example,

```
        READ(5,22)X,Y
22      FORMAT(E14.5,E10.2)
```

If $X = 2.71736E+02$ and $Y = 31.0E-11$, the values entered would appear as

```
        1            2            3
1234567890.........0.........0
   2.71736E+02 31.00E-11
```

An important point to keep in mind when reading numbers with an E format is that the exponent *must* be placed at the extreme right of the field and of course the numbers must fit within the specified field width. If the format specification were

```
22      FORMAT(E13.5,E10.2)
```

the above line would assign

```
X  =  2.71736E + 0  =  2.71736×10⁰  =  2.71736
Y  =  2 31.00E - 1  =  23.1    (or 2031.00E–1 if a blank
                                 is interpreted as a zero)
```

which are incorrect assignments. As with F format, if a decimal point is included with the number, it overrides the d specification of the format.

Thus, if the above numbers were read with either of the formats below, correct assignments would still result.

```
22   FORMAT(E14.7,E10.3)
22   FORMAT(E14.0,E10.0)
```

Writing numbers with the latter format would result in only the exponent being printed.

In summary, you should remember the following points concerning the Ew.d format:

For output especially, $w \geq d + 7$. For input the decimal point included in the number overrides the d specification of the format.

For input, the exponent field must be right-adjusted, i.e. positioned to the extreme right of the field.

The mantissa (the fractional part of the number) should contain a decimal point.[1]

For input, the exponent should contain no more than two digits.

If the E field is missing on input, the E format is used as if it were an F format. So, if the number entered contains a decimal point, it will be correctly read and stored. For example, the number 12.34, read with a format E10.3, would be stored as 12.34.

The G Format (General-Purpose Numerical)

The G format combines features of both the F and E formats and is often found to be quite useful in constructing tables of numbers. The G format can be used for both input and output, but since it is particularly useful in output, only that application will be described here.

The form of the G-format specification is similar to the E and F formats, namely:

Gw.d

    The number of significant digits to be displayed

    The total number of spaces allotted to the number

---

[1]As with the F format, the decimal point is optional when reading real numbers. If the decimal point is omitted, the d specification in the format Ew.d will act as a negative power-of-ten scaling factor. Thus, reading the number 1234E+4 with a format E9.2 would result in the value

$$(1234.E+4) \times 10^{-2} = 0.1234E+6 \qquad d = 2$$

being stored.

A number printed with a G format may or may not be printed with an exponent according to the following rule:

If the number X is printed with a Gw.d format, then the number is displayed *without* an exponent if $|X|$ is in the range

$$10^d > |X| > 0.1$$

Numbers outside this range are printed with an exponent.

In either case, the number will occupy w spaces on the line. The exponent, if present, will be placed in four spaces on the extreme right. If no exponent is displayed, these spaces will be blank. A few examples are given below.

```
    X = 1.23456
    Y = 123.456              <Printed output>
    Z = 123456.7              1     1     2
12  FORMAT(G15.4)      1...5....0....5....0
    WRITE(*,12)Z             0.1234E+04
    WRITE(*,12)X             1.234
    WRITE(*,12)Y             123.4
```

Notice, the d specification (for example, the 4 in G15.4) denotes the total number of significant digits to be displayed, either with or without an exponent. This is to be contrasted with the same symbol in an Fw.d specification wherein the d designates the number of significant digits that follow the decimal point.

Frequently the size of computed numbers may not be known very accurately in advance. The G format permits the safety of an E format in accommodating large or small values along with the convenience of printing numbers in a more readable F format if the values turn out to be of suitable size.

Some compilers even permit INTEGER values to be printed using a G format. This is seldom useful and I would discourage the practice.

6.3.2 Format Specifications—Position Descriptors

The X Format (Skip a Space)

The form of The X format specification is

        nX

        └────Number of spaces to be skipped

For example,

```
                    I = 12
                    J = 34
                    K = 56
                    WRITE(6,6)I,J,K
            6       FORMAT(1X,I2,9X,I2,5X,I2)
```

would result in

```
              1                 2               3
         1234567890.........0.........0
         ─────────────────────────────────────────
           12              34      56
```

Recall, we are avoiding the printing of numbers in column 1 for the moment.

The Slash (/) Format (Skip to Next Line)

An example of the slash format is

```
                    WRITE(6,5)I,J,K
            5       FORMAT(1X,I2/2X,I2///3X,I2)
```

These statements would result in

```
              1               2
         1234567890.........0
           12
           34


             56
```

Note that

> Multiple line skips cannot be specified as n/. To skip five lines and begin on the sixth, you could use //////. An alternative is discussed in the next section.
>
> The line skip specification need not be separated from other format specifications by commas. In fact, format specifications may be separated by either commas or slashes. However, you may find it easier to read the format statement by explicitly separating off the slashes by commas. Thus format 5 above could also be written as

```
            5       FORMAT(1X,I2,/,2X,I2,///,3X,I2)
```

Immediately following a slash, the position of the format specification is in column 1, the carriage control position for printed output.

The T Format (Tab)

You can have the printed output (or input) skip to a particular position on the line in much the same way you use a tab key on a typewriter by means of the T format. This format has the forms

Tn   Tab to column number n moving either right or left

TRn   Skip forward (tab right) by n positions, measured from the current position.

TLn   Skip backward (tab left) by n positions, measured from the current position. If n is greater than the current column number, it will position in column 1.

These forms are particularly useful in constructing tables. For example,

```
      READ *,X,Y,Z
      WRITE(6,10)
      WRITE(6,11)X,Y,Z
 10   FORMAT(2X,'OUTPUT TABLE')
 11   FORMAT(T15,'VALUES',/,T17,F3.1,/,T17,F3.1,TL6,
     +  F3.1)
```

with input of X = 1.0, Y = 2.0, and Z = 3.0, will result in

```
        OUTPUT TABLE
                  VALUES
                    1.0
                  3.02.0
```

6.3.3 Repeatable Format Specifications

With the exception of the slash and T formats, each of the formats discussed thus far may be repeated by preceding the format specification by a multiplying integer factor. That is,

```
        3F5.2
```

is the same as

```
        F5.2,F5.2,F5.2
```

It is also possible to have multiples of combinations of format specifications by enclosing groups in parentheses and preceding the parentheses by a multiplicative factor. For example,

```
      WRITE(6,33)X,IA,A,IB,B,IC,C
 33   FORMAT(5(/),3X,F10.5,3(/,3X,I2,2X,F5.2))
```

will print a total of nine lines.

6.4 ELEMENTARY INPUT AND OUTPUT OF TEXT

6.4.1 Introduction

The output of most programs will consist of more than just numbers. Tables will have labels and most computed results will require some identification and explanation. The printing of text (also called Hollerith or character strings) is quite simple in Fortran 77. More elaborate procedures for handling character strings will be addressed in Chapter 8, but for the moment the elementary methods outlined below will be sufficient for most applications.

6.4.2 The Apostrophe as a String Delimiter

Up to this point, textual material and character strings have been printed by using the list-directed output statement

```
PRINT *, <character strings and/or variables>
```

in which the character strings are enclosed in apostrophes. A character string constant can also be included in a FORMAT statement and the procedure is similar: the string is simply enclosed in apostrophes and separated from other format specifications by commas or slashes. The string may contain any of the valid symbols in Fortran (see page 24) and will be printed exactly as it appears, including blanks. For example,

```
      IDAY=11
      IYEAR=86
      WRITE(6,11)IDAY,IYEAR
   11 FORMAT(5X,'TODAY IS SEPT.',I2,', 19',I2)
```

First string
14 characters

Second string
4 characters
(includes the
comma)

will produce the output

```
            1         2         3
  1234567890.........0.........0.....
      TODAY IS SEPT.11, 1986
  5X                I2    I2
```

Also, the statements

```
            WRITE(6,12)
    12   FORMAT(10X,60('*'))
```

will result in a line of 60 asterisks.

As a somewhat more complicated example, the statements below will produce a table of temperature and resistance values. Obviously, some care was used in designing the FORMAT statement so that the column headings and numerical values are properly aligned and centered.

```
      WRITE(6,13)T1,R1,T2,R1,T3,R3
   13  FORMAT(5X,'A TABLE OF ELECTRICAL RESISTANCE',
      +              ' VS,TEMPERATURE',
      +//,9X,'TEMPERATURE      RESISTANCE',/,
      +   9X,'  (DEG-C)          (OHMS)',/,
      +3(10X,F10,3,5X,F10,1,/))
```

Assuming that the variables have been assigned values, these statements would produce the following output:

```
A TABLE OF ELECTRICAL RESISTANCE VS, TEMPERATURE

      TEMPERATURE      RESISTANCE
       (DEG-C)          (OHMS)
       124,300         10344,3
        93,200         10113,7
        21,700          9958,4
```

If the string itself contains an apostrophe, it can be printed by typing two consecutive apostrophes.

```
      PRINT 14
   14  FORMAT(2X,'DON''T FORGET TODAY''S ASSIGNMENT')
```

results in

```
      DON'T FORGET TODAY'S ASSIGNMENT
```

Note, on some printing devices, the apostrophe is represented as ↑.

6.4.3 Replacing the ∗ with Format Specifications

Unless the formatting of a PRINT or WRITE statement is quite complicated, it is usually more convenient to use the list-directed form of the output statements, i.e., PRINT ∗, or WRITE(6, ∗), especially for preliminary results. However, there are frequently situations that fall somewhere between simple and complicated. For example, if you wished to print A = 5.0 and B = 3.0 on separate lines and centered on the page, this could easily be done with a FORMAT statement or you may simply replace the '*' with the required format specification. That is,

```
        PRINT 11, A,IA
        WRITE(6,11)B,IB
  11    FORMAT(25X,F5.2,/,27X,I2)
```

could be replaced by

```
      PRINT '(25X,F5.2,/,27X,I2)',A,IA
      WRITE(6,'(25X,F5.2,/,27X,I2)')B,IB
```

or

```
      PRINT '(2(25X,F5.2,/,27X,I2))',A,IA,B,IB
```

Notice, when replacing the asterisk with format specifications, they must be enclosed in parentheses and delimited fore and aft by apostrophes. Thus, if the format specifications themselves contain an apostrophe, an error will result. To print labels in addition to numbers using this option, a slightly different approach is required and is explained in the next section.

6.4.4   The A Format for Printing Text

The A Format

A variable that is of type CHARACTER is printed by means of the alphanumeric format specification, Aw, where w is the length of the character expression to be printed. For example, to print a character variable of length 24, the format specification would be

```
       CHARACTER PHRASE*24
       PHRASE = 'CHARACTER VARIABLE OF LENGTH 24'
       WRITE(6,21)PHRASE
  21   FORMAT(5X,A24)
```

Fortran 77 also permits a very convenient variation of the above A format specification. In place of 'A24' in the format statement we could simply use 'A' and the computer will count the length of the variable and allot the proper number of spaces.

```
      CHARACTER NAME*10,STATE*2
      NAME = 'ANDERSON  '
      STATE = 'PA'
      WRITE(6,22)NAME,STATE
22    FORMAT(2X,'MR. ',A,' LIVES IN ',A)
```

In summary, the form of the A format specification is

Aw   or   A

And it is used to print character-type variables or expressions. Next consider how labels might be printed with the list-directed form.

The Form PRINT '(A)',⟨character string⟩

Text may be printed using list-directed output by employing the A format in place of the *.

```
      CHARACTER STAR*1
      STAR = '*'
      PRINT '(A)',STAR
```

Note that the format specification must be enclosed in parentheses. An even more concise procedure is

```
      PRINT '(A)','*'
```

Variations on this particular form are quite easy to construct and can make the chore of obtaining intermediate results from a program almost painless.

```
PRINT '(5X,A,I2,A,F6.2)','THE LOAD ON BEAM NO. ',I2,
+       ' IS ',LOAD
WRITE(6,'(5X,A,E9.2,A)')'THE FLOW RATE IS ',FLOW,
+       'CUBIC METER/SEC'
```

Input of Character Variables

If a variable has been designated as type CHARACTER, it may be assigned a value by means of assignment statements as in the previous examples, or by reading input data using the A format. Thus if the code below is used to read data entered at a terminal, an entire line of 80 characters will be stored in the variable LINE.

```
      CHARACTER LINE*80
      READ '(A)', LINE
```

Note that ⟨READ *, LINE⟩ will only work if the character string to be as-

---

signed to the variable LINE is enclosed in apostrophes. This is not required with the A format. Also, if the length of the character string being read is less than 80 characters, the rest of variable LINE is filled-in with blanks. The significant characters are left adjusted.

From the discussion thus far it should be apparent that facility in using character variables for manipulating text is an important and useful part of overall Fortran 77 fluency. However, it is not a skill that comes easily, but requires considerable practice. Character string manipulation is a feature that was added to Fortran to broaden the usefulness of the language to applications in fields outside its historical base of science and engineering. Alphabetizing a list or an inventory search is done using character variables. As important as these uses may be, our primary concern as engineers is plain old "number crunching," and so more detailed explanations of character variable features in Fortran 77 will be postponed to Chapter 8.

6.5 SOME TECHNICAL POINTS TO REMEMBER

6.5.1 Carriage Control

All of the programs and examples thus far have avoided printing in column 1. The reason for this is that Fortran uses the character placed in column 1 of each line as a vertical spacing or carriage-control command. This character is not printed. Carriage-control commands apply particularly to printed output, they do not generally affect the output at a terminal. The four characters that are used for vertical spacing are listed in Table 6-1.

Table 6-1 Vertical spacing control characters.

Character	Effect
Blank	Space down one line then print
0	Space down two lines then print
1	Advance to first line of next page
+	No advance before printing; allows overprinting

The list-directed forms of output (PRINT *, ⟨list ⟩) always insert a blank at the beginning of each line to ensure single spacing. The program that follows illustrates some uses of carriage control. In particular, notice the use of the tab format and how Mr. Smith's score is given a commendation. In addition, the program terminates by comparing each name read with the character string 'END OF SET'.

```
        PROGRAM SCORES
        CHARACTER NAME*10
        PRINT '(A)','1'    ⟨Ejects to next page⟩
        PRINT '(1X,A,/,T5,A,T20,A)',
       +'RESULTS OF FINAL EXAM','NAME','SCORE'
*
*                       CONTINUE TO READ NAMES UNTIL
*                       THE PHRASE 'END OF SET' IS READ
*
     1  READ(5,'(A10,F4.0)')NAME,SCORE
           PRINT '(T2,A10,T20,F4.0)',NAME,SCORE
           IF(SCORE .EQ. 100)THEN
               PRINT '(A1,T30,A)','+','EXCELLENT'
           END IF
           IF(NAME .EQ. 'END OF SET')THEN
               STOP
           ELSE
               GO TO 1
           END IF
*
        END
```

Input		Output		
JONES,A.	63.	RESULTS OF FINAL EXAM		
ADAMS,J.	71.	NAME	SCORE	
SMITH,C.	100.	JONES,A	63.	
BROWN,B.	82.	ADAMS,J.	71.	
END OF SET		SMITH,C.	100.	EXCELLENT
		BROWN,B.	82.	
		END OF SET	****	

6.5.2 What If the Computer Runs Out of Format?

If the number of elements to be printed or read by a READ/WRITE/PRINT statement exceeds the number allotted in the associated FORMAT statement, the computer will always complete the entire command by reusing the same FORMAT statement, each time starting a new line. Thus

```
        WRITE(6,17)A,B,C,D,E,F,G
     17 FORMAT(1X,2F5.1)
```

will result in three lines with two numbers and a fourth with the single value associated with G. After printing A and B, the WRITE statement runs out of format and continues to print the elements in the list until the list is exhausted. Notice in the final line the entire format is only partially used. The situation for READ statements is similar. For example, suppose the following input data file is stored on the disk and named DATA3.

```
                              Input  file
      column             1               2
                 1234567890.........0

                 1.0  2.0  3.0  4.0
                 5.0  6.0  7.0  8.0
                 9.0 10.  11.  12.
                 -1.
```

and is read and printed by the following code.

```
      OPEN(37,FILE = 'DATA3')
      REWIND 37
    1 READ(37,'(3F4.0)')A,B,C,D
      PRINT *,A,B,C,D
      IF(A .NE. -1.)GO TO 1
```

The result will be

```
1.000  2.000  3.000  5.000
9.000  10.00  11.00  -1.00
```
⟨*The value 4.0 was missed.*
The READ ran out of format
and went on to the next line.⟩

and the program will fatally terminate by reading the end of the file. The clever use of −1.0 as a flag for the end of the data was missed by the computer by assigning it to D rather than A. Apparently the READ format was intended to be '(4F4.0)'.

If the format specifications themselves contain parentheses, the situation is somewhat more complicated. The rule is

> **If the computer runs out of format, it scans the format specifications *right to left* and reuses the first paired set of parentheses it encounters, or its multiple.**

Thus, assuming the variables A to G have been assigned the values 1.0 to 7.0, respectively,

```
      PRINT 77,A,B,C,D,E,F,G
   77 FORMAT(X,'THE RESULTS ARE ',/,X,20(*),/,
  +          2(1X,(2F3.0)))
```

results in

```
THE RESULTS ARE
********************
  1.02.0 3.04.0
  5.06.0 7.0
```

The recycled specification was 2(1X,(2F3.0)).

6.5.3 Additional Features of READ/WRITE Statements

Very frequently a data file will consist of an unknown number of lines and the program will read some data, compute results, return to the READ statement, and repeat the calculation until the data file is exhausted. However, if the READ statement is used beyond the last data line an execution time error is generated. This has been avoided in previous examples by placing a trailer data line at the end of the file and after each READ checking for a known flag (See program SCORES on page 197 and the program on page 198.)

As we have seen, this procedure is not always successful. A preferred option is available in the READ statement.

$$\text{READ}(\text{unit}, \text{format}, \text{ERR} = sl_1, \text{END} = sl_2)$$

The option ERR = sl_1 directs the computer to statement number sl_1 if an error was encountered during the read (e.g., a real number with an integer format). The option END = sl_2 causes a transfer to statement number sl_2 only if an END OF FILE is encountered. An example follows:

```
     READ(5,11,ERR=100,END=99)X,Y
     READ(*,11,END=99)A,B
     READ(*,*,END=99)I,J
 11  FORMAT(1X,2F5.0)
     PRINT *,'X= ',X,' Y= ',Y,
   +           'A= ',A,' B= ',B,
   +           'I= ',I,' J= ',J,
     STOP
 99  PRINT *,'END OF DATA'
     STOP
100  PRINT *,'I/O ERROR'
     STOP
     END
```

The ERR = sl option may also be included in WRITE statements.

PROBLEMS

1. Formatted READ and WRITE
   a. Give two methods of specifying the input device as INPUT (i.e., default to the terminal for data input) in a READ statement.
   b. Give two methods of testing for the end of a data file.
   c. Using only one line of Fortran, print e^{π} with an F9.6 format.
   d. If your program has a FORMAT statement that is never used, will the computer detect this and inform you of a possible error? (Try it and see.)
   e. Which of the following format specifications can be repeated by preceding by a multiplying integer? (F,T,E,/,I,X,A,')
   f. Give an example of when replacing the word READ by WRITE in a correct Fortran statement will result in a compilation time error.
   g. Can you GO TO a FORMAT statement?

2. Identify the errors, if any, in the following:
   a. `READ(*)X,Y`   d. `READ(5,5)FIVE`
   b. `PRINT *,X,Y,Z`   e. `READ *,X + Y`
   c. `WRITE (6,*)X + Y`   f. `READ(5,6),X,Y`

3. Locate any errors in the following. If none write OK.
   a. `READ(5,1)X,IX,Y,IY`   d. `READ(5,4)X,Y,IX,IY`
      `1 FORMAT(2X,F6.6,I5,2F4.1)`   `4 FORMAT(150X,2F9.3,2I10)`
   b. `READ(*,2)X,Y,IX,IY`   e. `READ(5,5)X,Y,Z,W`
      `2 FORMAT(2X,2(F5.1,1X,I5))`   `5 FORMAT(1X,F3.0,4(1X,F7.1))`
   c. `READ(*,3)X,Y,Z,W`   f. `READ(5,6)X,Y,IX,IY`
      `3 FORMAT(2X,3F7.1)`   `6 FORMAT(2E9.6,2I9)`

4. Identify the errors, if any, in the following FORMAT statements assuming an appropriate READ statement of the form READ(*,10)(list of variables).
   a. `10 FORMAT(I1,F3.0,I1,F3.0)`
   b. `10 FORMAT(2X,I1,2X,F3.0,2X,F3.0,2X,I1)`
   c. `10 FORMAT(1X,2(I2,F3.1,I1))`
   d. `10 FORMAT(2X,I6,F6.7,I2,F3.2)`
   e. `10 FORMAT (5X,I10,F7.2E+02,I4,F3.0)`
   f. `10 FORMAT(T5,I5,E8.1,I4,E4.0)`
   g. `10 FORMAT(1X,'I=',I2,'X=',F3.0,'J=',I4,'Y=',F6.2)`
   h. `10 FORMAT(T20,I5,TR10,I5,TL10,F5.1,TR10,F5.1)`
   i. `10 FORMAT(///,2(I5,2/,F4.2))`
   j. `10 FORMAT(1X,(I1,F3.1,I1))`

5. Identify the errors, if any, in the following formatted WRITE statements and their associated FORMAT statements.
   a. `WRITE(*,10)X,Y,X+Y`
      `10 FORMAT(F5.0)`
   b. `WRITE(6,11)X,X**2,X**3,EXP(X)`
      `11 FORMAT(4E9.1)`

```
c.      WRITE(6,12)X,I,Y,J
     12 FORMAT(1X,2('R=',E8.1,'K=',I10))
d.      WRITE(*,13)A,B,C,D,E
     13 FORMAT(5X,F5.1,TL5,F5.1)
e.      WRITE(*,15)'A=',A,'B=',B
     15 FORMAT(5X,A2,F3.1,/,5X,A,F3.1)
f.      WRITE(6,16)A,B,IA,IB
     16 FORMAT('0',F3.1,'+',F3.1,/,'+',2I4)
g.      WRITE(6,17)A,'B=',B
     17 FORMAT(5X,'A=',A2,2F5.0)
h.      PRINT 18,A,B
     18 FORMAT('1','A=',F5.0,'B=',F5.0,'C=',F5.0)
```

6. Using the assignments X = 2., Y = 3., Z = 1./3., I = 2, J = 3, K = 4, determine the output of the following WRITE statements. Indicate blanks as ȸ.

```
a.   WRITE(6,1)X,Y,Z
   1 FORMAT(2X,F3.1,2X,F2.0,2X,F7.4)
b.   WRITE(6,2)X,Y,Z
   2 FORMAT(/,2X,F7.5,2(/,1X,F6.3,/))
c.   WRITE(*,3)X,I,K,J,Y,Z
   3 FORMAT(5X,F2.0,'*',I2,'=',I2,/,5X,I1,'/',F3.1,'=',F7.6)
d.   WRITE(*,4)Y*Z,Y,J,J,J,J
   4 FORMAT(/,4X,F9.6,'/',F2.0,'= ',5I1)
e.   WRITE(*,5)K,J,I
   5 FORMAT(T12,I1,TL2,I1,TL3,I1)
f.   WRITE(*,6)I,J,K
   6 FORMAT(T2,I1,T5,I1,T7,I1)
g.   WRITE(*,7)I,J,K
   7 FORMAT(T2,I1,TR5,I1,TR7,I1)
h.   WRITE(*,8)I,J,K
   8 FORMAT(1X,I1)
i.   WRITE(*,9)X,Y,Z
   9 FORMAT(2X,E7.0,1X,E9.2,1X,E12.5)
```

7. What is the output from the following, assuming the assignments, X = 1.0, Y = 2.0, Z = 1./3., I = 2, J = 3. Indicate blanks as ȸ.

```
a.      PRINT '(F4.0,2X,F4.2,F2.1,2I5)',X,Y,Z,I,J
b.      WRITE(*,10)X,Y,I,J
     10 FORMAT(T5,2E9.1,/,2I1)
c.      WRITE(*,11)X,I,Y,J,Z
     11 FORMAT(2X,E9.1,I5)
d.      WRITE(*,'(F5.1,A4)')X,' = X'
e.      WRITE(*,12)X,Y,Z
     12 FORMAT(/,F5.0)
f.      WRITE(6,13)I,X
     13 FORMAT(1X,F5.0,I5)
g.      WRITE(*,'(F4.1,A3,I4)')X,'***',I,Y,'---',J
h.      WRITE(*,14)Z,1./Z,1./Z*Z-1.
     14 FORMAT(5X,3E10.3)
```

8. Determine the output from the following. Use the assignments: I = 2, J = 3, K = 4, X = 4., Y = 5., Z = 1./6.

   a. `WRITE(*,1)`
   ` 1 FORMAT(5X,3(4('*'),2X))`

   b. `WRITE(*,2)I,J,K`
   ` 2 FORMAT(2X,'I=',I1,/,2X,'AND J=',I1,/,2X,'AND K=')`

   c. `WRITE(*,3)I,J,Z`
   ` 3 FORMAT(/,2X,'1/(',I1,'*',I1,') =',F9.6)`

   d. `WRITE(*,4)X,Y,X+Y,1./Z`
   ` 4 FORMAT(1X,'X=',F4.1,'Y=',F4.1,/,`
   ` +      1X,'X+Y=',F4.1,'1/Z=',F4.1)`

   e. `WRITE(*,'(1X,3F3.0,3I2)')X,Y,Z,I,J,K`
   f. `WRITE(*,'(1X,A,I1,A,I1)')'I = ',I,'AND K = ',K`
   g. `WRITE(*,'(1X,F5.1,/,1X,I2)')X,I,Y,J,Z,K`
   h. `WRITE(*,'(A)')'CANDY IS DANDY BUT LIQUOR IS ... '`
   i. `WRITE(*,5)X,Y,K,X*Y,X*Y*K`
   ` 5 FORMAT(///,5X,'DIMENSIONS OF ROOM',/,5X,`
   ` +      'LENGTH= ',F5.1,2X,'WIDTH= ',F5.1,'HEIGHT= ',I3,//,`
   ` +      3X,'SURFACE AREA = ',F12.7,//,9X,`
   ` +         'VOLUME = ',E10.2)`

   j. `WRITE(*,6)`
   ` 6 FORMAT(30X,70('X')/6(2X,'*',2X),3(T31,70('X')/),`
   ` +      3X,5(2X,'*',2X)///6(2X,'*',2X)//T31,70('X')/,`
   ` +      3X,5(2X,'*',2X),3(T31,70('X')/),6(2X,'*',2X)///,`
   ` +      3X,5(2X'*',2X)//,`
   ` +      30X,70('X')/6(2X,'*',2X),3(T31,70('X')/),`
   ` +      3X,5(2X,'*',2X)///6(2X,'*',2X)//,`
   ` +      3(4(100('X')/)//////),4(100('X')/))`

9. Write a single line of Fortran that will:
   a. Print the variables X, Y, X + Y on separate lines.
   b. Print 'X = ',X and 'Y = ',Y on separate lines.

10. Write a program to compute the height, $y = -9.8t^2 + 5t$ (m), and the velocity, $v = -9.8t + 5$ (m/sec), for a falling object for time $t = 0$ to $t = 10$ sec in steps of 0.2 sec. The results should be displayed in a neat table with column headings, including units. The numerical results should be centered in the columns and a counter printed for each time value.

11. Write a program to read and print the name of a metal and its melting temperature in the form of a table with column headings, including units. In addition, if the melting temperature is greater than 1400° C, print TOO HIGH on the same line, if less than 600° C, print TOO LOW on the same line. Use the carriage-control characters.

12. Write a program to read a data file stored on disk and named CLASS which has a student's last name (A10), gender (M or F, A1), hometown (A10), and class (I1) on each line. The program is to count the number of:
   a. Male juniors and seniors from Boston

b. All females from Cincinnati

c. Freshmen from Chicago

In addition, the program should print the names of all students from Detroit, with their name preceded by either Mr. or Miss.

13. Write a program to print, with appropriate headings, the integers from 1 to 100. If the number is a perfect square or perfect cube, print on the same line a statement like

```
27     IS THE CUBE OF     3
```

Use the carriage-control character +.

14. Write a program to read and print a Fortran program devoid of all comment lines. The program is to be read from a data file.

15. Design format statements to handle the following:

a. Print a "block" letter; for example,

```
MM              MM
MMM             MMM
MM M          M MM
MM   M      M   MM
MM     M M      MM
MMMM     M     MMMM
```

b. With one format, print a table heading for (i,x_i,y_i) plus positions for five entrees in the table.

c. Ask the terminal user if he or she wishes to compute the SUM. If the answer is YES, instruct the user to enter ten values, one at a time, in a specified format. The program should then add the stored numbers and use a single format to print the ten numbers (F9.4) and their SUM (E12.4) in the form

```
        ENTREES
               XX.XXXX
               XXX.XXXX
                ......
               X.XXXX
               --------
   TOTAL  =    X.XXXXE+XX
```

d. Read the time of day expressed as the total number of minutes past midnight and print the time in the form XX:XXAM (or PM).

PROGRAMMING ASSIGNMENT III

III.1   SAMPLE PROGRAM

The programming problems in this section are meant to illustrate the material covered to this point in the text, particularly the manipulation of data files and the creation of neat and attractive printed output by the use of format statements. These programs are quite demanding and will require a considerable amount of your time. I suggest that you start early and allow ample time for debugging the programs. In addition, you should attempt to structure the programs in block form as suggested in Section 3.5. The programs should contain safeguards with diagnostic PRINT statements to handle any potential problems. A carefully constructed, neat code may take somewhat longer to design, but it pays dividends when problems arise. Of course, as your programs grow longer and more complex, the role of the flowchart or its alternatives becomes more critical. Again, you will find that the effort that you put into a clear outline of the workings of a proposed program will be time well spent.

Sample Program: Gas Separation (Chemical Engineering)

A very important problem in chemical engineering is the separation of a gas mixture into its constituents by means of selective absorption of gases in liquids based on differences in absorption rates of the gases. For example, it is desired to separate a gas containing (percentages by moles) hydrogen (65%), methane (18%), ethane (14%), and propane (3%) by bringing the gas into contact with an oil. After each pass the gas is assumed to be in equilibrium with the oil. Each pass is called a *plate* and a series of plates will constitute a packed tower. The equation defining the properties of a set of plates is

(III.1)
$$E = 1 - \frac{A - 1}{A^{n+1} - 1}$$

where E = Desired absorption efficiency for a particular chemical
  A = Absorption factor for a particular chemical (see below)
  n = The number of plates

Of course, the liquid will absorb only as much of each gas required to reach a phase equilibrium, which depends upon a variety of thermodynamic properties. An empirical expression for the absorption factor is

$$A = \frac{1}{k}\frac{L}{G} \qquad \text{(III.2)}$$

where L = mole/sec of flowing liquid
  G = mole/sec of flowing gas
  k = experimentally determined equilibrium vaporization constant which is equal to the ratio of the equilibrium concentrations of the particular chemicals in the supply gas to that of the liquid and is dependent on pressure, temperature, and the types of gas and liquid.

Since A is not really constant, an effective value is used.

$$A = A_e = [A_b(A_t + 1) + \tfrac{1}{4}]^{1/2} - \tfrac{1}{2} \qquad \text{(III.3)}$$

where $A_b(A_t)$ is the absorption constant appropriate for conditions at the bottom (top) of the process.

Problem

For the given gas-oil combination, it is desired to have an absorption efficiency E for ethane. If the following conditions hold

	k	
Constituent	Bottom	Top
Ethane	1.470	1.439
Methane	9.500	8.000
Propane	0.510	0.480
Gas Flow	55.00	52.00
Oil Flow	95.00	98.00

The program to solve for n, the number of plates required, is constructed along the following lines and is given in Figure III-1. The output is given in Figure III-2.

1. Read the names of the gases and their vaporization constants k from a data file, along with the flow rates of the gas and the oil.
2. The program should have statement functions for Equations (III.3) and (III.1).
3. Use the bisection algorithm to solve Equation (III.1) for n, the number of

plates required for $E = 0.9, 0.91, \ldots, 0.98$. The number of plates should then be rounded to an integer and the resulting E (for $n =$ integer) printed.

4. For each value of n (integer), determined above, print the resulting absorption efficiencies for Methane and Propane.

Figure III-1   Fortran code for sample program two.

```
      PROGRAM DISTIL
*--
*-- THIS PROBLEM DETERMINES THE NUMBER OF THEORETICAL ABSORPTION
*-- PLATES REQUIRED TO ACHIEVE A GIVEN ABSORPTION EFFICIENCY FOR
*-- ETHANE GAS FLOWING THROUGH OIL. THE PROBLEM IS SOLVED FOR
*-- EFFICIENCIES E = 0.90 THROUGH 0.98 IN STEPS OF 0.02. ONCE
*-- THE NUMBER OF PLATES IS DETERMINED, THE ABSORPTION EFFICIENCY
*-- FOR PROPANE AND METHANE IS COMPUTED. AN EFFECTIVE ABSORPTION
*-- FACTOR A IS COMPUTED FOR EACH GAS FROM THE PROPERTIES AT THE
*-- TOP AND BOTTOM OF THE COLUMN.
*-------------------------------------------------------------------
* VARIABLES
*
      REAL EETH,EMTH,EPRP,KETHT,KETHB,KMTHT,KMTHB,KPRPT,KPRPB,
     +               AETHT,AETHB,AMTHT,AMTHB,APRPT,APRPB,
     +               GFLOWT,GFLOWB,LFLOWT,LFLOWB,
     +               N,A,N1,N2,N3,F1,F2,F3,AT,AB,E
      INTEGER I,PLATES
*
*                   IN THE VARIABLE LIST
*                    ETH => ETHANE, MTH => METHANE, PRP => PROPANE
*                    LAST LETTER T => TOP, B => BOTTOM
*
*                    E***   --  ABSORPTION EFFICIENCY
*                    K****  --  EXPERIMENTAL VAPORIZATION CONST.
*                    A****  --  ABSORPTION FACTOR
*                    GFLOW* --  GAS FLOW RATE
*                    LFLOW* --  OIL (LIQUID) FLOW RATE
*                    N      --  COMPUTED NUMBER OF PLATES
*                    PLATES --  NUMBER OF PLATES (INTEGER)
*-------------------------------------------------------------------
* STATEMENT FUNCTION FOR EFFECTIVE ABSORPTION FACTOR
*
      A(AT,AB) = SQRT(AB * (AT + 1.) + 0.25) - .5
*-------------------------------------------------------------------
* STATEMENT FUNCTION FOR EQ. 1.
*
      F(N,E,AT,AB) = E - 1. + (A(AT,AB) - 1.)/(A(AT,AB)**N - 1.)
*-------------------------------------------------------------------
* INITIALIZATION
*
      OPEN(9,FILE='GASDAT')
      REWIND 9
      READ(9,10) GFLOWT,GFLOWB,LFLOWT,LFLOWB,
     +           KETHT,KETHB,KMTHT,KMTHB,KPRPT,KPRPB
*
      AETHT = LFLOWT/KETHT/GFLOWT
      AETHB = LFLOWB/KETHB/GFLOWB
      X1 = 2.0
      X3 = 20.0
      EPS = 1.E-3
      IMAX = 25
      DO = (X3 - X1)
```

```
*
*                    PRINT OUT INPUT PARAMETERS AND TABLE HEADINGS
*
      WRITE(*,11)X1,X3,EPS,IMAX
      WRITE(*,12)GFLOWT,LFLOWT,KETHT,KMTHT,KPRPT,
     +           GFLOWB,LFLOWB,KETHB,KMTHB,KPRPB
      WRITE(*,13)
*
*                    START OF THE E - LOOP
*
      EETH = 0.9
*        --------------------------------------------------------
    1 N1 = X1
      N3 = X3
      F1 = F(N1,EETH,AETHT,AETHB)
      F3 = F(N3,EETH,AETHT,AETHB)
      D = 1.
*                    --------------------------------------------------
*                    THE BISECTION CODE TO FIND THE ROOT OF F(N)
*                    IS THE SAME AS FIG. 4.13
*
      IF(F1*F3 .GT. 0.)THEN
         WRITE(*,20)N1,N2
         STOP
      END IF
*
    2 N2 = (N1 + N3)/2.
      F2 = F(N2,EETH,AETHT,AETHB)
      IF(D .LT. EPS)THEN
         GO TO 3
*
*                    STMT 3 IS THE SUCCESS PATH
*
*     ELSE IF(I.GT. IMAX)THEN
         WRITE(*,*)'--ERROR-- EXCESSIVE ITERATIONS = ',I
         STOP
      END IF
*
      IF(F1*F2 .LT. 0.)THEN
         D = (N3 - N2)/D0
         F3 = F2
         N3 = N2
      ELSE IF(F2*F3 .LT. 0.)THEN
         D = (N2 - N1)/D0
         N1 = N2
         F1 = F2
      ELSE IF(F2 .EQ. 0.)THEN
         GO TO 3
      ELSE
         WRITE(*,22)I,N1,N3
         STOP
      END IF
*
      I = I + 1
      GO TO 2
```

Continued

```
*                -----------------------------------------------------------------
*
*                    SUCCESS PATH - PLATES IS THE NEAREST INTEGER
*                               TO THE ROOT N2
*
    3 PLATES = N2
      II = N2 - 0.5
      IF(II .EQ. PLATES)PLATES = PLATES + 1
*
*                    PRINT THE RESULTS FOR ETHANE
*
      WRITE(*,23)EETH,PLATES
*
*                    COMPUTE AND PRINT THE RESULTS FOR METHANE AND
*                    PROPANE.
*
* METHANE-------------------------------
      AMTHT = LFLOWT/KMTHT/GFLOWT
      AMTHB = LFLOWB/KMTHB/GFLOWB
      AMTH = A(AMTHT,AMTHB)
      EMTH = 1. - (AMTH - 1.)/(AMTH**PLATES - 1.)
      WRITE(*,24)EMTH
*
* PROPANE-------------------------------
      APRPT = LFLOWT/KPRPT/GFLOWT
      APRPB = LFLOWB/KPRPB/GFLOWB
      APRP = A(APRPT,APRPB)
      EPRP = 1. - (APRP - 1.)/(APRP**PLATES - 1.)
      WRITE(*,25)EPRP
*
*                    INCREMENT E AND REPEAT
*
      EETH = EETH + 0.02
      IF(EETH .LE. 0.98)GO TO 1
      STOP
*-----------------------------------------------------------------------------------
* FORMATS
*
   10 FORMAT(10F5.2)
   11 FORMAT(///,10X,'THE BISECTION ALGORITHM PARAMETERS ARE',//,
     +        10X,'INTERVAL: FROM ',F6.2,' TO ',F6.2,//,
     +        10X,'EPS = ',E7.1,' IMAX = ',I4)
*
   12 FORMAT(///,5X,'INPUT PARAMETERS',//,43X,'ABSORPTION',//,
     +     10X,'   GAS         OIL                CONSTANTS',//,
     +     10X,'   FLOW        FLOW     ETHANE    METHANE    PROPANE'
     +     ,//,2X,'TOP',5X,5(2X,F7.3,1X)
     +     ,//,2X,'BOTTOM',2X,5(2X,F7.3,1X))
*
   13 FORMAT(///,20X,'COMPUTER RESULTS',//,
     + T10,'ABSORPTION',T25,'REQUIRED',T45,'ABSORPTION',//,
     + T10,'EFFICIENCY',T25,'NUMBER',  T45,'EFFICIENCY',//,
     + T10,' (INPUT) ',T25,' OF ',    T45,'(COMPUTED)',//,
     + T10,'  ETHANE',T25,'PLATES',T40,'  METHANE     PROPANE',//,
     + T10,50(1H-))
*
```

```
20 FORMAT(2X,'--ERROR-- NO ROOT IN INTERVAL ',2F8.4)
22 FORMAT(2X,'--ERROR-- AFTER ',I4,' ITERATIONS',/,15X,
   +        'NO ROOT IN SUB-INTERVAL ',2F8.4)
23 FORMAT(11X,F8.6,T26,I3)
24 FORMAT(1H ,T40,F8.6)
25 FORMAT(1H ,T52,E8.2)
*
   END
```

```
THE BISECTION ALGORITHM PARAMETERS ARE

INTERVAL: FROM   2.00 TO  20.00
EPS =  .1E-02 IMAX =   25
```

INPUT PARAMETERS

	GAS FLOW	OIL FLOW	ETHANE	ABSORPTION CONSTANTS METHANE	PROPANE
TOP	52.000	98.000	1.439	8.000	.480
BOTTOM	55.000	95.000	1.470	9.500	.510

COMPUTER RESULTS

ABSORPTION EFFICIENCY (INPUT) ETHANE	REQUIRED NUMBER OF PLATES	ABSORPTION EFFICIENCY (COMPUTED) METHANE	PROPANE
.900000	6	.188912	.28E-13
.920000	7	.188942	.28E-13
.940000	8	.188947	.28E-13
.960000	9	.188948	.28E-13
.980000	12	.188949	.28E-13

III.2   PROGRAMMING PROBLEMS

Programming Problem A: File Manipulation and Merging

A common task involving the manipulation of large data files is concerned
with the merging of two lists and finding the combined information pertaining to individual items in either list. For example, one list could be a
large data file including the physical constants such as molecular weight,
melting and boiling point, specific gravity, ignition temperature, and so on,

of a great many chemical compounds listed by chemical name. The second and smaller data file could be a list of hazardous materials by chemical name. This file might also contain antidotes, toxic dosages, solubility in water, as so on. The purpose of the merging of the files might then be to obtain more complete information on all toxic materials.

Consider an analogous problem. The athletic department of a large university would like to check up on the academic performance of its student athletes. The department has a file that includes the following information on each line:

Athlete's name, (last, first MI)	Sport,	Class, (Fr, So, . . .)	Sex, (M, F)	Years of, eligibility	Amount of Scholarship

and the department has access to the registrar's data file, which contains information on all students in the university.[1] The data on each line includes:

Student's name, (last, first MI)	Campus, address	College,	Major,	Sex,	Credits, earned to date
	Credits, attempted this sem.	Current, G.P.A.	Last Sem. G.P.A		

The athletic department would like a list of all the information on both files that pertains to its athletes. Additionally, it would like a separate list of the names of the current male freshmen on a football scholarship who were on probation last semester. Note that the registrar declares class status based on credits completed as

Freshman < 30cr ≤ Sophomore < 60cr < Junior < 90cr ≤ Senior

Also, a student is placed on probation if his or her current G.P.A. falls below 2.00 on an A = 4.00 scale.

An outline of a program to accomplish this merging follows.

1. OPEN files associated with the two existing data files in question and for the new files that are to be created.

---

[1] This is very likely illegal without the permission of all students involved. The more sensitive the file, the more limited will be the access to that file and the more elaborate will be the security precautions protecting it. Recently, a student "hacker" presented me with a neat list of all students and staff connected with a computer course I was teaching, along with their secret passwords. I hope the registrar's security precautions are more elaborate and clever than were my own.

2. READ a line of information from the longer (registrar's) file. Use the END = option to terminate this section of the program.
3. READ a line of information from the shorter (athletic department) file and
   a. Check for a matching name (or ID number if given).

```
IF(a match is found)THEN
     write the information from both files
     to the new file
ELSE
     GO TO 3 above to read the next athlete's name
END IF
```

The END = option for this read will handle the case of no match in the entire athlete's file.

4. REWIND the new combined file.
5. READ each line of the new file and, if the student athlete is a male freshman football player on probation, write this line to a separate file.
6. Neatly print the contents of both merged files.

The two files given in Figures III-3 and III-4 may be used to test your program. Of course, before the program can be executed, both of these or equivalent files must be entered and stored on the disk. (A kind instructor may have already done this for you.) The files are arranged assuming that they will be read using a list-directed READ.

```
'Altman, James   ',','123A East Quad',','A+S ',','Math',','M',000,16,2.77,2.91
'Ambler, Susan   ',','55 N. Main     ',','Engr',','Civl',','F',102,14,3.09,2.83
'Anderson, Tom   ',','6 Logan Crt.   ',','A+S ',','Hist',','M',061,16,2.02,2.74
'Armstrong, Jack',','17A Stadium Rd',','A+S ',','Engl',','M',032,17,3.87,3.64
'Bates, Sarah    ',','731B West Quad',','Bus ',','Acct',','F',111,16,1.88,2.04
'Becker, Adam    ',','111 Market St.',','Educ',','Elem',','M',045,17,2.24,3.31
'Bradshaw, Terri',','43 Miller Hall',','Engr',','Elec',','F',012,12,0.95,1.43
'Corning, Jesse ',','12C Stadium Rd',','Educ',','PhEd',','M',037,14,1.98,2.33
'Decker, Mary    ',','21 Miller Hall',','Educ',','Spec',','F',008,17,3.01,2.22
'Donchez, Juan   ',','414A East Quad',','A+S ',','Germ',','M',092,15,2.87,3.12
'Freeman, Frank ',','123B East Quad',','Bus ',','Finc',','M',027,15,2.25,2.13
'Goodman, Doug   ',','12 Logan Crt. ',','Bus ',','Acct',','M',088,16,2.44,3.12
'Huge, Bubba     ',','17C Stadium Rd',','A+S ',','Drma',','M',009,16,2.35,2.01
'Kilpatrick, Ed ',','612 Market St.',','Engr',','Mech',','M',030,13,3.15,2.66
'Kingston, Bill ',','Rd4, Box 446   ',','Agrc',','Polt',','M',116,16,2.75,3.21
'Levy, Ann       ',','42B Smith Hall',','Engr',','Elec',','F',052,14,2.11,2.37
'Lewis, Jane     ',','55 N. Main     ',','A+S ',','Chem',','F',077,15,3.42,3.21
'Mental, Mickey ',','322C East Quad',','Educ',','PhEd',','M',016,17,3.65,3.42
'ONeil, Sam      ',','111 Market St.',','Educ',','Elem',','M',042,15,2.14,2.39
'Roberts, Joel   ',','772 Center St.',','Engr',','Chme',','M',021,13,1.73,3.44
'Root, Babe      ',','17B Stadium Rd',','Bus ',','Mgmt',','M',000,12,3.02,0.00
'Smith, John     ',','55 N. Main     ',','A+S ',','Soci',','M',011,13,2.54,1.75
```

Figure III-3
The registrar's student information file (named *roster*).

Continued

```
'Tall, Tina     ',,'32 Miller Hall','A+S ','Geol','F',071,15,2,33,2,98
'Trapezoid, O.J.',,'12A Stadium Rd','Engr','Mech','M',008,12,1,53,1,99
'Thomas, Karen  ',,'115 River Rd  ','A+S ','Phil','F',104,16,3,95,3,97
'Weber, Willy   ',,'8 Country Lane','Educ','Art ','M',112,17,2,61,2,88
```

Figure III-4
The athletic department's directory file (named *athlts*).

```
'Armstrong, Jack','Football','So','M',2,2500,
'Corning, Jesse ','Track    ','Fr','M',2,3200,
'Decker, Mary    ','Track    ','So','F',3, 450,
'Huge, Bubba    ','Wrestlng','Fr','M',1,3000,
'Kingston, Billy','Tennis   ','Sr','F',1, 500,
'Mental, Mickey ','Baseball','Fr','M',3,1200,
'Tall, Tina      ','BsKtball','Jr','F',2,1250,
'Trapezoid, O.J.','Football','Fr','M',3,4000,
'Root, Babe     ','Baseball','Fr','M',3, 250,
```

Programming Problem B: Civil Engineering and Mechanics: The Buckling of a Tall Mast

A standard, but complex, problem in civil engineering and mechanics is the determination of how tall a mast can be before it will begin to buckle under its own weight. The solution of this problem, which is usually covered in an advanced mechanics course,[2] yields the following result:

Defining the quantities

L = mast length (m)

Y = Young's modulus of the material (N/m²). Young's modulus is an experimental value of the ratio of the size of a deformation to the applied force.

λ = mass per unit length of the mast (kg/m)
(λ = density × cross-sectional area of the mast)

I_2 = the second area moment of inertia of the mast given as $\pi r^4/2$, where r is the radius of the round mast

x = a dimensionless parameter that is related to the above quantities by

$$x = \frac{4}{9} g \frac{\lambda L^3}{Y I_2} \qquad \langle g = 9.8 \rangle$$

---

[2] See, for example, S. Timoshenko, *Strength of Materials,* Part I, Van Nostrand, Princeton, N.J., 1955.

it is found that the mast will just begin to buckle when x has the value corresponding to the smallest positive root of the function

$$F(x) = a_0 + a_1 x + a_2 x^2 + a_3 x^3 + a_4 x^4 \qquad \text{(III.4)}$$

where $a_0 = 1$

$$a_1 = -\frac{3}{8}$$

$$a_2 = +\frac{9}{320} \qquad \text{(III.5)}$$

$$a_3 = -\frac{9}{10,240}$$

$$a_4 = +\frac{27}{1,802,240}$$

The problem is then to find the first root of Equation (III.4), say x_1, and once this value is determined, to specify the maximum lengths of a mast for a variety of materials. The root-solving technique to be used is Newton's method, and the program will have to step in small increments of x, starting at zero, to first determine the vicinity of the root.

Problem Specifics

1. The program should have statement functions for the function in Equation (III.4) and for its derivative, DFDX(X).
2. The parameters associated with Newton's method such as convergence criterion, EPS, and maximum number of iterations, IMAX, should be read from the terminal. The data for the various materials given in Table III-1 should be read from a data file. Note that, for some of the materials, there is a range of Young's moduli. A corresponding range of mast heights is then to be calculated. Assume that all masts have a radius of $r = 0.10$ meters.

Material	Young's Modulus, Y (N/m²)(× 10^{10})	Density, ρ (kg/m³)(× 10^3)
Aluminum (cast)	5.6 to 7.7	2.70
Brass	9.02	8.44
Gold	7.85	19.3
Iron (cast)	8.4 to 9.8	7.86
Lead	1.5 to 1.67	11.0
Platinum	16.7	21.4
Steel	20.0	7.83
Tin	3.9 to 5.39	7.29
Tungsten	35.5	18.8

Table III-1 Young's modulus and density values for various materials.

3. The output consists of:
    a. The initial guess for x_0.
    b. The computed value of the root, the value of the function at the root, and the number of iterations required.
    c. For each of the materials in Table III-1 list the following
        i. The name of the material
        ii. The density (ρ)
        iii. The mass per unit length (λ).
        iv. Young's modulus (Y)
        v. The second area moment (I_2).
        vi. The maximum height (or range of heights) of a mast of this material.

Programming Problem C: Nuclear Engineering: Shielding of a Nuclear Reactor

In a nuclear reactor, the fuel, uranium with enriched amounts of a fissionable isotope ^{235}U, is burned by capturing a free neutron on a ^{235}U nucleus, whereupon it splits roughly in half, releasing considerable energy in the form of kinetic energy of the fragments. In addition, high-energy neutrons and gamma rays are also produced. Some of these neutrons are used to keep the reaction going, but most will escape the reactor core. Since this intense and penetrating radiation can cause considerable damage to personnel and equipment, it is vital to surround the reactor core with adequate material called *shielding* to prevent the escape of radiation. The shielding material is usually concrete or water.

To design a protective shield we must have a measure of the total number of "prompt" neutrons released per second during the reactor's operation.[3] The source strength of the reactor core will be labeled S (neutrons/sec). The approximate thickness of the concrete shield can then be computed from the equation

(III.6)
$$D = \frac{Se^{-\mu t}}{4\pi R^2}F$$

where D = amount of radiation that penetrates the shield, called the dose rate (units are rem/hr). The maximum dose is about 0.1 rem over a 24-hour period.

R = distance of observer from the reactor

F = dose rate (converts neutrons/m²-sec to rem/hr)

μ = effective removal coefficient for concrete. The units are m^{-1}.

t = the thickness of the concrete shield

---

[3] Since the fission fragments are frequently radioactive, additional "delayed" neutrons are often released in their decay. Delayed neutrons represent only about 1/2 percent of the total neutrons released. Moreover, their energy is always much less than the prompt neutrons and therefore they are not of great significance in the design of the shielding of the core.

This equation can be solved for the concrete thickness and yields

$$t = \frac{1}{\mu} \ln\left(\frac{SF}{4\pi R^2 D}\right) \qquad \text{(III.7)}$$

The problem is to determine the source strength S experimentally and then to use reasonable values for the other parameters to compute the shielding thickness t. The problem will consist of two parts:

1. The creation of an "experimental" data file containing measurements of number of neutrons emitted as a function of energy, $N(E)$, for equal steps of energy from $E = 0$ to $E = E_{max}$.
2. A program to read the data file, line by line, and to evaluate simultaneously a Simpson's rule approximation to the integral

$$S \simeq \int_0^{E_{max}} N(E)\ dE \qquad \text{(III.8)}$$

Once the integral is computed, the shield thickness can be evaluated by using Equation (III.7).

Creating the Data File

The number of prompt neutrons of energy E emitted per second by a reactor is given approximately by the relation

$$\begin{aligned} N(E) &= 0.453 \times 10^{12} e^{-1.04E} \sinh(/\sqrt{2.47E}) \qquad \text{for } 0 < E < 4 \text{ MeV} \\ &= 1.75 \times 10^{12} e^{-0.766E} \qquad \text{for } E > 4 \text{ MeV} \end{aligned} \qquad \text{(III.9)}$$

The units of MeV (million electron volts) are standard in nuclear physics and are related to more common SI units by 1 MeV $= 1.602 \times 10^{-25}$ Joule.

The experiment consists in evaluating $N(E)$ for 101 values of energy from $E = 0$ to $E = E_{max} = 14$ MeV. (That is, 100 intervals for the Simpson integration.) To simulate actual experimental data, it would be preferable if the data did not fall precisely on a given curve such as Equation (III.9).[4] To accomplish this we will assume that each energy value is known only to an accuracy of $\Delta E = \pm 0.01$MeV so that the "actual" energy can be anywhere within the band

---

[4] This part of the problem may be made optional.

We will next *randomly* choose a point within this band to represent the actual energy. To do this we make use of a nonstandard intrinsic function RANDOM.[5] The function will return a random number in the range $0 \leftrightarrow 1$. Thus, the procedure we will use to create simulated data is to replace each energy value greater than zero by the "actual" energy, defined as

(III.10)
$$E_{actual} = E + (\text{RANDOM}(\ ) - 0.5) \times \Delta E$$

The program that creates the simulated data file will then proceed as follows:

1. OPEN a file for the computed data with STATUS = 'NEW'.
2. Step through the 101 energy values E from 0 to E_{max}. For each value of $E > 0$, compute E_{actual} from Equation (III.10), and from this value compute the number of prompt neutrons using Equation (III.9).
3. WRITE I, E_{actual}, N onto the data file (*not E*) using a suitable FORMAT.
4. WRITE an END-OF-FILE mark on the data file.
5. After the program executes correctly, save the data file on the disc.

Integrating the Data and Computing the Shielding Thickness

The second program to read and integrate the simulated data file will be constructed along the following lines:

1. READ the additional parameters required in Equation (III.7) entered at the terminal and neatly echo print them using formatted output.
2. OPEN and REWIND the data file.
3. Initialize $\Delta E = E_{max}/100.$, SUModd = 0, SUMeven = 0.
4. The Simpson rule integration of the data can be effected by writing Equation (4.14) in the following form:

(III.11)
$$S = \int_0^{E_{max}} N(E)\ dE \simeq \frac{\Delta E}{3}\left[N_0 + \left(2 \sum_{even} N_i + 4 \sum_{odd} N_i \right) + N_{101} \right]$$

That is, read a line (I, E, N) from the data file and then

$$\begin{aligned}
&\text{IF}(i = 0)\text{THEN}\\
&\qquad N_0 = N\\
&\text{ELSE IF}(i = even)\text{THEN}\\
&\qquad\qquad \text{SUMevn} = \text{SUMevn} + N\\
&\text{ELSE IF}(i = odd)\text{THEN}\\
&\qquad\qquad \text{SUModd} = \text{SUModd} + N
\end{aligned}$$

---

[5] Unfortunately the form and name of this function is different for each computer operating system, and you will have to determine the details of using the random number function from your computing center.

```
        ELSE
            N_{101} = N
        END IF
```

5. Once S is evaluated compute the shielding thickness t.

The additional parameters required in this problem are listed as follows:

penetration dose limit (rem/hr)	dose rate conversion factor	distance from core	concrete removal rate
D	F	R	μ
0.004	1.43	10m	8.6m^{-1}

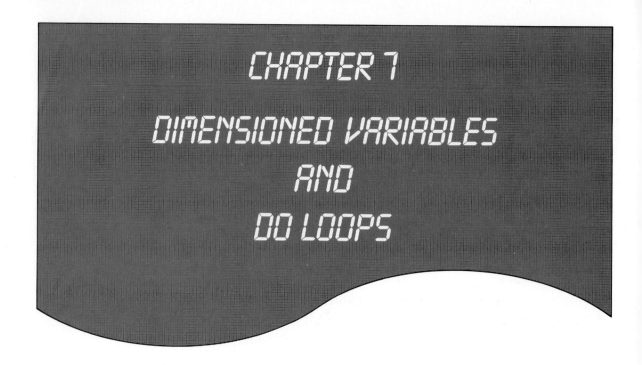

CHAPTER 7
DIMENSIONED VARIABLES
AND
DO LOOPS

7.1 INTRODUCTION

The Fortran described thus far can be used to solve a great many types of problems. However, there is as yet an entire class of problems that is not easily handled without introducing some new Fortran features. For example, if we wish to alter the program in Section 4.4.3 that computed the average of a list of exam scores to calculate in addition the deviation from the average, we face a formidable problem. The data set will have to be scanned twice, once to obtain the average score, and then again to compute the deviations of each score from this average. However, the individual scores were never stored. A single variable was used for the current value of the exam score and was repeatedly updated with each READ. Thus to alter the program we will have to assign each score as it is read to a separate variable which then can be reused in the second calculation. If 400 students took the examination, the READ statement might then appear

```
READ(*,*)STU1,STU2,STU3, ... ,STU400
```

It will probably take longer to type this single READ statement than it took to enter the entire previous program.

The algebraic notation for this problem suggests a remedy.

$$\text{avg} = \frac{1}{n} \sum g_i$$

The set of exam scores is represented by the variable g_i, and individual scores of each student are distinguished from each other by the subscript i. Subscripted variables are introduced into Fortran by means of the DIMENSION statement.

7.2   THE DIMENSION STATEMENT

As we saw in Chapter 1, the Fortran feature most responsible for the success of Fortran and other higher-level computer languages is the method of automatically assigning an address to variable names. Whenever the compiler encounters a new variable name it associates the name with a location in memory. The contents of that memory location are then accessible by simply using the variable name. Each variable is allotted one word of memory.

It is also possible to instruct the compiler to reserve more than one memory word to a variable, say an entire block of words. Each word in this block would have the same variable name, but of course the address of each word would be different, analogous to a subscripted variable. The Fortran instruction to accomplish this is the DIMENSION statement. The form of the DIMENSION statement is

DIMENSION A(subscript),X(subscript$_1$,subscript$_2$),NUMBR(subscript$_1$)

The names A, X, and NUMBR could be any valid integer, real, or character Fortran names. Any number of variables, separated by commas, may be listed after DIMENSION. The quantities in the parentheses specify the range of the subscript associated with the variable. A subscripted variable in Fortran may have up to seven subscripts. The simplest form for specifying the subscript range merely uses the number of words to be reserved for a particular variable.

```
DIMENSION INDEX(50),STU(400,4)
```

The integer variable INDEX occupies 50 words of memory. The first element of the block is accessed by INDEX(1) and the last by INDEX(50). The number appearing in the parentheses in this form *must* be a *positive integer*. The real variable STU occupies 1600 words of memory (i.e., STU$_{ij}$, $i = 1–400, j = 1–4$). Variable names may never appear in the specification for the subscript range in a Fortran program. The compiler must know, prior to the program's execution, how much space to reserve for each variable.

The DIMENSION statement is a nonexecutable statement and should be placed before any executable statement.

Subscripted variables in Fortran are called *arrays*. As mentioned, the individual elements of an array are accessed by specifying the subscript or index.

```
X(3) = 13.4 + EXP(-5.2*T)
```

or

```
A = X(16) + 4.*Y(4)**2
```

or

```
I = 12
RADIUS = SQRT(X(3*I - 2)**2 + Y(3*I - 2)**2)
```

When referencing an array element, the index may be any integer expression. All of the elements of an array are of the same data type, which is specified by the name of the array.

7.2.1 Combining Type and DIMENSION Statements

Type declarations and DIMENSION statements may be combined as follows:

```
REAL KLASS
DIMENSION KLASS(50)
```

has exactly the same effect as

```
REAL KLASS(50)
```

The combining of type INTEGER or CHARACTER declarations with a DIMENSION statement is similar.

```
REAL X,JUNIOR        is the    REAL X(50),JUNIOR(20)
INTEGER C            same      INTEGER C(512)
CHARACTER NAME*8      as       CHARACTER NAME(100)*8
DIMENSION X(50),
+    JUNIOR(20),
+    C(512),NAME(100)
```

I suggested earlier that *all* variables that appear in a program be type-declared at the beginning of the program. Applying this to dimensioned variables as well means that the shorter combined form of a dimension declaration is preferred. That is, modern programming style demands that the word DIMENSION not appear in your program. The Fortran DIMENSION statement was introduced in this chapter only to facilitate your understanding of the concept of arrays, and all subsequent program examples will employ the preferred combination of type and DIMENSION statements to declare a variable to be an array.

7.2.2 Example Program for the All-University GPA

Again assuming we have a data file containing the grade point average (GPA) of all 8000 students in the university with each line of the file containing the following information:

1. Student's name (A10)
2. ID (I9)
3. College (I1) (1 = arts and science, 2 = engineering, 3 = education,
           4 = architecture, 5 = business, 6 = nursing,
           7 = night, 8 = other)
4. Class (I1)   (1 = freshman, 2 = sophomore,
           3 = junior, 4 = senior)
5. Sex (A1)   (M = male, F = female)
6. Current GPA (F4.2)
7. Last semester's GPA (F4.2)
8. Credit hours attempted thus far (F6.2)

The code to compute the average GPA would be

```
      PROGRAM AVERAGE
      REAL SUM,STU(8000),XN,GPA
      INTEGER I

      OPEN(8,FILE='GRADES')
      REWIND(8)

      I = 1
      SUM = 0.0

    1 READ(8,10,END=2)STU(I)
   10 FORMAT(22X,F4.2)
```

```
            SUM = SUM + STU(I)
            I = I + 1
            GO TO 1

     2   XN = I - 1
         GPA = SUM/XN
         WRITE(*,*)'THE ALL-UNIVERSITY GPA = ',GPA
         STOP
         END
```

If we also need to compute the standard deviation, given by the expression

$$\sigma^2 = \frac{1}{n-1} \sum_i |s_i - s_{avg}|^2$$

a second pass through the list of students will be required. This could be accomplished by inserting the following code in the above program just before the STOP.

```
         I = 1
         DEV = 0.
     3   DEV = DEV + (STU(I) - GPA)**2
             I = I + 1
             IF(I .LE. XN)GO TO 3
         DEV = SQRT(DEV/(XN - 1.))
         WRITE(*,*)'AND THE AVERAGE DEVIATION IS ',DEV
```

The index for the array STU must not be zero or negative and must not exceed the limit specified in the DIMENSION statement, 8000. If the index is outside the proscribed range (here 1–8000), most compilers will print an execution time error message, but many will not. In those cases the computer may simply use some element stored in memory near the location of the variable. For example, STU(8007) would access the memory location seven words beyond the end of the block called STU. This location may contain a number, or a format, or some executable instruction, and the result will be unpredictable. This type of error, since it is often extremely difficult to track down, is one of the most grievous that a programmer can make. If your computer does not check whether the subscript is within the limits given in the DIMENSION statement, you must always do so yourself. Thus the program above should contain a check of the form

```
     IF(I .GT. 8000)THEN
             WRITE(*,*)'INDEX OUT OF RANGE, I = ',I
             STOP
     END IF
```

The previous program could be easily adapted to compute the individual averages and deviations for each class, for each college, etc. To accomplish this we could separately dimension four arrays to hold the grades for freshmen, sophomores, juniors, and seniors.

```
REAL FRSH(2000),SOPH(2000),JUNR(2000),SENR(2000)
```

Or we could accommodate all four classes in a single variable STU(I,J), where I (from 1 to 2000) identifies a student in a class, and J (from 1 to 4) identifies the class. There will now be four separate averages and four counts of the total number in a class. Thus we will need two addtional arrays, GPA(4) and N(4). The algebraic statement of the problem is

$$c_k = \frac{1}{n_k} \sum_{i=1}^{n} g_{ik} \qquad \text{Class averages for each } k = 1 \text{ to } 4$$

$$t = \frac{1}{4} \sum_{k=1}^{4} c_k \qquad \text{Average of the four averages}$$

The program to accomplish this is shown in Figure 7-1.

```
* PROGRAM AVERAGE
*--
*--   THIS PROGRAM READS THE DATA FILE <GRADES> AND COMPUTES THE
*--   AVERAGE GRADE BY CLASS (FRESHMAN, SOPHOMORE, ETC.) AND THE
*--   STANDARD DEVIATION FOR EACH CLASS.
*------------------------------------------------------------------
* VARIABLES
*--
      REAL SUM(4),STU(1000,4),STDEV(4),GPA(4),DELTA,AVG
      INTEGER I,N(4)
      CHARACTER*9 CLASS(4)
*--
*        STU(I,K)    --   CURRENT GPA OF STUDENT I IN CLASS K
*        SUM(K)      --   SUM OF ALL GPA'S IN CLASS K
*        GPA(K)      --   AVERAGE GPA FOR CLASS K
*        STDEV(K)    --   STANDARD DEVIATION FROM THE AVERAGE
*                         FOR CLASS K
*        N(K)        --   NUMBER OF STUDENTS IN CLASS K
*        AVG         --   ALL UNIVERSITY AVERAGE GPA
*        DELTA       --   AVERAGE OF THE FOUR STANDARD DEVIA-
*                         TIONS
*        XXX         --   TEMPORARY VALUE FOR CURRENT STUDENT
*                         GPA
*------------------------------------------------------------------
* INITIALIZATION
*--
      CLASS(1) = 'FRESHMAN'
      CLASS(2) = 'SOPHOMORE'
      CLASS(3) = 'JUNIOR   '
      CLASS(4) = 'SENIOR   '
```

Continued

Figure 7-1 A program to compute individual class averages.

```
              N(1) = 0
              N(2) = 0
              N(3) = 0
              N(4) = 0
              AVG = 0.0
              DELTA = 0.0
              OPEN (UNIT=66,FILE='GRADES')
              REWIND 66
*-------------------------------------------------------------
*--
*--              READ THE GRADES FILE AND COUNT THE NUMBER
*--              IN EACH CLASS
*--
    1 READ(66,10,END=99)K,XXX
              N(K) = N(K) + 1
              STU(N(K),K) = XXX
              GO TO 1
*-------------------------------------------------------------
*--
*--              FOR CLASS K = 1 TO 4, SUM STU(I,K) FROM
*--              I = 1 TO N(K)
*--
   99 K = 0
    2 K = K + 1
      IF(K .LE. 4)THEN
          SUM(K) = 0
          I = 0
    3     I = I + 1
          IF(I .LE. N(K))THEN
              SUM(K) = SUM(K) + STU(I,K)
              GO TO 3
          ELSE
*--
*--
*--              THE SUM OVER I IS COMPLETE, COMPUTE GPA(K)
*--
              GPA(K) = SUM(K)/N(K)
          END IF
          AVG = AVG + GPA(K)
*--
*--              RETURN TO START OF LOOP AND SUM NEXT CLASS
*--
          GO TO 2
      ELSE
*--
*--              ALL CLASSES HAVE BEEN SUMMED
*--              COMPUTE GRAND AVERAGE
*--
          AVG = AVG/4.
      END IF
*-------------------------------------------------------------
*--              RESCAN THE ENTIRE LIST TO COMPUTE THE STANDARD
*--              DEVIATIONS FOR EACH CLASS
*--
      K = 0
    4 K = K + 1
      IF(K .LE. 4)THEN
          STDEV(K) = 0.0
          I = 0
    5     I = I + 1
          IF(I .LE. N(K))THEN
              STDEV(K) = STDEV(K) + (STU(I,K) - GPA(K))**2
              GO TO 5
```

```
          ELSE
               STDEV(K) = SQRT(STDEV(K)/(N(K) - 1.))
          END IF
          DELTA = DELTA + STDEV(K)
          GO TO 4
      ELSE
          DELTA = DELTA/4.
      END IF
*------------------------------------------------------------
*-- PRINT RESULTS
*--
      K = 0
      NTOT = 0
      WRITE(*,11)
    6 K = K + 1
      IF(K .LE. 4)THEN
          NTOT = NTOT + N(K)
          WRITE(*,12)CLASS(K),N(K),GPA(K),STDEV(K)
          GO TO 6
      ELSE
          WRITE(*,12)'TOTALS',NTOT,AVG,DELTA
          STOP
      END IF
*------------------------------------------------------------
*--FORMATS
*--
   10 FORMAT(20X,I1,1X,F4.2)
   11 FORMAT(////,10X,'    RESULTS -- ALL UNIVERSITY GPA',//,
     +        T20,'NUMBER',/,
     +        T5,'CLASS',T20,' OF',T30,'AVERAGE',T40,'AVERAGE',
     +        T20,'STUDENTS',T30,' GPA',T40,'STD. DEV.',//,45('-'))
*--
   12 FORMAT(T5,A9,T23,I4,T33,F5.2,T42,F7.4)
      END
```

This program could be adapted further to calculate the averages classified not only by class, but by gender as well. For example, the average of junior women. In this case the array STU would be dimensioned as STU(1000,4,2), the last index designating male or female. You should have no trouble in constructing the code to handle this case and so it will not be given here.

7.2.3 Upper and Lower Subscript Bounds in a DIMENSION Statement

The simplest form of the DIMENSION statement

```
          DIMENSION X(10) ,K(6,4)
```

or the equivalent form

```
          REAL X(10)
          INTEGER K(6,4)
```

allow for an index range on the variable x_i from $i = 1$ to $i = 10$, and the range of the two indices of the variable k_{mn} from 1 to 6 and 1 to 4, respectively. Frequently it is desirable to have the index assume negative values or zero. This is permitted in Fortran, provided the compiler is informed of the precise range of the index when the variable is first allotted space in a type or DIMENSION statement.

$$\text{REAL } A(I_{lower}\!:\!I_{upper})$$

where I_{lower} is the lower limit of the index range and I_{upper} the upper limit. Thus,

```
REAL PROFIT(1963:1984)
```

would allot 22 words of memory ($I_{upper} - I_{lower} + 1$) to the variable PROFIT. Other examples are

```
REAL X(0:20),TIME(-50:50)
INTEGER B(1:10)     ⟨This is the same as B(10).⟩
```

7.3   INTERNAL STORAGE OF ARRAYS

After a program has been compiled, the entire program is stored sequentially in memory. That is, variables, FORMAT statements, arithmetic instructions, etc. are assigned addresses in a string. Thus, we can imagine an array B(5) to be stored as

$$
\begin{vmatrix}
B_1 \\
B_2 \\
B_3 \\
B_4 \\
B_5
\end{vmatrix}
$$

An array with two subscripts is usually visualized as a two-dimensional rectangular block; i.e., B(2,4) would be pictured as

$$
\begin{array}{cccc}
b_{11} & b_{12} & b_{13} & b_{14} \\
b_{21} & b_{22} & b_{23} & b_{24}
\end{array}
$$

where the first index characterizes the row and the second the column in the block. But since the computer must store all arrays in a string, the array is actually stored as

$$b_{11}$$
$$b_{21}$$
$$b_{12}$$
$$b_{22}$$
$$b_{13}$$
$$b_{23}$$
$$b_{14}$$
$$b_{24}$$

That is, the first index increments first.

If an array is dimensioned as $X(k_1, k_2)$, then $X(i_1, i_2)$, with $i_1 \leq k_1$, $i_2 \leq k_2$, is the nth element in the block, where

$$n = i_1 + (i_2 - 1)k_1$$

Thus if two arrays are dimensioned as $X(10,10)$ and $Y(11,9)$, then $X(3,7)$ is the 63rd element of X, while $Y(3,7)$ is the 69th element of Y. This has important consequences in connection with the input and output of arrays. The Fortran statement

WRITE(*,*)B   ⟨*B was dimensioned as B(2,4)*⟩

will print *all* of the variables with the name B, i.e., eight elements. The order in which the elements are printed is the same as the order in which the elements are stored in memory—i.e., the same as the string given above. Similarly, a statement like

READ(*,*)B

will read eight numbers and store them in the sequence given.

However, a Fortran statement like

B = 35.   ⟨*Error if B has been dimensioned*⟩

will result in a compilation time error. Since a unique memory location has not been specified to the left of the assignment operator, this statement cannot be executed. Similarly, the arithmetic operations in the statement

Z = 2. * B + 4.0   ⟨*Error if B has been dimensioned*⟩

will likewise result in a compilation time error. No unique value has been specified to be used in place of the symbol B.

In short, in all I/O operations the appearance of an array name only, without specifying the indices, causes the entire contents of the array to be used. In other Fortran statements the use of the variable name only will cause an execution time error.

7.4   THE DO LOOP

The primary goal of structured programming is to construct code that is easy to read and that employs logic that is easy to follow. This usually translates into a total banishment of GO TO statements. The block IF structures of Section 3.3 are a major step in this direction. Yet there remain cases for which GO TOs are unavoidable: namely any algorithm that involves looping.

```
    I = 1
    SUM = 0.0
1   SUM = SUM + X(I)
    I = I + 1
    IF(I .LT. IMAX)GO TO 1
```

Frequently the loop will consist of a *known* number of cycles (IMAX) which is monitored by a loop index (I). The arithmetic in each cycle will often involve subscripted variables identified by the loop index X(I). A special structure in Fortran to handle situations like this without employing GO TOs is the DO loop.

7.4.1   The Structure of the DO Loop

The DO loop structure is a block of statements with an entry point (the top) and a normal exit point (the bottom). The beginning line of a DO loop has the general form

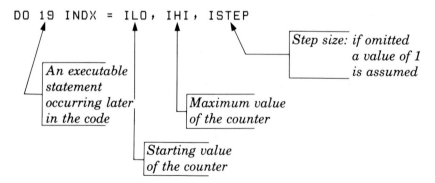

All Fortran statements beginning with the DO statement down to and including the DO terminator (here statement 19) are executed first with the index INDX = ILO. The loop is then repeated with INDX = ILO + ISTEP, etc. until the maximum value is reached. The terminal statement of the DO loop must be an executable statement, but it cannot be any of the following:

```
GO TO      (unconditional)
IF         (arithmetic)
ELSE
ELSE IF
STOP
END
DO
```

If the terminator is a logical IF, it must not contain a DO, another IF (block or logical), ELSE, ELSE IF, END, or END IF. Rather than attempt to remember these rules, we reintroduce a Fortran statement whose primary purpose is to serve as the terminator of DO loops. This is the CONTINUE statement, which has the simple form

⟨stmt number⟩ CONTINUE

The CONTINUE statement is defined to be an executable (though do-nothing) statement.

The index range parameters, ILO, IHI, and ISTEP, may be either integer or real constants, variables, or expressions. The indexing variable INDEX may also be integer or real, however it is suggested that whenever possible integer values be used exclusively. The DO-loop structure has a built-in test for completion of the loop.[1] After completion of the DO loop, the index parameter retains its most recent value. A few examples are given in Table 7-1.

Table 7-1   Examples of DO statements.

DO statement	No. of cycles executed	Value of index after loop	Comments
`DO 44 I = 1,5`	5	6	
`DO 73 K = 5,1`	0(or 1)	5	
`DO 73 K = 5,1,-1`	5	0	Negative steps
`DO 11 M = 1,9,3`	3	10	
`DO 15 X = 1.0,4.0,0.8`	4	4.2	
`DO 15 M = 1.0,4.0,0.8`	None (execution time error—zero step not allowed)	Undefined	The loop limits are first converted to the type of M, so this is the same as DO 15 M = 1,4,0
`DO 91 R = SQRT(2.),5.`	4	$4. + \sqrt{2}.$	

---

[1] The test for completion may be at the beginning or at the end of the loop, depending on the local installation of Fortran. Thus on some machines, a DO loop is always executed at least once, while on other computers the same loop may be executed zero times. See the second example in Table 7-1.

7.4.2 Examples of DO Loops

Several of the previous programs that have employed a loop structure may now be rewritten using a DO loop.

1. Summations

```
SUM = 0.0
DO 1 I = 1,N
    SUM = SUM + X(I)
1   CONTINUE
```

2. Zero-fill an array

```
REAL X(50),Y(50)
DO 3 I = 1,50
    X(I) = 0.0
    Y(I) = 0.0
3   CONTINUE
```

3. Factorials, $n! = n(n - 1)(n - 2) \ldots 2 \times 1$

```
INTEGER N,FACT
READ(*,*)N
FACT = N
DO 1 I = N - 1,1,-1
    FACT = FACT * I
1   CONTINUE
```

7.4.3 Abnormal Exits from a DO loop

Frequently, you may wish to terminate a loop structure before the total number of cycles are executed. This is usually done with a logical IF and a GO TO; such a situation arises in iterative calculations. For example, in searching for the first positive root of a function of x, $f(x)$, you might start at $x = 0.0$ and proceed along the x axis in steps of 0.1 and monitor the relative sign of two successive values of $f(x)$. If you are fairly sure that the root is less than, say, $x = 15$, the code would be

```
PROGRAM ROOT
F(X) =                          ⟨Statement function for f(x)⟩
X = 0.0
DX = 0.1
F1 = F(X)
```

```
       DO 1 I = 1,150
           X = X + DX
           F2 = F(X)
           IF(F1 * F2 .LE. 0.0)THEN

                     GO TO 99
           ELSE
                     F1 = F2
           END IF
     1 CONTINUE

       WRITE(*,*)'NO ROOT'
       STOP
    99 X3 = X
       X1 = X - DX
       F1 = F(X1)
       F3 = F2
       X2 = .5 * (X1 + X3)
              ⟨etc.⟩
```
⟨*Leave the loop before it completes 150 cycles*⟩

⟨*No root found in 150 steps, print and stop*⟩

⟨*An axis crossing was found before x = 15, so start bisection from here*⟩

When exiting a DO loop before completion, the value of the index parameter is the last value assigned. You should never branch from outside a DO loop to statements within a DO loop. The counter will most likely not have an assigned value. In addition, the DO loop index and the limits should not be redefined within the loop.[2]

To improve the readability of DO loops it is suggested that the body of the loop be indented in a manner similar to block IF statements.

7.4.4 Nested DO loops

When one DO loop is entirely contained within another DO loop, the grouping is called a *DO loop nesting*. Nesting of DO loops to any depth is permitted, provided each inner loop is entirely within the range of the next-level outer loop. Basically this means DO loops may never overlap. An inner loop and an outer loop may, however, share a terminal statement. A few examples of allowable nestings are shown in Figure 7-2. For each cycle of an outer loop, the inner loops are completely executed. Thus

---

[2] The potential number of cycles of the DO loop is ordinarily computed by the compiler at the very beginning of the loop from the index range parameters in the DO statement. And, except for abnormal exits from the loop, that many cycles are executed *even if* the index and index range parameters are changed within the loop.

Figure 7-2
Examples of nested DO loops.

```
DO 4 I1 = 1,10
   ...
   ...
      DO 3 I2 = 3,7
         ...
         ...
            DO 2 I3 = 1,8
               ...
               ...
2           CONTINUE
         ...
         ...
3        CONTINUE
   ...
   ...
4 CONTINUE
```

```
        DO 300 K1 = 1,12
           ...
           ...
              DO 100 K2 = 3,1,-1
                 ...
                 ...
100           CONTINUE
           ...
           ...
              DO 200 K3 = 5,9
                 ...
                 ...
200           CONTINUE
           ...
           ...
300     CONTINUE
```

```
DO 11 I1 = 1,10
   DO 11 I2 = 1,12
      DO 11 I3 = 1,20
         ...
         ...
11 CONTINUE
```

```
     DO 2 I = 0,9
        DO 2 J = 0,9
           WRITE(*,'(5X,I2)')J + 10*I
  2 CONTINUE
```

will print the integers from 0 to 99 in order.

The program on page 223 may be rewritten using nested DO loops as follows:

```
...

...
STNDEV = 0.0
GPA = 0.0
DO 3 K = 1,4
    AVG(K) = 0.0
    DO 1 I = 1,N(K)
        AVG(K) = AVG(K) + STU(I,K)
1       CONTINUE

    AVG(K) = AVG(K)/N(K)

    DEV(K) = 0.0
    DO 2 I = 1,N(K)
        DEV(K) = DEV(K) + (AVG(K) - STU(I,K))**2
2       CONTINUE

    DEV(K) = DEV(K)/(N(K)-1)
    DEV(K) = SQRT(DEV(K))
```

$\langle n_k$ = the number per class;

STU_{ik} = individual student

GPA's (read in)$\rangle$

$\langle a_k = \dfrac{1}{n_k} \sum\limits_{i=1}^{n} STU_{ik} \rangle$

$\langle \sigma_k^2 = \dfrac{1}{n_k - 1} \sum\limits_{i=1}^{n} (a_k - STU_{ik})^2 \rangle$

```
      GPA = GPA + AVG(K)
      STNDEV = STNDEV + DEV(K)
3 CONTINUE

   GPA = GPA/4.                              ⟨GPA = 1/4 Σ(k=1 to 4) a_k⟩



   STNDEV = STNDEV/4.                        ⟨σ_avg = 1/4 Σ(k=1 to 4) σ_k⟩
```

$$\langle GPA = \frac{1}{4} \sum_{k=1}^{4} a_k \rangle$$

$$\langle \sigma_{\text{avg}} = \frac{1}{4} \sum_{k=1}^{4} \sigma_k \rangle$$

7.5   THE IMPLIED DO LOOP

A special form of DO loop is available exclusively for I/O: the implied DO loop. The form of the implied DO loop for input is

```
READ( ) (<list of variables> , INDX = ILO,IHI,ISTEP)
```

and for output is

```
WRITE( ) (<list of variables> , INDX = ILO,IHI,ISTEP)
PRINT sl,(<list of variables> , INDX = ILO,IHI,ISTEP)
```

In either case, the list of variables is read or printed first with the index, INDX = ILO, then again with INDX = ILO + ISTEP, etc., until the loop is completed. The rules concerning the index and range parameters are the same as those of an ordinary DO loop. Note, the loop is enclosed in parentheses and the index specifications are separated from the I/O list by a comma. For example,

```
        WRITE(*,'(1X,6F5.2)') (X(I),I=4,9)
```

has exactly the same effect as

```
 WRITE(*,'(1X,6F5.2)')X(4),X(5),X(6),X(7),X(8),X(9)
```

and is distinctly different from

```
        DO 1 I = 4,9
            WRITE(*,'(1X,6F5.2)')X(I)
      1 CONTINUE
```

The first two examples print all six numbers on a single line, while the last will result in six lines of output from six separate WRITEs. An implied READ is interpreted as a *single* READ of a repeated list of variables rather than a repeated READ. Similar considerations apply to WRITE/PRINT statements. Thus,

```
REAL X(50)
PRINT *,(X(I),I=1,50)
```

is the same as

```
REAL X(50)
PRINT *,X
```

More than one variable may be included in the I/O list.

Fortran	Output

```
     REAL X(6)
     DO 1 R = 1.,6.
        I = R
1       X(I) = R
     PRINT *,(X(J),2*J,J = 3,5)     ⟨3.000 6  4.000 8  5.000 10⟩

     WRITE(*,11)(X(I),X(I + 1),I = 1,5,2)   ⟨1. 2.00
11 FORMAT(1X,F3.0,2X,F4.2)                    3. 4.00
                                              5. 6.00⟩
     WRITE(*,'(5F4.1)')(X(3),I = 1,4)     ⟨3.0  3.0  3.0  3.0⟩
```

Implied DO loops are also useful for reading data, as illustrated in the following:

```
     REAL X(50),Y(50)
10 FORMAT(10F5.1)
11 FORMAT( 2F5.1)

     READ(*,11)(X(I),Y(I),I=1,50)      ⟨Requires 50 lines of
                                        input, 2 numbers per
                                        line⟩

     READ(*,10)(X(I),Y(I),I=1,50)      ⟨Requires 10 lines of
                                        input, 5 pairs xᵢ, yᵢ
                                        per line⟩

     DO 1 I = 1,50                     ⟨Requires 50 lines of
        READ(*,10)X(I),Y(I)            input, 2 numbers per
1 CONTINUE                             line⟩

     READ 10,X,Y                       ⟨Requires 10 lines of
                                        input, first 5 lines
                                        for Xs, last 5 for Ys⟩
```

Implied DO loops may also be nested as illustrated in the following example.

```
    REAL A(4,5)
10  FORMAT(5F6.2)
    WRITE(*,10)((A(I,J),J = 1,5),I = 1,4)
```

Inner
loop

Outer
loop

will print the elements of A in the arrangement

$$a_{ij}$$

j $\backslash$ i	1	2	3	4	5
1	a_{11}	a_{12}	a_{13}	a_{14}	a_{15}
2	a_{21}	a_{22}	a_{23}	a_{24}	a_{25}
3	a_{31}	a_{32}	a_{33}	a_{34}	a_{35}
4	a_{41}	a_{42}	a_{43}	a_{44}	a_{45}

7.6 MISCELLANEOUS EXAMPLES OF LOOP STRUCTURES AND ARRAYS

7.6.1 Infinite Series

In any book of mathematical tables you can find examples of series representations of various common functions. One such series that we have already encountered in Section 4.4.3 is

$$e^x = 1 + x + \frac{x^2}{2!} + \frac{x^3}{3!} + \cdot \cdot \cdot + \frac{x^n}{n!} + \cdot \cdot \cdot$$

If you attempt to sum this series by computing the individual terms as

```
TERM = X**N/FACTN
```

where FACTN is $n!$ and is evaluated on page 230, you may get the correct answer, but the program would be extremely inefficient. The evaluation of a single term would require exponentiation, which is relatively slow, and the use of a DO loop to compute $n!$. If a large number of terms is required to obtain an accurate result, serious problems involving computation time (i.e., money!) may result, even leading to failure of the code due to round-off error.

Frequently, when a series expansion involves factorials, it is more efficient to relate successive terms by means of their ratio, as was done in Section 4.4.3. For the series above we would once again obtain

$$R = \frac{(\text{term})_{n+1}}{(\text{term})_n} = \frac{x}{n+1}$$

The code to sum the series without using DO loops was given in Section 4.4.3. Rewriting this code using a DO loop yields the following:

```
READ(*,*)X
SUM = 1.
TERM = X
DO 2 I = 1,1000
    SUM = SUM + TERM
    IF(ABS(TERM) .LT. 1.E-6)THEN
        WRITE(*,*)'EXP(',X,') = ',SUM
        STOP
    ELSE
        TERM = TERM * X/(I + 1.0)
    END IF
2 CONTINUE
WRITE(*,*)'DID NOT CONVERGE AFTER 1000 TERMS'
STOP
```

This procedure for summing an infinite series, either with a DO loop or without, is a very important computational technique. You should attempt to code the following series expansions.

1. The $\sin(x)$

$$\sin(x) = x - \frac{x^3}{3!} + \frac{x^5}{5!} - \frac{x^7}{7!} + \cdots \qquad \text{only odd terms}$$

where you should verify that

$$(\text{term})_{n+2} = (\text{term})_n \times \left[-\frac{x^2}{(n+1)(n+2)} \right]$$

and SUM and TERM should be initialized as

$$\begin{array}{c} \text{SUM} = 0.0 \\ \text{TERM} = X \end{array}$$

2. The $\cos(x)$

$$\cos(x) = 1 - \frac{x^2}{2!} + \frac{x^4}{4!} - \frac{x^6}{6!} + \cdots \qquad \text{only even terms}$$

where

$$(\text{term})_{n+2} = (\text{term})_n \times \left[-\frac{x^2}{(n+1)(n+2)} \right]$$

3. The natural logarithm, $\ln(1 + x)$

$$\ln(1 + x) = x - \frac{x^2}{2} + \frac{x^3}{3} - \frac{x^4}{4} + \cdots \qquad \text{for } |x| \le 1$$

where

$$(\text{term})_{n+1} = (\text{term})_n \left(-x \frac{n}{n+1} \right)$$

4. The binomial expansion

$$(1 + x)^p = 1 + px + \frac{p(p-1)}{2!} x^2 + \frac{p(p-1)(p-2)}{3!} x^3$$

$$+ \cdots + \binom{p}{n} x^n + \cdots$$

where the notation $\binom{p}{n}$ is called the combinatorial and is defined as

$$\binom{p}{n} = \frac{p!}{n!(p-n)!}$$

For example,

$$\binom{p}{2} = \frac{p!}{2\,(p-2)!} = \frac{p(p-1)}{2}$$

The ratio of successive terms in the expansion may then be evaluated as

$$R = (\text{term})_{n+1}/(\text{term})_n$$

$$= \frac{\binom{p}{n+1} x^{n+1}}{\binom{p}{n} x^n}$$

$$= \frac{p!}{(n+1)!(p-n-1)!} \frac{n!\,(p-n)!}{p!} x$$

and since

$$\frac{n!}{(n+1)!} = \frac{1}{n+1}$$

and

$$\frac{(p-n)!}{(p-n-1)!} = \frac{(p-n)(p-n-1)!}{(p-n-1)!} = (p-n)$$

we obtain

$$(\text{term})_{n+1} = \left(x\frac{p-n}{n+1}\right)(\text{term})_n$$

For example, using $p = 4$ we obtain from the above:

$$\binom{4}{0} = 1 \quad \binom{4}{1} = 4 \quad \binom{4}{2} = 6 \quad \binom{4}{3} = 4 \quad \binom{4}{4} = 1 \quad \binom{4}{5} = 0$$

so

$$(1+x)^4 = 1 + 4x + 6x^2 + 4x^3 + x^4$$

If p is a positive integer, the infinite series terminates and the expression reduces to a polynomial. However, if p is not a positive integer, the binomial expansion results in an infinite number of terms.[3] What are the expansions for $p = -1$?, for $p = \frac{1}{2}$?

7.6.2 Maximum/Minimum of an Array

In Section 4.3 I described a simple method for finding the largest or smallest element of a list of numbers. This procedure can be easily adapted to finding the minimum and/or maximum element of an array. However, we are frequently interested in finding not only what the maximum element is, but where it is—that is, which element in the array is the largest (or smallest). This is accomplished by using what is called a *pointer,* in the following manner.

```
REAL X(100)
XMAX = X(1)
IMAX = 1
DO 1 I = 2,100
```

---

[3] The factorial for half-integers has not been defined, so the combinatorial expression cannot be used. However, the equation for the ratio of successive terms remains valid. For example,

$$(1+x)^{-1} = 1 + (-1)x + \frac{(-1)(-2)}{2}x^2 + \frac{(-1)(-2)(-3)}{6}x^3 + \cdots$$

so that

$$R = (\text{term})_{n+1}/(\text{term})_n = x\frac{(-1-n)}{(n+1)} = -x$$

```
        IF(X(I) .GT. XMAX)THEN
             XMAX = X(I)
             IMAX = I
        END IF
1 CONTINUE
  WRITE(*,*)'THE LARGEST ELEMENT IS THE I = ',IMAX,
 +    ' ELEMENT'
  WRITE(*,*)'WHICH HAS A VALUE OF',X(IMAX)
  STOP
  END
```

The position of the largest element of an array with more than one index is found in a similar manner.

7.6.3 Printing Two-Dimensional Arrays

We saw in Section 7.5 that a two-dimensional array is usually pictured as a rectangular block of numbers labeled by rows and columns. Such a grouping is also called a *matrix*. When printing such an object, the output should also include the row and column labels. The Fortran to do this is not difficult, but occasionally takes some trial and error to get everything lined up correctly. The code below will print a square 8 by 8 array with F6.2 formats.

```
      REAL B(8,8)
      WRITE(*,11)(I,I = 1,8)
      DO 1 IROW = 1,8
          WRITE(*,12)IROW,(B(IROW,ICOL),ICOL = 1,8)
    1 CONTINUE
   11 FORMAT(5X,8(3X,I1,3X))
   12 FORMAT(2X,I1,2X,8(F6.2,1X))
      . . .
```

PROBLEMS

1. Find any errors in the following. If none write OK.
    a. DIMENSION A(50),B(5,000)
    b. DIMENSION A(9,9,9,9,9,9,9)
    c. M = 5
        N = 3
        DIMENSION X(M,N)
    d. DIMENSION INTEGER I(50), REAL X(25)

---

e. DIMENSION A(5,5,5,5)
f. DIMENSION A(B(5))
g. DIMENSION A(5)(1:2)
h. REAL A(6,6),B(2,2,0:1)
i. CHARACTER*4 FOUR(-4:4)
j. INTEGER A(0:0)
k. CHARACTER*2 REAL(4)
l. INTEGER REAL(2:3)
m. INTEGER X(-5:-7)

2. Write a program to count the number of times each vowel is used in a text of several hundred lines. The text is contained on a data file. Store all the counters in a single array.

3. Starting with a function of two variables, for example,

$$z = f(x, y) = e^{x-y} \sin(5x) \cos(2y)$$

for

$$0 \le x \le 2 \qquad 0 \le y \le 2$$

a. Write a Fortran program employing DO loops to compute and store a square grid (21 by 21) of values of Z(I, J) for equally spaced values of x and y within the specified range.

b. Determine the maximum and minimum of the array Z. Copy the array Z into an integer array IZ(0:20,0:20) defined by

```
IZ(I,J) = 10*(Z(I,J) - ZMIN)/(ZMAX - ZMIN)
```

c. Print the array IZ as a 20 by 20 square block of integers. Interpret the result.

4. Write a segment of a Fortran program that determines and prints which element of a two-dimensional array A(50, 50) is largest. That is, the output should be something like

```
THE LARGEST ELEMENT IS IN ROW  III, COLUMN  III
```

5. Redo the solutions to Problem 4.4, using DO loops.

6. To determine whether an integer k is a prime number or not you must test whether k is divisible by all primes less than $\sqrt{k}$. To do this you must have a table of primes. Write a program to determine and print the first 100 prime numbers. A pseudocode outline of the program is given below.

INTEGER p_i, $i = 1,100$
$p_1 = 2$, $p_2 = 3$
For $m = 2,100$
    let $k = p_m + 2$
    test whether k is divisible by $p_1, p_2, \ldots, p_m$

if yes → increment k by 2 and retest

if no → $p_{m+1} = k$

The test to determine whether a number is evenly divisible by another number is given in Problem 3.13.

7. Translate the following into Fortran using DO loops:

a. $s = \sum\limits_{i=1}^{10} a_i x_i$

b. For all $i = 1,10$

$t_i = \sum\limits_{j=1}^{10} a_{ij} x_j$

c. For all $i, j = 1,10$

$c_{ij} = \sum\limits_{k=1}^{10} a_{ik} b_{kj}$

8. Write a segment of a Fortran program that:

a. Interchanges the entire rows K and J of a square array Z(10, 10).

b. Scans column K of the array, determines the row containing the largest element, and then interchanges this row with the first row of the array.

9. Determine the output from the following:

```
      INTEGER K(0:4,0:4)
      DO 1 I = 0,4
          DO 1 J = 0,4
              K(I,J) = 10*I + J
    1 CONTINUE
```

a.
```
      WRITE(*,1)(K(0,J),J=1,4)
    1 FORMAT(1X,10I2)
```
b.
```
      WRITE(*,2)((K(I,J),J=1,4),I=1,4)
    2 FORMAT(2X,5I3)
```
c.
```
      WRITE(*,3)(I,I=0,4),(M,(K(M,J),J=0,4),M=0,4)
    3 FORMAT (5X,5(2X,I1,2X),/,
     +             5(2X,I1,2X,5(2X,I2,1X),/))
```
d.
```
      WRITE(*,4)((K(I,J),J=1,4),I=0,4)
    4 FORMAT(2X,5I3)
```
e.
```
      WRITE(*,5)((K(I,J),I=4,0,-1),J=0,4)
    5 FORMAT(2X,5I3)
```
f.
```
      DO 3 I=0,4
          WRITE(*,6)(K(I,J),J=I,4)
    3 CONTINUE
    6 FORMAT(2X,5I4)
```
g.
```
      DO 4 I=0,4
          WRITE(*,7)(K(I,J),J=4,I,-1)
    4 CONTINUE
    7 FORMAT(20X,5(I3,TL8))
```

PROGRAMMING ASSIGNMENT IV

The goal of this programming assignment is to construct moderately complicated programs in modular form, employing numerous subprograms, and to illustrate the use of DATA, IMPLICIT, SAVE, and PARAMETER statements.

IV.1 SAMPLE PROGRAM

Industrial Engineering

Each of the traditional engineering disciplines (civil, mechanical, electrical, chemical, and metallurgical/mining) relies on a particular area of natural science for its foundation. Industrial engineering, however, seeks to incorporate the knowledge of the social sciences into designing improvements in the overall human-machine systems. The industrial engineer has responsibility for design, installation, and evaluation of not merely machines or systems but also for their interfacing with people to effect an overall productivity improvement. This may involve an understanding of human behavioral characteristics and their effect on the design of machines or the workplace, or on demands and services from outside from customers or clients. The industrial engineer will draw heavily on knowledge in economics, business management, and finance, as well as in the natural sciences. The areas of specialization of the industrial engineer may be classified as

Operations Research: The application of analytical techniques and mathematical models to phenomena such as inventory control, simulation, decision theory, and queuing theory to optimize the total systems necessary for the production of goods.

Management or Administrative Engineering: The increasingly complex interplay of management and production skills in modern industrial

242

operations have resulted in a need for technically trained managers. They will evaluate and plan all manner of corporate ventures, interact with labor, engineering departments, and subcontractors. In addition, a management engineer may participate in the financial operations of a company, drawing on knowledge in economics, business management, and law.

Manufacturing and Production Engineers: Before a product is produced, the complete manufacturing process must be designed and set-up to optimize the economics involved and the final quality of the item. This requires a broad knowledge of process design, plant lay-outs, tool design, robotics, and man-machine interactions.

Information Systems: The use of computers to gather and analyze data for decision making, planning, and to improve man-machine activity.

In a recent survey by the American Institute of Industrial Engineers the following list includes the most common responsibilities of those industrial engineers that responded.

> Facilities planning and design
> Methods engineering
> Work systems design
> Production engineering
> Management information and control systems
> Organization analysis and design
> Work measurement
> Wage administration
> Quality control
> Project management
> Cost control
> Inventory control
> Energy conservation
> Computerized process control
> Product packaging, handling, and testing
> Tool and equipment selection
> Production control
> Product improvement studies
> Preventive maintenance programs
> Safety programs
> Training programs

Sample Program: Minimizing Repair Costs

A large manufacturing plant has many identical machines, all of which are subject to failure at random times. Repairmen are hired to patrol and service the machines. The determination of the appropriate number of repairmen to hire is dependent on the following considerations:

1. Good repairmen are expensive. Assume an hourly wage of W (dollars/hr). The number of repairmen is R.
2. A repairman can work on only one machine at a time, so several down machines may have to wait for repair resulting in a productivity loss to the company. We will assume the loss per machine while it is inoperative is L (dollars/hr).
3. Too many repairmen will reduce the number of malfunctioning machines but may result in excessive idle time for these workers.

In addition we will make the following assumptions regarding the servicing of these machines:

1. The failure rate is known and is characterized by the average time between failures for an individual machine.

$$\lambda = \text{failures per machine per unit time (no./hr)}$$

$$1/\lambda = \text{time between failures per machine (hr)}$$

2. The average repair time is known and is characterized by

$$\mu = \text{repairs per hour per repairman (no./hr)}$$

$$1/\mu = \text{average repair time per machine (hr)}$$

The quantity

$$\rho = \frac{\lambda}{\mu}$$

is called the *traffic intensity* and $1/\rho$ represents roughly the number of machines that one repairman can handle [(time between machine failures)/(repair time per machine)].

The basic problem is to minimize the cost of the wages to the repairmen plus the downtime costs of inoperative machines; i.e.,

(IV.1) $$\text{Cost}(R) = WR + \langle N_d \rangle L$$

where $\langle N_d \rangle$ is the average number of inoperative machines at any one time. This will of course depend on how many repairmen (R) are hired. The main complexity of this problem is in calculating $\langle N_d \rangle$.

Let us assume that the likelihood that all machines are working can be represented by a number P_0 between 0 and 1 with $P_0 \sim 0$ representing a very small possibility that all are working and $P_0 \sim 1$ representing near certainty that all are working. Thus P_0 is the probability that zero machines are down.

Similarly, P_1 is the probability that only one machine is down, P_n the probability that n machines are down. The following result may then be obtained from queuing theory to relate these probabilities:
If $n < R$

$$P_{n+1} = \rho \left(\frac{N - n}{n + 1} \right) P_n \qquad \text{(IV.2)}$$

If $R \le n < N$

$$P_{n+1} = \rho \left(\frac{N - n}{R} \right) P_n$$

where N is the total number of machines. For example if $N = 50$ and $R = 5$, then

$$P_1 = 50 \rho P_0$$

$$P_2 = \frac{49}{2} \rho P_1 = 49(25) \rho^2 P_0$$

$$P_3 = \frac{48}{3} \rho P_2 = 16(49)(25) \rho^3 P_0$$

$$\cdots \quad \cdots$$

$$P_6 = \frac{45}{6} \rho P_5 = 105938 \rho^6 P_0$$

Thus every P_n can be related to P_0.
　　Finally, P_0 is determined from the condition

$$\sum_{n=0}^{N} P_n = 1 \qquad \text{(IV.3)}$$

Since each P_n is proportional to P_0, this is accomplished by

First setting $P_0 = 1$
Computing all the P_n on this basis
Evaluating the sum in Equation (IV.3)
Rescale each of the P_n by replacing P_n by P_n/sum

The average number of down machines may then be expressed as

$$\langle N_d \rangle = \sum_{n=0}^{N} n P_n \qquad \text{(IV.4)}$$

That is, the number down times the likelihood of that many down, summed over all possibilities.

Once this expression is computed, Equation (IV.1) may be evaluated for a variety of values of R and the value that results in a minimum cost determined.

Problem Specifics

Write a program using the parameters of Table IV-1 to accomplish the following:

1. Read the values of N, R_{max}, λ, μ, W, and L for the first case (fast, expensive repairmen and good, reliable machines). Print with appropriate headings. Use character variables for "fast, expensive" and "good, reliable."
2. For $R = 1$ to R_{max}, execute the following calculation:
   a. For $n = 1$ to N, compute and store the values of P_n assuming a value of 1 for P_0. Use a DO loop and store the values in an array $P(0: 75)$.
   b. Sum all the P_n (including P_0) and then rescale so that the sum of all the P_n is now 1.
   c. Compute the average number of down machines, $\langle N_d \rangle$, from Equation (III.4) and store in an array DOWN(R,ICASE) where ICASE is the case number.
   d. Compute and store the total repair costs for this case and for this number of repairmen in a similar array.
3. Determine and print the results for the minimum repair costs for this case.
4. Repeat for the three subsequent cases.

The Fortran code for this problem is shown in Figure IV-1 and the output from the program is shown in Figure IV-2.

Table IV-1   Input parameters for repair costs problem.

Case, i	Repair-men	Machine	No. of machines, N	Maximum No. of repair-men R_{max}	Failure rate, λ	Repair rate, μ	Repair-man wages, W	Loss per machine, λ
1	fast, expensive	good, reliable	65	15	0.03	0.66	16.0	250.0
2	fast, expensive	poor, cheap	70	15	0.04	0.66	16.0	215.0
3	slow, cheap	good, reliable	65	15	0.03	0.58	11.0	250.0
4	slow, cheap	poor, cheap	70	15	0.04	0.58	11.0	215.0

Figure IV-1 The Fortran code for the minimum repairs cost problem.

```fortran
      PROGRAM REPAIR
*--
*-- Computes the optimum number of repairmen to be hired
*-- to service a large number of machines that break
*-- down randomly. Several situations are considered
*-- and the overall best case is found.
*--
* Variables
*--
      CHARACTER*4 MEN,MACHINE
      INTEGER N,RMAX,ROPT(4),IC,IBEST,R,NCASE
      REAL  FRATE,RRATE,W,LOSS,CSTMIN(4),RHO,P(0:75),
     +      DOWN(15,4),COST(15,4),SUM,CBEST
*--
*-- IC      -- Case number (1 to 4)
*-- N       -- Number of machines
*-- RMAX    -- Maximum number of repairmen
*-- R       -- Current number of repairmen
*-- NCASE   -- Total number of cases considered
*-- FRATE   -- Failure rate of the machines
*-- RRATE   -- Repair rate of the repairmen
*-- RHO     -- Traffic intensity (Frate/Rrate)
*-- W       -- Repairmen's hourly wage
*-- LOSS    -- Down time loss per machine
*-- MEN     -- Type of repairmen (fast - slow)
*-- MACHINE -- Type of machine (good - poor)
*-- P(K)    -- Probability that K machines are down
*-- DOWN(R,I)-- Avg. no. down for case I and R
*--             repairmen
*-- COST(R,I)-- Total cost for the same conditions
*-- CSTMIN(I)-- Minimum cost for case I
*-- ROPT(I)  -- Optimum no. of repairmen for case I
*-- IBEST   -- The optimum case number
*-- SUM     -- Sum of the P(K)
*-- CBEST   -- Overall minimum cost
*--
* Initialization
*--
      OPEN(UNIT=36,FILE='RPRCST')
      REWIND 36
*--
*--        Print table headings
*--
      WRITE(*,21)
      IC = 1
1     READ(36,*,END=10)MEN,MACHNE,N,RMAX,FRATE,RRATE,
     +                 W,LOSS
      WRITE(*,22)      MEN,MACHNE,N,RMAX,FRATE,RRATE,
     +                 W,LOSS


      RHO = FRATE/RRATE
*--
* Computation
*--
*--      Determine the probabilities (unscaled) for
*--      R = 1 to Rmax.
*--
      DO 4 R = 1,RMAX
        P(0) = 1.0
        SUM = 0.0
        DO 2 I = 0,N - 1
          IF(I .LT. R)THEN
            P(I + 1) = RHO*(N - I)/(I + 1.)*P(I)
          ELSE
            P(I + 1) = RHO*(N - I)/R*P(I)
          END IF
          SUM = SUM + P(I + 1)
2       CONTINUE
*--
*--      Next scale the P's and determine the avg.
*--      number down.
*--
        DOWN(R,IC) = 0.0
        DO 3 I = 0,N
          P(I) = P(I)/SUM
          DOWN(R,IC) = DOWN(I,IC) + I*P(I)
3       CONTINUE
*--
*--      Once the average down is known, compute costs.
*--
        COST(R,IC) = W*R + DOWN(R,IC)*LOSS
4     CONTINUE
*--
*--      Find the minimum costs and no. of
*--      repairmen for this case.
*--
      CSTMIN(IC) = COST(1,IC)
      ROPT(IC) = 1
      DO 5 R = 1,RMAX
        IF(COST(R,IC) .LT. CSTMIN(IC))THEN
          CSTMIN(IC) = COST(R,IC)
          ROPT(IC)   = R
        END IF
5     CONTINUE
*--
*--      This case is complete, return and read next
*--      data set.
*--
```

Continued

```
*--
        IC = IC + 1
        GO TO 1
*--
*--     All cases considered, determine optimum case.
*--
10      NCASE = IC - 1
        IBEST = 1
        CBEST = CSTMIN(1)
        DO 11 IC = 2,NCASE
        IF(CSTMIN(IC) .LT. CBEST)IBEST = IC
11      CONTINUE
*--
* Print final results
*--
        WRITE(*,23)
        DO 12 IC = 1,NCASE
        WRITE(*,24)IC,ROPT(IC),CSTMIN(IC)
        IF(IC .EQ. IBEST)THEN
        WRITE(*,25)'OPTIMUM SITUATION'
        END IF
12      CONTINUE
        STOP
*--
* Formats
*--
21      FORMAT(///,13X,'RESULTS OF THE REPAIR COST ',
       +          'PROBLEM',//,
       +       5X,'INPUT PARAMETERS',//,
       +       6X,'NUMBER',T31,'MAX',//,
       +       7X,'OF',5X,'TYPE',3X,'NUMBER',4X,'OF',T57,
       +          'DOWN',//,
       +       6X,'REPAIR',9X,'OF',9X,'REPAIR FAIL REPAIR'
       +          ,' WORKER TIME',//,
       +       ' CASE MEN MACHINE MACHINE MEN RATE ',
       +          'RATE WAGES LOSS',//,
       +       1X,60('-'))
*--
22      FORMAT(3X,I1,2(3X,A4),5X,I2,6X,I2,3(2X,F5.2),
       +       2X,F5.1)
23      FORMAT(//,5X,'COMPUTED RESULTS',//,
       +       T20,'OPTIMUM',T32,'OPTIMUM',//,
       +       T20,'NUMBER ',T32,' COSTS',//,
       +       11X,'CASE',T22,'OF',T31,'FOR THIS',//,
       +       10X,'NUMBER REPAIRMEN',5X,'CASE',//,
       +       10X,6('-'),2X,9('-'),3X,8('-'))
*--
24      FORMAT(12X,I1,9X,I2,6X,F8.2)
25      FORMAT('-',T40,A)
*--
        END
```

Figure IV-2 The output from the minimum repair costs problem.

```
                 RESULTS OF THE REPAIR COSTS PROBLEM
```

	NUMBER OF REPAIR- MEN	TYPE OF MACHINE	NUMBER OF MACHINE	MAX OF REPAIR- MEN	FAIL RATE	REPAIR RATE	WORKER WAGES	DOWN TIME LOSS
CASE								
1	FAST	GOOD	65	15	.03	.66	16.00	250.0
2	FAST	POOR	70	15	.04	.66	16.00	215.0
3	SLOW	GOOD	65	15	.03	.58	11.00	250.0
4	SLOW	POOR	70	15	.04	.58	11.00	215.0

INPUT PARAMETERS

COMPUTED RESULTS

CASE NUMBER	OPTIMUM NUMBER OF REPAIRMEN	OPTIMUM COSTS FOR THIS CASE	
1	6	815.77	OPTIMUM SITUATION
2	8	996.84	
3	7	883.84	
4	9	1076.65	

```
                STOP
```

Programming Problem A: Cooling Curve for Transfer Ladle Cars

The transfer of molten pig iron from a blast furnace to the steelmaking facilities is a seemingly simple procedure. The molten iron is placed in an elongated tilting ladle railroad car as pictured in Figure IV-3. If the distance to be transferred is considerable or if for other reasons the molten iron will remain in the ladle car, it is very important that the cooling rate of the car and its contents be continually monitored. If the molten iron begins to solidify in the transfer car, you can imagine the resulting problems.

The purpose of this problem is to estimate and plot the temperature of the molten iron vs. time based on two assumptions:

1. The heat (Q) flowing out from the iron through the car walls (J/sec = watts) is equal to the heat transferred from the car surface to the air by convection and radiation.

$$Q_{cond} = Q_{conv} + Q_{rad} \qquad \text{(IV.5)}$$

The conduction term depends upon the geometry of the ladle car. If we approximate the car as a cylinder of length L, outer radius a, and wall thickness δ, and if the car material has a heat conductivity k, the heat flow Q_{cond} can then be expressed approximately as

$$Q_{cond} = k \frac{A_\lambda}{\delta}[T(t) - T_s] \qquad \text{(IV.6)}$$

where A_λ is the log-mean area involved in the heat conduction, $T(t)$ is the temperature of the molten iron at time t, and T_s is temperature of the outside surface of the car.

The log-mean area A_λ is expressed in terms of the areas of the outside cylinder surface ($A_O = 2\pi aL$) and the inside surface ($A_i = 2\pi(a - \delta)L$) as

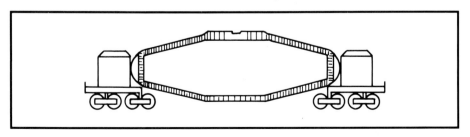

Figure IV-3
Sketch of a 150-ton transfer ladle car.

(IV.7)
$$A_\lambda = \frac{A_0 - A_i}{\ln (A_0/A_i)}$$

which can be written as

(IV.8)
$$A_\lambda = -\frac{2\pi\delta L}{\ln (1 - \delta/a)}$$

(Note: The logarithm is negative so A_λ is positive.)

2. If the temperature of the surrounding air is T_a, the heat loss from the car to the air is

(IV.9)
$$Q_{conv} + Q_{rad} = h_c A_\lambda (T_s - T_a)^{5/4} + \sigma\epsilon A_\lambda (T_s^4 - T_a^4)$$

where h_c = coefficient for convective heat transfer from a cylinder

σ = coefficient for radiative heat transfer, Stefan-Boltzman constant

ϵ = emissivity of the car's surface

Thus Equation (IV.5) may be written

(IV.10)
$$T(t) = T_s - \frac{a}{k \ln(1 - \delta/a)}[h_c(T_s - T_a)^{5/4} + \sigma\epsilon(T_s^4 - T_a^4)]$$

Knowing the temperature of the iron at a time t, $T(t)$, and the air temperature T_a, we will use this equation to determine the temperature of the car surface, T_s.

The rate of heat loss by the molten iron is related to its temperature through its specific heat C by

(IV.11)
$$Q_{cond} = \frac{\Delta Q}{\Delta t} = -mC\frac{\Delta T}{\Delta t}$$
$$= -mC\frac{T(t + \Delta t) - T(t)}{\Delta t}$$

where m is the mass of the molten iron and C its specific heat. This equation may be written

(IV.12)
$$T(t + \Delta t) = T(t) - \frac{\Delta t Q_{cond}}{mC}$$

This equation will be used to compute T at a later time $t + \Delta t$, knowing $T(t)$ and Q_{cond}.

The basic procedure is thus

1. Start at $t = 0$
   a. Knowing $T(t)$ and T_a, use the bisection algorithm to solve Equation (IV.10) for T_s
   b. Use Equation (IV.6) to compute Q_{cond}
   c. Use Equation (IV.12) to compute T at the next time step—i.e., $T(t + \Delta t)$
2. Increment $t \rightarrow t + \Delta t$ and repeat
3. Print a neat table of t and $T(t)$ for each case.

Problem Specifics

The input parameters for this problem are listed in Table IV-2. These should be stored in a data file that will be read by your program. The calculation should proceed as follows:

1. Read all the input data from a data file and neatly print them.
2. For the initial calculation ($t = 0$) of T_s use the interval

$$T_a < T_s < T(0)$$

while for all subsequent steps use an interval based on the previous calculation. For example, if T_s is the latest computed surface temperature, the next computed value should be only slightly lower. Thus a reasonable interval to use for the next calculation might be $T_s - \frac{1}{25}(T_s - T_a) \leftrightarrow T_s$.

Table IV-2 Input parameters for the cooling curve problem.

a	= 1.50	= outer radius of ladle car (m)
L	= 8.00	= length of ladle car (m)
δ	= 0.50	= thickness of car walls (m)
k	= 4.20	= heat conductivity of walls (W/K-m)
h_c	= 1.70	= heat convection coefficient for a cylinder (W/K-m)
ε	= 0.80	= emissivity of ladle car surface
m	= 1.35E5	= mass of molten iron (kg)
C	= 1172.	= effective specific heat of molten iron (J/kg-K)
T_a	= 298	= air temperature (K)
σ	= 5.67E-9	= radiative heat coefficient
Δt	= 1200	= time step (sec)
t_{max}	= 9.0E4	= maximum time (sec), i.e., 25 hr
T_{solid}	= 1430.	= solidification temperature of molten iron (K)
eps	= 10^{-3}	= convergence criterion for bisection (K)
I_{max}	= 40	= maximum number of iterations in bisection algorithm
$T(0)$	= case 1 1800.0	= initial temperature of molten iron (K)
	case 2 1700.0	

3. For the first-case initial temperature compute the 76 temperature values for the 75 time steps and store in an array $T(0:75, 2)$.
4. Repeat for the second-case initial temperature.
5. Print a table of the results for the two initial temperatures.
6. All internal values of temperature are in Kelvin but all your printed values should be in degrees Celsius $(K - 273 = °C)$. Likewise, the internal time is in seconds while the output should be in hours.
7. Indent DO loops and IF blocks, segment the code into blocks.

Programming Problem B: The Minimum of a Multi-dimensional Array or Function

The procedure given in Section (4.3) for finding the minimum of a function $f(x)$ or a list is easily implemented provided the search is only in one direction (for example, along the x-direction or down the list). However, the problem is considerably more complicated if there are several independent variables involved. A moderately simple, though somewhat inefficient, procedure is to construct an algorithm along the following lines:

Consider a set of numbers that are a function of two variables or counters. The set can be thought of as a rectangular array, a_{ij}. We pick a starting point anywhere within the array and examine a_{ij} and the eight neighboring points in the array. If the minimum of these nine values is a_{ij}, we have found a local minimum. If not, we replace a_{ij} by the minimum of the set and repeat the procedure for this new point in the array. This is continued until a minimum is found or until an edge of the array is reached.

The idea is illustrated on the rectangular set of numbers below.

```
9  9  8  7  6  5  5  6  9
      start
9  8  7 (7) 6  5  6  8  9
8  7  6  6 (5) 4  4  5  8
7  6  5  4 (3) 5  6  6  7
6  5  4  3 (2) 3  5  6  8
5  4  3  2 (1) 2  4  5  7
6  5  4  3  2  2  3  4  6
```

The algorithm to find the maximum of a function of two (or more) variables proceeds in a similar manner.

Start with an initial guess for the minimum.
(x_0, y_0) = initial guess
$f_0 = f(x_0, y_0)$

$$f_{min} = f_0$$

and specify the step sizes, Δx, Δy.

1   for $i_x = -1, 0, +1$ and for $i_y = -1, 0, +1$

      $x_1 = x_0 + i_x \Delta x$

      $y_1 = y_0 + i_y \Delta y$

      IF$(f(x_1, y_1) < f_{min})$THEN

            redefine the minimum of the set

            $x = x_1$

            $y = y_1$

            $f_{min} = f(x_1, y_1)$

      END IF

      IF$(x \neq x_0$ or $y \neq y_0)$THEN

            redefine the minimum point of the set

            $x_0 = x$

            $y_0 = y$

            GO TO 1 to continue the search.

      ELSE

            the point (x_0, y_0) is a local minimum,

            so, either reduce the step size and

            continue, or stop the search.

            IF$(\Delta x < EPSx$ and $\Delta y < EPSy)$THEN

                  PRINT x, y, f_{min}, Δx, Δy

                  STOP

            ELSE

                  $\Delta x = \Delta x/5.$

                  $\Delta y = \Delta y/5.$

            END IF

      END IF

Problem Specifics

Use the pseudocode outline just given to construct a program to find the optimum design of a water-carrying trough. The trough is to be made from a strip of sheet metal of width $d = 30$cm wide. The edges will be bent up to angles θ to form sides of length t. Both sides are identical. (See Figure IV-4.)

To maximize the capacity of the trough, we wish to find the value of both t and θ that will maximize the cross-sectional area $A(t, \theta)$, which is given by

$$A(t, \theta) = (d - 2t)t \sin \theta + t^2 \sin \theta \cos \theta \qquad \text{(IV.13)}$$

Your program should use the starting values

$$t_0 = d/5$$

$$\theta_0 = \pi/4$$

Figure IV-4 A
water-carrying
trough formed
from sheet metal.

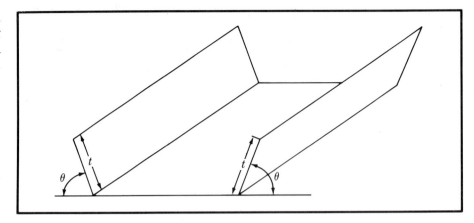

$$\Delta t = 0.1$$

$$\Delta \theta = 0.1$$

The program should complete a maximum of eight searches. After
each search the step size is diminished by a factor of 5.
The program should terminate with a diagnostic print if the edge of
the physical region is reached; that is

$$[t \le 0, t \ge d/2 \quad \text{or} \quad \theta \le 0, \theta \ge \pi/2]$$

If successful, the program should print the optimum values of t, θ,
and the current step sizes along with the optimum area.

By the way, the analytical solution, easily obtained by methods of advanced calculus, is $t_{\text{best}} = d/3$, $\theta_{\text{best}} = \pi/3$.

Programming Problem C: Calculation of a Trajectory with Air Drag

The calculation of the trajectory of a projectile requires that we integrate the
differential equation $F = ma$. This is ordinarily a formidable task, but, for
some simple forces, the problem can be stated in a form understandable to
the novice. First of all, we need to have formulas for numerically evaluating
the derivatives of functions.
The first derivative of a function $y(t)$ can be approximated by

(**IV.14**)
$$\left. \frac{dy}{dt} \right|_{t_1} = y'(t_1) \approx \frac{y(t_1 + \Delta t) - y(t_1 - \Delta t)}{2 \, \Delta t}$$

where Δt is assumed to be very small. If we designate the equally spaced
points along the t-direction as t_0, $t_1 = t_0 + \Delta t$, $t_2 = t_0 + 2 \, \Delta t$, and so on, and

the values of the function at these points as $y_2 = y(t_2)$, this equation can be written more succinctly as

$$y'(t_n) \simeq \frac{y_{n+1} - y_{n-1}}{2\,\Delta t} \tag{IV.15}$$

Similarly, it can be shown that an approximation for the second derivative of a function is

$$y''(t_n) \simeq \frac{y_{n+1} - 2y_n + y_{n-1}}{(\Delta t)^2} \tag{IV.16}$$

We will use these two equations to solve for the trajectory of a projectile shot straight up.

The equation describing the motion of a rocket shot straight up in air is Newton's second law: $F = ma$, or

$$ma = m\frac{d^2 y}{dt^2} = my'' = -mg - \lambda v \tag{IV.17}$$

where $v = dy/dt = y'$ is the velocity and λ is the air-drag coefficient. The drag force is always directed opposite to the direction of the velocity. (This expression is a rather poor approximation and so the results of the problem may be correspondingly unreliable.)

Replacing the first and second derivatives in Equation (IV.17) by the approximate finite difference equations, we obtain

$$y_{n+1} - 2y_n + y_{n-1} = -g(\Delta t)^2 - \frac{\lambda}{2m}(y_{n+1} - y_{n-1})\Delta t \tag{IV.18}$$

This is really a series of equations relating the set of unknowns, $y_0, y_1, \ldots, y_n$. This set of simultaneous equations can be solved by a procedure known as Gauss-Jacobi iteration. We first formally solve the equation with $n = 1$ for y_1, the equation with $n = 2$ for y_2, and so on:

$$y_1 = \frac{1}{2}g\,\Delta t^2 + \frac{\lambda}{4m}\Delta t\,(y_2 - y_0) + \frac{1}{2}(y_2 + y_0)$$

$$y_2 = \frac{1}{2}g\,\Delta t^2 + \frac{\lambda}{4m}\Delta t\,(y_3 - y_1) + \frac{1}{2}(y_3 + y_1) \tag{IV.19}$$

$$\cdots \quad \cdots \quad \cdots \quad \cdots$$

$$y_n = \frac{1}{2}g\,\Delta t^2 + \frac{\lambda}{4m}\Delta t\,(y_{n+1} - y_{n-1}) + \frac{1}{2}(y_{n+1} + y_{n-1})$$

We will start with the information that at $t = 0$, the projectile starts at $y_0 = y(t = 0) = 0.0$ and, that after 2 minutes ($t_{n+1} = 120$ seconds), the rocket returns to earth ($y_{n+1} = 0.0$). We next need to compute *all* the intervening points, y_n. To accomplish this we *guess* values for $y_1, y_2, \ldots, y_n$; we insert these values in the right side of Equation (IV.19) and compute improved estimates. This is repeated until the average change in the y-values from one iteration to the next is less than some prescribed tolerance EPS.

Details

1. Your program should read values for

$$t_{\text{hit}} = \text{the time (seconds) the rocket returns}$$
$$\text{to earth (use } t_{\text{hit}} = 120 \text{ seconds)}$$
$$\lambda_0 = \text{the air drag coefficient (use } \lambda = 0.008)$$
$$g = \text{the gravitational acceleration } (9.8\text{m/sec}^2)$$

2. The number of intermediate y-values should be $n = 100$. Thus $y_0 = y_{101} = 0.0$, and $\Delta t = t_{\text{hit}}/(101)$. Use the equation

(IV.20)
$$y_f(t) = \frac{1}{2} gt(t_{\text{hit}} - t)$$

to obtain the original set of estimates for the y-values; that is, $y_1^{\text{old}} = y_f(t = \Delta t)$, $y_2^{\text{old}} = y_f(t = 2\,\Delta t)$, and so on. This is the equation for the trajectory of the projectile with zero air drag.

3. Compute a new set of y-values using Equations (IV.19). It is important to note that the air drag must be directed opposite to the direction of travel.

4. Compute a running sum of the changes in each of the y-values:

$$\delta^2 = \frac{1}{101} \sum_i \delta_i^2 = \frac{1}{101} \sum_i |y_i^{\text{new}} - y_i^{\text{old}}|^2$$

If, after a complete iteration, $\delta < $ EPS, the calculation is successfully terminated. Use a value of EPS = 1.50 and limit the number of iterations to IMAX = 100.

5. After a successful calculation, find the maximum y-value and the value of the time a maximum height is attained. Print these values and compare with the same quantities for zero drag. (For zero drag, the maximum height is reached when $t = t_{\text{hit}}/2$.)

6. Repeat the calculation with a drag force twice as large. Does the maximum height decrease by half?

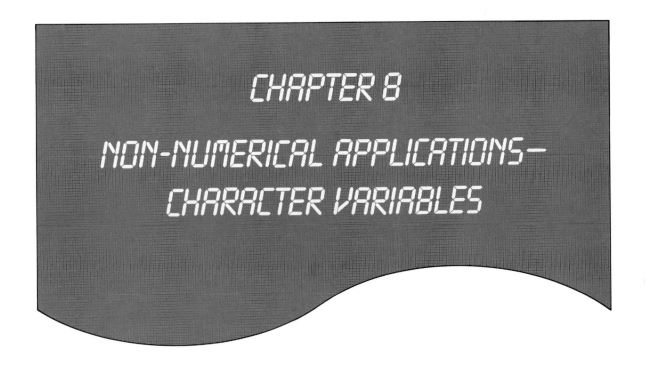

CHAPTER 8
NON-NUMERICAL APPLICATIONS— CHARACTER VARIABLES

8.1 REVIEW OF CHARACTER VARIABLES

Non-numerical applications of computers like graphics or word processing are the areas of computer science experiencing the most rapid growth. And even though an engineer will be using a computer primarily to solve mathematical problems or analyze numerical data, the character manipulation abilities of the computer will frequently be found to be quite useful. This chapter deals with applications of character variables in a variety of situations, such as sorting a list of names, or plotting a graph on the printer or terminal screen.

First, let us review the properties of character-type variables introduced in Section 2.2.3. A variable is defined to be of type CHARACTER with a length of ⟨integer⟩ characters by a statement

```
CHARACTER [name]*<integer>
```

or

```
CHARACTER*<integer> [name₁], [name₂], ...
```

For example,

```
CHARACTER*5 METAL,OXIDE,ACID(6)      ⟨all have length five⟩
CHARACTER GAS(0:5)*8, SYMBOL(6)*1    ⟨Each of the six ele-
                                       ments of GAS is of
                                       length eight⟩
```

These variables may then be assigned "values," which must be a sequence of characters.

Assignment statement	Value stored
`METAL = 'STEEL'`	S | T | E | E | L
`OXIDE = 'AL302'`	A | L | 3 | 0 | 2
`ACID(1) = 'H2CO3'`	H | 2 | C | O | 3
`ACID(2) = 'H2S'`	H | 2 | S | |

The remainder of the word is filled with blanks.

```
GAS(0) = 'NEON'     N | E | O | N |   |   |   |   |
SYMBOL(2) = 'CL'    C |
```

The value is truncated from the right until it fits the length of the variable.

```
SYMBOL(1) = GAS(0)    N |

GAS(1) = SYMBOL(2)    C |   |   |   |   |   |   |   |
```

and may be compared in IF tests

Comparison	Value of comparison
`IF(METAL .EQ. 'STEEL')`	⟨true⟩
`IF(METAL .EQ. 'STEAL')`	⟨false⟩
`IF('NEON' .EQ. GAS(0))`	⟨true⟩

⟨*The shorter variable ⟨'NEON'⟩ is extended with blanks to the same length as the longer variable before the comparison.*⟩

`IF(SYMBOL(2) .EQ. 'CL')`	⟨false⟩

⟨*Compares* ⟨C\| \|⟩ *with* ⟨C\|L⟩⟩

The input and output of character variables is accomplished with the A format.

```
        WRITE(*,12)METAL
     12 FORMAT(5X,A5)
```

or

```
12 FORMAT(5X,A)
```

As a reminder: Character variables contain coded values for symbols and can never be used in any arithmetic expressions.

```
CHARACTER M*5
INTEGER N
N  =  3 + 4          ⟨N contains the numerical value 7⟩
M = '3 + 4'          ⟨M contains the symbols '3 + 4'⟩
```

8.2 CHARACTER SUBSTRINGS

The result of an assignment statement involving a character variable is to store in the variable a set of symbols in a so-called string. A character string may be any length from one to several thousand characters. Clearly, it will frequently be necessary to access parts of a long string. The parts of a complete string are called substrings, and the form of a reference to a character substring is shown in Figure 8-1. For example, if a character variable ABCS is defined as

```
CHARACTER ABCS*26
ABCS = 'ABCDEFGHIJKLMNOPQRSTUVWXYZ'
```

then references to substrings might be

$$\text{CHNAME}(p_l : p_r)$$

where   CHNAME   is the name of the character variable.

p_l is an integer or integer expression designating the position of the first or leftmost character in the substring. If omitted, a value of 1 is assumed.

p_r is an integer or integer expression designating the position of the last or rightmost character in the substring. If omitted, the last position in the string is used.

If the entire string is of length n characters, then $1 \leq p_l \leq p_r \leq n$.

Figure 8-1
Character substring reference.

Substring reference	Value
ABCS(1:2)	'AB'
ABCS(3:6)	'CDEF'
ABCS(:4)	'ABCD'
ABCS(20:)	'TUVWXYZ'
ABCS(4:4)	'D'
ABCS(:)	same as ABCS—i.e., entire string

If the character variable itself is a dimensioned array, then substrings of each array element may be referenced in a similar manner. First a particular element in the array is given, followed by the specification of the substring in that element. For example,

```
CHARACTER NAME(6)*8, ADDRESS(0:5)*25, A*2
NAME(2) = 'JONES'
ADDRESS(0) = '1442 STATE, BOSTON, MA'
```

A = NAME(2)(2:3) ⟨A is assigned the value of the second and third characters in the second element of the array NAME⟩

ADDRESS(0)(6:10) = 'MAIN' ⟨STATE is replaced by MAIN. Note, MAIN is extended by a blank.⟩

Just as with an array, you must be very careful that the substring specifications are within the bounds of the complete character string. If p_1 is less than 1 or p_r is greater than the length of the string, an execution error will result.

8.3 INTRINSIC FUNCTIONS RELATED TO CHARACTER VARIABLES

Many of the applications of character variables involve searching a string for a particular substring and then performing a replacement. There are several Fortran intrinsic functions that have been designed to help in the coding of such tasks.

8.3.1 The Length of a String—Function LEN

The Fortran function LEN simply returns the length of the character string that is the argument of the function.

```
LEN(string)
```

Some examples are

```
CHARACTER A*4,B*5
INTEGER I,K,M,N
A = '1234'
B = 'NAME'
```

I = LEN(A)	⟨value assigned to I is 4⟩
K = LEN(B)	⟨value assigned to K is 5, the value stored in B is ⌊N⌋A⌊M⌋E⌊ ⌋⟩
M = LEN(A(2:4))	⟨value assigned to M is 3, the length of the substring is $(4-2)+1$⟩
N = LEN(A)/LEN('ABC')	⟨LEN('ABC') is 3, so N is assigned the value 1⟩

8.3.2 The Location of a Substring—Function INDEX

The Fortran function

$$\text{INDEX(string,substring)}$$

will return the position of the substring within the string. Both entrees in the argument list must be of type CHARACTER. If the second character string in the argument list occurs as a substring in the first, the result is an integer corresponding to the starting position of the substring within the first named string. If a match is not found, including the case where the substring is larger than the string, the value returned is zero. If there is more than one match within the string, only the starting position of the first occurrence is given. For example,

```
CHARACTER ABCS*26,A*1,B*2
ABCS = 'ABCDEFGHIJKLMNOPQRSTUVWXYZ'
A = 'A'
```

B = 'B'	⟨note, stored in B is ⌊B⌋⌊ ⌋⟩
I = INDEX(ABCS,A)	⟨value assigned to I is 1⟩
J = INDEX(ABCS,'C')	⟨value assigned to J is 3⟩
K = INDEX(ABCS,'DEF')	⟨value assigned to K is 4⟩
L = INDEX(ABCS,B)	⟨value assigned to L is 0, the substring ⌊B⌋⌊ ⌋ is not contained in ABCS⟩
ABCS(INDEX(ABCS,'P'):INDEX(ABCS,'R')) = '**********'	⟨the positions P through R are replaced with asterisks. Note the string on the right is truncated⟩

8.4 CHARACTER EXPRESSIONS—CONCATENATION

The only character string operation provided in Fortran is called *concatenation,* which means joining together. If S_1 and S_2 are two character strings of length n_1 and n_2, respectively, then the concatenation of S_1 and S_2 is effected by the operator $//$ (two slashes, interpreted as a single symbol). Thus

$$S_1//S_2$$

has a value of a string of length $n_1 + n_2$, consisting of the two individual strings joined into one. For example,

```
CHARACTER*10 NAME1,NAME2
NAME1 = 'JOHN SMITH'
NAME2 = NAME1(6:)//','//NAME(:4)
```

Successive concatenations proceed from left to right, so

```
NAME2 => | S | M | I | T | H | , | J | O | H | N |
```

The code to read a list of names in the form first name, middle initial, last name and store them, last name first, would then be

```
   CHARACTER*30 XX,NAME(100)
   INTEGER PERIOD,LAST,I

   I = 1
1  READ(*,'(A)',END=99)XX
   PERIOD = INDEX(XX,'.')        (Finds the location of the period after
                                 the middle initial)
   LAST = INDEX(XX,'  ') - 1     (The end of the name is the occurrence of
                                 two successive blanks)
   NAME(I) = XX(PERIOD+2:LAST)//','//XX(:PERIOD)
   I = I + 1
   IF(I .LT. 100) GO TO 1
99 CONTINUE
   ...
   ...
```

8.5 COMPARISON OF CHARACTER STRINGS

The binary code for symbols 'A', 'B', . . . , 'Z' has been set up so that the value of 'A' is less than the value of 'B', is less than the value of 'C', etc. Furthermore, the value of a blank is less than the value of 'A'. Thus character strings may be compared in logical IF tests. A few examples are given below

IF test	Value of argument	
IF('A' .LE. 'G')	⟨*true*⟩	
IF('AA' .LT. 'A ')	⟨*false*⟩	blank < 'A' so
		$\boxed{A}\boxed{A} > \boxed{A}\boxed{\ }$
IF('ABCDE' .LT.'ABCDZ')	⟨*true*⟩	

Once again, if the character variables being compared are of unequal length, the shorter variable is extended by adding blanks to the right before the comparison.

Unfortunately, the ordering of the remaining characters in Fortran is not standard and will vary from site to site. There is a Fortran intrinsic function that will tell you what ordering of characters is employed at your computing center: the function CHAR(I). The range of I values is also system-dependent and may take some experimenting on your part to determine. On my computer the allowed range of I is from 0 to 63 and the function CHAR(I) returns a single character for each I in this range. Thus CHAR(0) on my computer has a value of , i.e., ⟨blank space⟩, meaning that this character has the smallest binary value of all the allowed Fortran symbols. Printing out all the remaining 63 allowed Fortran symbols by means of this function then establishes their relative ordering

```
      DO 1 I = 1,63
      WRITE(*,'(2X,I2,2X,A)')I,CHAR(I)
    1 CONTINUE
```

and, for my computer, gives the following result:

I	CHAR(I)	I	CHAR(I)	I	CHAR(I)	I	CHAR(I)	I	CHAR(I)
1	!	12	'	25	9	38	F	51	S
2	''	13	-	26	:	39	G	52	T
3	#	14	.	27	;	40	H	53	U
4	$	15	/	28	<	41	I	54	V
There is no		16	0	29	=	42	J	55	W
symbol defined		17	1	30	>	43	K	56	X
for I = 5		18	2	31	?	44	L	57	Y
6	&	19	3	32	@	45	M	58	Z
7	'	20	4	33	A	46	N	59	[
8	(	21	5	34	B	47	O	60	\
9	)	22	6	35	C	48	P	61	]
10	*	23	7	36	D	49	Q	62	^
11	+	24	8	37	E	50	R	63	_

You might wish to try this on your machine.

The inverse of the function CHAR is the intrinsic function ICHAR( ). The argument is a *single* character and the function returns the local sequencing number associated with that character on your computer. Thus, from the list given above, the value of ICHAR('A') on my computer is the integer 33.

In Section 4.3 we discussed a procedure for finding the minimum or maximum of a list of numbers by repeatedly comparing pairs of numbers using an IF test. Since character variables may also be compared in an IF test, the same ideas may be used to alphabetize a list of names. The procedure is simply to scan the list of names and find the name with the minimum "value," for example, AARDVARK, and put this name at the top of the list. The remainder of the list is then scanned once more for the minimum in the remaining list of names, and this name is placed in the second position, and so on. This is an example of what is called sorting and is described more completely in the next section.

8.6   SORTING ALGORITHMS

The arranging of the elements of a list or set into some sort of ordered sequence is called *sorting* and the most common example is the alphabetizing of a list of names. In principle, the basic ideas involved in sorting are rather trivial, as we have seen; however, the difficulty increases dramatically with the size of the list being sorted, roughly speaking increasing with the square of the size of the list. So while one algorithm may be quite suitable to sort the names of the students in a university (about 10,000 names), it might be next to useless in sorting the names in the New York City phone book (about 3,000,000 names or 300^2 times more difficult). To a professional programmer, writing program code which will be used perhaps thousands of times, obviously efficiency of code is extremely important and the programmer will spend many hours or days rewriting a perfectly good code to optimize the speed and minimize the memory requirements of the program. However, for most of the rest of us such meticulous care would not be cost-effective. For programs that will only be executed a few times and from which we want an answer quickly, it would be foolish to spend an extra week reducing the run time from 73 seconds to 47 seconds. (Of course, a reduction of run time from 73 hours to 47 hours would be a different matter.) For this reason, only the simplest sorting algorithms will be discussed here, neither of which would be suitable for excessively long lists for which extremely sophisticated procedures have been devised.

8.6.1   The Selection Sort

The *selection sort* algorithm (also called *exchange sort*) is the most obvious method of arranging a list and was described at the end of the previous section.

Find the *minimum* of the list of n elements.
Place this element on top.

Find the *minimum* of the remaining list of $n - 1$ elements.
Place this element next.
Continue until the remaining list contains only one element.

The Fortran to accomplish a selection sort is given in Figure 8-2. After the first pass through the outer loop, NAME(1) = overall minimum in the list, after the second pass NAME(2) = second smallest value, etc. Also, notice that when the two values, MIN and NAME(I) were exchanged, three lines of code and the use of a temporary variable, TEMP, were required. If we had used simply

```
NAME(I) = MIN
MIN = NAME(I)
```

both variables would have been assigned the same value, namely MIN.
During execution of the program, a total of

$$(n - 1) + (n - 2) + (n - 3) + \cdots + 2 + 1 = \tfrac{1}{2}n(n - 1)$$

Figure 8-2 The Fortran code for a selection sort alphabetization.

```
*     SELECTION SORT PROGRAM
*
      PROGRAM SELECT
*
      CHARACTER*20 NAME(1000),MIN,TEMP
      INTEGER I,N,TOP

      READ(*,'(A20)',END=99)(NAME(I),I=1,1000)
99    N = I - 1

      DO 2 TOP = 1,N                        TOP is the number
         MIN = NAME(TOP)                     of the first element
         DO 1 I = TOP + 1,N                  in the remaining
                                             list.
            IF(NAME(I) .LT. MIN)THEN         If the current name
               TEMP = NAME(I)                is less than the cur-
               NAME(I) = MIN                 rent minimum, we
               MIN = TEMP                    redefine the mini-
                                             mum by switching
            END IF                           positions.

1        CONTINUE
         NAME(TOP) = MIN                     The minimum is
2     CONTINUE                               now placed at the
                                             top of the remain-
                                             ing list.
```

comparisons are made and, assuming a random initial ordering, the number of interchanges should be somewhat less than half the number of comparisons.

There are two further considerations concerning the selection sort algorithm: First, even if the original list were already in order, exactly the same number of comparison tests would have to be made, although of course there would be no replacements. Second, if the original, unsorted list consisted of not just names but also additional information such as addresses, ID numbers, and bank balances, during each exchange, all of these items would have to be interchanged as well. This could seriously affect the efficiency of the algorithm. These shortcomings of the selection sort are addressed in the next two sections.

8.6.2 The Bubble Sort

Another simple sorting algorithm, the bubble sort, is especially useful when the original list is partially sorted to begin with. The idea in the bubble sort algorithm is illustrated in the sequence below.

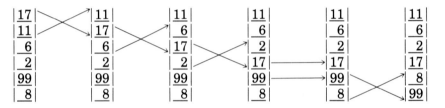

1. Compare the first two elements (17 and 11) and if out of order exchange.
2. Compare the next pair (now 17 and 6) and if out of order, exchange.
3. Continue this through the entire list. Notice that after the first pass the largest element in the list (99) has "bubbled" down to the bottom.
4. In the second pass through the list only compare the remaining elements, since the largest element is already in the proper position. The arrangement of the numbers after the second pass will be

$$
\begin{array}{|c|}
\hline 6 \\
\hline 2 \\
\hline 11 \\
\hline 8 \\
\hline 17 \\
\hline 99 \\
\hline
\end{array}
$$

The Fortran code to accomplish a bubble sort is given in Figure 8-3. The purpose of the variable FLAG is to signal when the entire list has been compared and *no exchanges* have been required. If this is the case, the list is already in order and the sorting can stop.

Figure 8-3 The Fortran code for a bubble sort.

```
*
*  THE BUBBLE SORT
*
      CHARACTER FLAG*3
      INTEGER I,N,J
      REAL A(1000)

      READ(*,'(F10.5)',END=99)(A(I),I=1,1000)
 99   N = I - 1
*
      DO 2 I = N,2,-1                        I is the bottom element in
                                             the current comparison set.
         FLAG = 'OFF'
         DO 1 J = 1,I - 1                    Compare successive pairs
                                             of the remaining set.
            IF(A(J) .GT. A(J + 1))THEN
               TEMP = A(J)                   If out of order, exchange.
               A(J) = A(J + 1)
               A(J + 1) = TEMP
               FLAG = 'ON'
            END IF
 1       CONTINUE
         IF(FLAG .EQ. 'OFF')THEN             If there were no exchanges
            GO TO 3                          at all in this pass, the
         END IF                              list is sorted. So ter-
                                             minate the sort.
 2    CONTINUE
 3    CONTINUE
         ...
```

8.6.3 Sorting with a Pointer

As mentioned above, a sorting problem is frequently complicated by the fact that associated with each name in a list are many other additional items. Each time two names are switched positions in the list, all the corresponding data must be exchanged as well. In such cases it is much more convenient to simply keep track of the rearrangements by means of an indexing array, called a *pointer*. The idea is to leave the original list, NAME(I), intact, while the alphabetic order of the elements is determined and stored in INDX(K). Thus the first name in the sorted list would be given by NAME(INDX(1)). Perhaps AARDVARK is the twelfth name in the original list and AARON is the seventh, so the result of the sorting would give:

```
                 INDX(1) = 12
                 INDX(2) =  7
                 NAME(INDX(1)) => 'AARDVARK'
                 NAME(INDX(2)) => 'AARON'
                     etc.
```

Either the exchange sort or the bubble sort can be rewritten using a pointing index array. The code for the exchange sort with a pointer is given in Figure 8-4.

The output from this code for a sample list of 10 names is given below:

Figure 8-4 The Fortran code for a selection sort with a pointer.

```
* EXCHANGE SORT WITH A POINTER ARRAY
*
      CHARACTER*5 NAME(1000),MIN
      INTEGER INDX(1000),I,N,TOP,TEMP
*
      READ(*,'(A)',END=99)(NAME(I),I=1,1000)
 99   N = I - 1

      DO 1 I = 1,1000                           Initialize the pointer
          INDX(I) = I                           array to the order 1, 2, . . .
 1    CONTINUE
*
      DO 3 TOP = 1,N - 1                         INDX(TOP) refers to the
                                                 name at the top of the re-
                                                 maining unsorted list, i.e.,
                                                 NAME(INDX(1)) = ⟨first
          MIN = NAME(INDX(TOP))                  name⟩. Initialize MIN to
                                                 be the top name.
          DO 2 I = TOP,N

              IF(NAME(INDX(I)) .LT. MIN)THEN
                  TEMP = INDX(I)                 If a smaller value is
                  INDX(I) = INDX(TOP)            found, redefine pointer;
                  INDX(TOP) = TEMP               i.e., INDX(TOP) and the
                  MIN = NAME(INDX(TOP))          value of MIN.
              END IF

 2        CONTINUE
 3    CONTINUE

      WRITE(*,11)
      WRITE(*,12)(I,NAME(I),NAME(INDX(I)),INDX(I),I = 1,N)
      STOP

 11   FORMAT(///,5X,'ORIGINAL SORTED ORIGINAL',/,
     +       2X,'I',4X,'LIST',5X,'LIST',3X,'POSITION',/)
 12   FORMAT(1X,I2,3X,A5,4X,A5,6X,I2)

      END
```

I	ORIGINAL LIST	SORTED LIST	ORIGINAL POSITION
1	ZEKE	ALICE	7
2	WILL	CARL	8
3	FRANK	CAROL	9
4	PAUL	DON	6
5	KEN	FRANK	3
6	DON	KEN	5
7	ALICE	PAUL	4
8	CARL	WILL	2
9	CAROL	ZEKE	1
10	ZELDA	ZELDA	10

To print out all the data associated with each name with the names in alphabetical order, we need only the information contained in the pointer array.

```
    ...
DO 37 I = 1,N
     K = INDX(I)
       WRITE(*,14)NAME(K),ADDRESS(K),ID(K), ...
37 CONTINUE
    ...
```

8.7 PLOTTING A GRAPH ON THE LINE PRINTER OR TERMINAL SCREEN

8.7.1 Plotting with a Typewriter

As with any problem, plotting a graph on the terminal screen requires that we first carefully analyze the steps involved in plotting a graph on ordinary graph paper and attempt to construct an algorithm understandable to the computer. Plotting a graph consists of the following steps:

1. Generate a table of number pairs, (x_i, y_i, for $i = 1,n$)

i	x_i	y_i
1	0.0	0.0
2	2.0	32.0
3	3.0	72.0
...	...	...
...	...	...

These numbers may represent experimental data, say position vs. time for a falling object, or may be generated from a particular functional relation between x and y as $y(x) = 8x^2$. Once the data set is complete we can begin to represent the set by points on a graph.

2. Determine the range of both x_i and y_i. This will require a determination of both the minimum and maximum values of x and y in the data set and then

$$(\text{Range})_x = (x_{\max} - x_{\min})$$

$$(\text{Range})_y = (y_{\max} - y_{\min})$$

These values are then used for scaling the x and y axes. That is, adjusting the scales of the axes so that the graph fits neatly on the graph paper. This is the most difficult aspect of plotting and I will return to the details in Section 9.5.

3. Step through the points and graph them one by one. You might not think that such a trivial step is worth mentioning. However, graphing a function in this manner is one of the surest ways to appreciate the meaning of the term "function." Since this is so critical to further understanding of mathematics, it warrants possibly insulting your intelligence. Thus, for each value of x_i (the independent variable) there is a corresponding y_i (the dependent variable). Or, for each value of i there is a pair (x_i, y_i).

Since a typewriter or line printer executes one horizontal line at a time, then proceeds to the next line, stepping along the x axis and graphing y values will mean that the graph will have to be constructed moving down the page, not across.

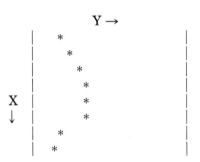

When one line is displayed, 80 or so characters will be printed (depending on the length of the y axis) for a particular value of x (say $x = 1.22$). *All* of these characters will be blanks except one. The position corresponding to $y(x = 1.22)$ will contain some symbol—e.g., an asterisk. To determine the proper placement of the asterisk, consider the following:

Suppose we have a function $y = f(x)$ and wish to graph the function y vs. x for $x = 0$ to $x = 5$. Thus $(\text{Range})_x = 5 - 0 = 5$. We also need the range in y values. We might determine the following:

$$y_{max} = 16.38$$

$$y_{min} = -7.21$$

so

$$(Range)_y = 16.38 - (-7.21) = 23.59$$

Next, if at $x = 3.0$, $y(3.0) = 12.2$, where in the horizontal line is the asterisk to be placed? If the y axis is to be 81 columns wide, we could first define

$$Ratio = \frac{y(3) - y_{min}}{y_{max} - y_{min}}$$

Notice that ratio is between 0.0 and 1.0. The appropriate column for the asterisk is

```
IY = 80. * RATIO
```

For the particular choice of numbers above, $y(3) = 12.2$, we obtain

```
RATIO = .82281
IY = 66
```

The Fortran code to execute this single line of the graph might then be:

```
CHARACTER*1 LINE(0:80),HIT,MISS
INTEGER IY,I
REAL X(0:80),Y(0:80)
      . . .
      . . .
X(27) = 3.0
Y(27) = 12.2
      . . .
      . . .
YMAX = 16.38
YMIN = -7.21
      . . .
      . . .
RATIO = (Y(27) - YMIN)/(YMAX - YMIN)
IY = 80.0 * RATIO
HIT = '*'
MISS = ' '
DO 13 I = 0,80
     LINE(I) = MISS     Initialize the entire line to be blanks.
13 CONTINUE
   LINE(IY) = HIT          Place a * in column IY.
   WRITE(*,'(2X,81A1)')(LINE(I),I = 0,80)
```

The idea then for graphing the complete set of values x_i, y_i, is to use this algorithm repeatedly for each line of the graph. The complete Fortran code is illustrated in Figure 8-5. The output from this program for the function

$$f(x) = \sin^2 \left(\pi \frac{x}{10} \right) \qquad 0 \le x \le 10$$

is shown in Figure 8-6.

8.7.2 Contour Plots

To represent a function of two variables on a graph, a contour plot is often used. You have probably seen contour maps that give the elevation of the land above sea level as closed paths drawn on a normal map. If x represents the east-west position on the map (longitude) and y represents the north-south position (latitude), then the elevation of any point of land can be written as a function of x and y: elevation = $E(x, y)$. If we next connect all those points with an elevation of 100 feet, we would get a curve of constant (100 feet) elevation. This is then repeated for elevations of 200 feet, 300 feet, etc. This is shown in Figure 8-7. The idea can be applied to any function of two variables.

Figure 8-5 The Fortran code to graph a function at the terminal.

```
PROGRAM PLOT
CHARACTER*1 LINE(0:60)
REAL X(0:25),Y(0:25),YMAX,YMIN,XHI,XLO,XSTEP,F,Z,

    F(Z) = ...                          A statement function for
                                        f(x); note I have used Z,
                                        since X is a dimensioned
                                        variable.

    N = 25
    READ(*,*)XLO,XHI                     The limits on the x's
    XSTEP = (XHI - XLO)/(N - 1.)
    DO 1 I = 0,N
        X(I) = I * XSTEP                 Fill in the table of xi, yi.
        Y(I) = F(X(I))
  1 CONTINUE

    YMAX = Y(0)                          Determine the min/max
    YMIN = Y(0)                          y values.
    DO 2 I = 1,N
        IF(Y(I) .LT. YMIN)YMIN = Y(I)
        IF(Y(I) .GT. YMAX)YMAX = Y(I)
  2 CONTINUE
```

```
      SCALE = YMAX - YMIN

      DO 3 I = 0,60
          LINE(I) = ' '                          Initialize the line to be
 3    CONTINUE                                    all blanks.

      WRITE(*,12)
      DO 4 IX = 0,25                              Step through the x values.
          ISTAR = (Y(IX) - YMIN)/SCALE * 60.      ISTAR is the position of
          LINE(ISTAR) = '*'                       the asterisk in the line.
          WRITE(*,11)X(IX),Y(IX)(LINE(I),I = 0,60)
          LINE(ISTAR) = ' '                       Blank out the star to
 4    CONTINUE                                    set up for next line.

      STOP
11    FORMAT(1X,2F6.2,')',61A1,')')
12    FORMAT(4X,'X',5X,'Y ',63('-'))
      END
```

Figure 8-6 A printer plot of $f(x) = \sin^2\left(\pi\dfrac{x}{10}\right)$.

```
INPUT  XLO,XHI 0., 12.5
   X      Y  -------------------------------------------------------
 0.00  0.00)*                                                        )
  .50   .02)  *                                                      )
 1.00   .10)        *                                                )
 1.50   .21)            *                                            )
 2.00   .35)                 *                                       )
 2.50   .50)                      *                                  )
 3.00   .65)                           *                             )
 3.50   .79)                               *                         )
 4.00   .90)                                  *                      )
 4.50   .98)                                    *    )
 5.00  1.00)                                        *)
 5.50   .98)                                    *    )
 6.00   .90)                                  *                      )
 6.50   .79)                               *                         )
 7.00   .65)                           *                             )
 7.50   .50)                      *                                  )
 8.00   .35)                 *                                       )
 8.50   .21)            *                                            )
 9.00   .10)        *                                                )
 9.50   .02)  *                                                      )
10.00   .00)*                                                        )
10.50   .02)  *                                                      )
11.00   .10)        *                                                )
11.50   .21)            *                                            )
12.00   .35)                 *                                       )
12.50   .50)                      *                                  )

       -------------------------------------------------------
```

Figure 8-7 A contour map.

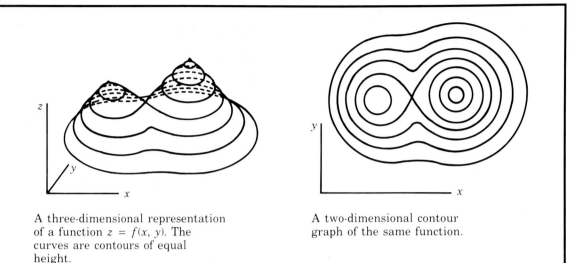

A three-dimensional representation of a function $z = f(x, y)$. The curves are contours of equal height.

A two-dimensional contour graph of the same function.

1. Given a function of two variables, $z = f(x, y)$, with known limits on x and y such that $a \leq x \leq b$, $c \leq y \leq d$, compute a two-dimensional table of z values for all combinations of x_i, $i = 1, n_x$ and y_j, $j = 1, n_y$ in this range. That is Z(I, J).
2. Compute the overall maximum and minimum values of Z(I, J) and thus determine the range of z values.
3. Print out a square array of symbols corresponding to the values of Z(I, J) defined in the following manner:
   a. Let $(\text{Range})_z = z_{\max} - z_{\min}$
   b. Replace $Z(I, J) \rightarrow 50 * (Z(I, J) - ZMIN)/(\text{Range})_z$. All the z values are now between 0.0 and 50.0.
   c. Construct an integer array IZ(I, J) = Z(I, J).
   d. For those points (I, J) that have an IZ(I, J) that is divisible by 5, assign a symbol from '0', '1', . . . , to '9', by the relation

$$\text{CONTUR(I,J)} = \text{NUMBER(IZ(I,J)/5)}$$

   where NUMBER(0) = '0'

   NUMBER(1) = '1'

   NUMBER(2) = '2'

   ＋＋＋   ＋＋＋

   NUMBER(9) = '9'

   e. If IZ(I, J) does not divide evenly by five, assign a blank.

   IF(IZ(I,J)/5 * 5 .NE. IZ(I,J))CONTUR(I,J) = ' '

The Fortran code for such a program is illustrated in Figure 8-8 and the output for the function

$$z = f(x,y) = \sin^2\left(\pi\frac{x}{5}\right)\sinh\left(\pi\frac{y}{5}\right)$$

is given in Figure 8-9.

Figure 8-8 Fortran code for a contour plot of a two-dimensional function.

```
      PROGRAM SQUARE
*--
*--   WILL PRODUCE A 20 BY 20 CONTOUR PLOT OF THE FUNCTION F(X,Y)
*--   FOR XMIN < X < XMAX, YMIN < Y < YMAX. THE MIN/MAX VALUES
*--   FOR THE FUNCTION ARE DETERMINED AND THE COUNTOURS ARE REP-
*--   RESENTED BY INTEGERS, O FOR MINIMUM, 9 FOR MAXIMUM.
*--------------------------------------------------------------------
* VARIABLES
      CHARACTER*1 NUMBER(0:9),CONTUR(20,20),BLANK
      REAL X,Y,XMIN,XMAX,YMIN,YMAX,ZMIN,ZMAX,RANGEZ,
     +     Z(20,20)
      INTEGER IZ(20,20)
*
*               Z(I,J)        - CONTAINS THE VALUES OF THE FUNCTION
*                               FOR X,Y VALUES ON THE GRID.
*               IZ(I,J)       - CONTAINS THE VALUES OF Z SCALED TO
*                               THE RANGE 0-50 AND TRUNCATED.
*               CONTUR(I,J)   - CONTAINS A SYMBOL '0' THROUGH '9'
*                               OR A BLANK AT EACH GRID POINT.
*--------------------------------------------------------------------
* STATEMENT FUNCTION FOR F(X,Y)
*
      F(X,Y) = ((SIN(3.1415926*X/5.))**2)*SINH(3.1415926*Y/5.)
*--------------------------------------------------------------------
* INITIALIZATION
*
      NUMBER(0) = '0'
      NUMBER(1) = '1'
      NUMBER(2) = '2'
      NUMBER(3) = '3'
      NUMBER(4) = '4'
      NUMBER(5) = '5'
      NUMBER(6) = '6'
      NUMBER(7) = '7'
      NUMBER(8) = '8'
      NUMBER(9) = '9'
*
```

Continued

```
*                              THIS COULD BE SHORTENED WITH A
*                              DO LOOP AND THE FUNCTION CHAR( ).
*
      BLANK = ' '
      PRINT *,'INPUT LIMITS ON THE X AXIS'
      READ *,XMIN,XMAX
      PRINT *,'INPUT LIMITS ON THE Y AXIS'
      READ *,YMIN,YMAX
      DX = (XMAX - XMIN)/19.
      DY = (YMAX - YMIN)/19.
*
*                              THERE ARE 20 POINTS BUT ONLY 19
*                              STEPS. NEXT FILL IN Z ARRAY.
*-------------------------------------------------------------------
*
      Y = YMIN
      DO 2 IY = 1,20
         X = XMIN
         DO 1 IX = 1,20
            Z(IX,IY) = F(X,Y)
            X = X + DX
1     CONTINUE
         Y = Y + DY
2 CONTINUE
*
*                              FIND MIN/MAX OF Z VALUES.
*
      ZMAX = Z(1,1)
      ZMIN = Z(1,1)
      DO 3 IX = 1,20
      DO 3 IY = 1,20
         IF(Z(IX,IY) .LT. ZMIN)ZMIN = Z(IX,IY)
         IF(Z(IX,IY) .GT. ZMAX)ZMAX = Z(IX,IY)
3 CONTINUE
*
*                              SCALE THE Z'S TO BE BETWEEN 0-50 AND
*                              CONSTRUCT THE CONTUR ARRAY.
*
      RANGEZ = (ZMAX - ZMIN)
      DO 4 IX = 1,20
      DO 4 IY = 1,20
         TEMP = (Z(IX,IY) - ZMIN)/RANGEZ * 50.0
         IZ(IX,IY) = TEMP
*
*                              SET CONTUR = BLANK UNLESS IZ IS
*                              DIVISIBLE BY 5.
*
      CONTUR(IX,IY) = BLANK
```

```
          IF(IZ(IX,IY)/5 * 5 .EQ. IZ(IX,IY)THEN
             CONTUR(IX,IY) = NUMBER(IZ(IX,IY)/5)
          END IF
    4 CONTINUE
*-------------------------------------------------------------------
* PRINT THE CONTOUR PLOT
*
      WRITE(*,8)XMIN,XMAX
      WRITE(*,9)YMIN
      WRITE(*,'(2X,62A1)')('*',I=1,62)
      DO 6 I = 1,20
         WRITE(*,10)'*',(CONTUR(J,I),J = 1,20),'*'
    6 CONTINUE
      WRITE(*,'(2X,62A1)')('*',I = 1,62)
      WRITE(*,9)YMAX
      STOP
*
    8 FORMAT(///,2X,F4.1,25X,'X',25X,F4.1)
    9 FORMAT(1X,'Y = ',F4.1)
   10 FORMAT(2X,A1,20(1X,A1,1X),A1)
      END
```

```
-1.0                          X                          2.0
Y = -2.0
  **************************************************************
  *    5              7  7        6        4           1     0 *
  *                   7  7        6     5        3             *
  *       6           7  7                    4                *
  *       6           7  7                 5                 2 *
  *                   7  7           6                    3    *
  *    6              7  7                    5                *
  *    6           7  7  7              6        5        4    *
  * 6              7  7  7                          5          *
  *                7  7  7  7              6              5    *
  *                7  7  7  7                    6             *
  *                7  7  7  7                       6  6       *
  *             7  7  7  7  7  7                               *
  *       7  7  7  7  7  7  7  7  7  7                         *
  * 7  7  7  7  7  7  7  7  7  7  7  7  7  7  7  7  7  7  7    *
  *       7  7  7  7  7  7  7  7  7  7                         *
  *          7  7  7  7  7  7  7                          8  8 *
  *             7  7  7  7  7  7                 8  8          *
  *             7  7  7  7  7                 8              9 *
  *                7  7  7  7              8           9       *
  * 8              7  7  7  7                                  *
  **************************************************************
Y = 1.0
```

Figure 8-9 Contour plot of $f(x, y) = \sin^2\left(\pi\frac{x}{5}\right)\sinh\left(\pi\frac{y}{5}\right)$.

PROBLEMS

1. What is the output from the following:

```
CHARACTER*5 B,C,D(0:3),E(-2:2)
CHARACTER F*1,G*2,A*1
A = 'A'
C = A
F = C
B = '12345'
G = B(4:)
E(-1)(3:4) = B(2:3)
E(-1) = A
D(0)(:) = B(:)
E(-2) = A//G//B(:2)
E(0) = B(5:5)//B(4:4)//B(3:3)//B(2:2)//B(1:1)
E(1) = A//B
```

   a. WRITE(*,'(1X,A)')A
   b. WRITE(*,'(1X,A)')B
   c. WRITE(*,'(1X,A)')C
   d. WRITE(*,'(1X,A)')F
   e. WRITE(*,'(1X,A)')G
   f. WRITE(*,'(1X,A)')E(-1)
   g. WRITE(*,'(1X,A)')D(0)
   h. WRITE(*,'(1X,A)')E(-2)
   i. WRITE(*,'(1X,A)')(E(0)(I:I),I = 5,1,-1)
   j. WRITE(*,'(1X,A)')E(1)
   k. WRITE(*,'(1X,A)')B//B

2. Write a program to read a character string ending with a period and print the string in reverse order. Assume the length of the string is less than 80 characters.

3. Design a program that will read a paragraph consisting of less than 25 lines and count the number of words. Assume that each line is 80 characters long, each line ends with two blank spaces, sentences end with a period followed by a single blank space, and no words are hyphenated. Also, the paragraph has fewer than 25 sentences and the first line is not indented.

4. Write a program to read a document of unknown length and everywhere replace the word "under" with "below." Watch out for compound words.

5. Write the code to read the phone book for the name 'JONES, JAMES' who lives on JENNINGS Street and prints out the phone number. The phone number in the listing is always in the form XXX-XXXX.

6. Execute the program on page 263 and determine the ordering of the Fortran symbols on your computer.

7. Rewrite the selection sort program to arrange the set in *decreasing* order.

8. Rewrite the selection sort program to also count the number of duplicate values in the list.

9. In the version of the bubble sort algorithm given in section 7.6.2, the largest element of the array "bubbles" down to the bottom of the array after one pass. Rewrite the algorithm so that the smallest element of the remaining list will "bubble" up to the top of the list.

10. Let each digit of a huge integer (less than 200 digits) be stored, right-adjusted, in the integer array, L(200). Write the Fortran code to accomplish the following:

   a. Form a copy of this number in the character array DIGIT (200)*1. Each element of DIGIT is a numerical symbol from "0" to "9" in place of the numerical values, 0 to 9, stored in L.

   b. Since the number is originally right-adjusted, it will likely contain a great many leading zeros. Replace all the leading zeros in DIGIT by blanks.

   c. If all of the digits of L are multiplied by an integer k, it would be useful to incorporate the normal arithmetic concept of carrying into the multiplication. Thus if

L_5	L_4	L_3	L_2	L_1
0	6	7	1	5

is multiplied by 5, the elements of L would be

L_5	L_4	L_3	L_2	L_1
0	30	35	5	25

which we would prefer to write as

L_5	L_4	L_3	L_2	L_1
0	0	5	5	5
+ 3	3	0	2	0
3	3	5	7	5

By considering several numerical examples, show that the code below accomplishes the task of carrying.

```
      NTERMS = 200
      ICARRY = 0
      DO 1 I = 1,NTERMS
          L(I) = L(I) * K + ICARRY
          ICARRY = L(I)/10
          L(I) = L(I) - 10 * ICARRY
    1 CONTINUE
```

d. Construct a program to compute 75! *exactly*. This number has more than 100 digits.

11. Rewrite the bubble sort program to include a pointer array. The original list should be left intact. Test your program on a sample list.

12. Alter the plotting program to include a border of asterisks around the graph. Also print four y values along the y axis along with equally spaced tic marks. For example,

In addition, along the x axis, print a value of x only every tenth line.

13. Alter the plotting program to graph two functions simultaneously, using different symbols. Note you will have to determine an overall minimum/maximum.

14. Execute contour plots for the following functions:

a. $f(x,y) = \sin(x^2 + y^2)$ $\quad -1 \le x \le 1$
$\qquad\qquad\qquad\qquad\qquad -1 \le y \le 1$

b. $g(x,y) = e^{-xy}$ $\quad -1 \le x \le 1$
$\qquad\qquad\qquad\quad -1 \le y \le 1$

c. $h(x,y) = (x - 1)^2 + 2(y - 1)^2$ $\quad 0 \le x \le 2$
$\qquad\qquad\qquad\qquad\qquad\qquad\qquad 0 \le y \le 2$

15. To produce a graph with the x axis horizontal and the y axis vertical, a different procedure must be employed. Once again a table of (x_i, y_i) is first obtained and the range of the x and y values determined. This time the entire plot will be stored in the character array

$$\texttt{CHARACTER PAGE(0:60,0:50)*1}$$

After initializing the entire array as all blanks, step through the tabulated values and determine the positioning of the points on the page by

$$I_x = 60\frac{(x_i - x_{min})}{(\text{Range})_x}$$

$$I_y = 50\frac{(y_i - y_{min})}{(\text{Range})_y}$$

and assign a symbol, say *, to this point,

$$\texttt{PAGE(IX,IY) = '*'}$$

Finally the array PAGE is printed

```
   DO 7 IY = 50,0,-1
      WRITE(*,'(2X,61A1)')(PAGE(IX,IY)IX = 0,60)
 7 CONTINUE
```

Follow this procedure to obtain a graph of $y = e^{-x/3} \sin(\pi x/2)$ for $0 \le x \le 4$.

PROGRAMMING ASSIGNMENT V

V.1 SAMPLE PROGRAM

Cooling Fins on a Steam Pipe (Mechanical Engineering)

The heat from a steam pipe is more effectively transferred to its surroundings by the addition of metal radiator fins. The fins are heated by the
pipe, and because of their large surface area can efficiently transfer heat by
radiation and convection to the surrounding air. To simplify the analysis, I
have assumed that the steam pipe is square and that the fin is rectangular,
as shown in Figure V-1.

The first step in determining the heat transfer is to compute the temperature distribution $T(x, y)$ across the area of the fin. This is ordinarily a
rather complex mathematical problem, requiring sophisticated techniques.
However, the algorithm for a numerical solution is quite transparent and is
based on some simple ideas concerning temperature and heat flow.

A fundamental property of heat flow is that of diffusion. A hot or a cold
spot will smooth or average out in time and the final temperature distribution will be the smoothest possible distribution consistent with the
constraints of the problem. For example, if a metal bar has one end in ice
water (0 °C) and the other end in boiling water (100 °C), and the rest of the

Figure V-1 A
square steam
pipe surrounded
by a radiator fin.

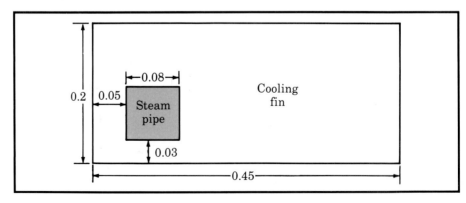

282

bar insulated, the final temperature distribution will be a simple linear decrease across the length of the bar. (See Figure V-2.)

Another way of stating this idea of smoothing is that the temperature at any point will ultimately be equal to the *average* of the temperatures of its immediate surroundings. This algorithm can be easily adapted to our fin problem. First we superimpose a two-dimensional grid over the fin and pipe and require that the initial guess for the temperature at each point on the fin, T_{ij}, be replaced by the average of the temperatures of its four nearest neighboring points. (See Figure V-3.)

$$T_{ij,\,\text{new}} \rightarrow \tfrac{1}{4}(T_{i+1,j} + T_{i,j+1} + T_{i-1,j} + T_{i,j-1}) \qquad \textbf{(V.1)}$$

This replacement is done for all the points on the fin and is repeated until there is very little change from one pass to the next. The same algorithm can be used to find the temperature distribution on an airplane wing or on a thin semiconductor element of a transistor.

Program Specifics

Write a program to solve for the temperature distribution on the radiator fin and to execute a contour printer plot. (Also see Section 8.7.2.)

The value of the temperature at the edges of the fin and inside the pipe should be initialized via a DATA statement. (Use $T_{\text{steam}} = 180\ °C$, $T_{\text{air}} = 20\ °C$.)

For all *interior* fin grid points, start with an initial guess of TOLD(I, J) = 50.

Use Equation (V.1) to obtain a new set of interior temperature values, T_{ij}^{new}. Also, evaluate

$$\Delta = \frac{1}{N} \sum_{ij} |T_{ij}^{\text{new}} - T_{ij}^{\text{old}}|$$

where N is the total number of interior points. If this quantity is greater than 0.01, replace $T_{ij}^{\text{old}} = T_{ij}^{\text{new}}$ and repeat the calculation.

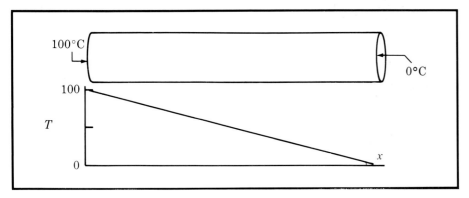

Figure V-2
Equilibrium temperature distribution in a bar.

Figure V-3   A grid superimposed on the fin pipe problem.

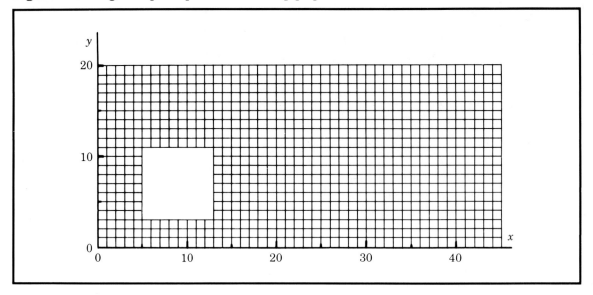

The number of iterations should be limited to IMAX = 75.

Once the equilibrium temperature distribution has been obtained, the program is to execute a contour printer plot of the temperature distribution in the fin.

An outline of the section of the code for the contour printer plot is:

First, initialize an array GRID(0:45, 0:20) of type CHARACTER*1 as

```
GRID(I,J) = '*'      around the edge of the fin
GRID(I,J) = 'X'      inside and on the edge of the pipe
GRID(I,J) = ' '      elsewhere
```

Initialize a CHARACTER*1 array, LETTER(0:27) to contain letters and blanks in the sequence

```
'A', ' ', ' ', 'B', ' ', ' ', 'C', ... , ' ', ' ', 'I'
```

This could be done with several assignment statements or with a DO loop and the functions CHAR and ICHAR in the following manner:

Use ICHAR to find the position of 'A' in the collating sequence on your computer.
```
IA = ICHAR('A')      <so now 'A' = CHAR(IA)>
Then
LETTER(0) = CHAR(IA)
```

```
       DO 7 I = 1,9
          IA = IA + 1
          LETTER(3*I - 2) = ' '
          LETTER(3*I - 1) = ' '
          LETTER(3*I)     = CHAR(IA)
    7  CONTINUE
```

For all interior points, assign the elements of GRID to one of the
  elements of LETTER by

```
IJ = (T(I,J) - TMIN)/(TMAX - TMIN)*27
```
⟨*Note: IJ is between 0 and 27.*⟩
```
GRID(I,J) = LETTER(IJ)
```

Print the GRID
Repeat the calculation with the bottom edge of the fin in ice water.

Computer Solution

The output of the program constructed on the basis of the above outline is
given in Figure V-4.

```
*************************************************
*AAA                   AAAAAAAAAAAAAAAAAAAAAAAAA*
*A     BBB    BBBBB                        AAAAA*
*A BB     CCCC        BBB                    AAA*
*      C           C      BBB                 AA*
* B    D          D   C    BBBBBBBBBBBBBB     AA*
*      D      FF       D C   BBBBBBBBBBBBBBB   AA*
*BCDEF     GGGG    E D C       BBBBBBBBBB      A*
*B    F  H           FE    C       BBBBB       A*
*        XXXXXXXXXH     D   C        BBBB      A*
* D      XXXXXXXXX GFE      C        BBBB      A*
* DF  XXXXXXXXX GFE       C        BBBB       A*
*    FHXXXXXXXXX    FE     C         BBBBB     A*
*    FHXXXXXXXXX    FE     C    BBBBBBBBBB     A*
* DFHXXXXXXXXX GFED C     BBBBBBBBBBBBBB      AA*
* D      XXXXXXXXX     D C   BBBBBBBBBBBB     AA*
*        XXXXXXXXX   E  C  BBB                AA*
*B       XXXXXXXXXG     C  B                 AAA*
*        F           ED    B               AAAAA*
*                   C B     AAAAAAAAAAAAAAAAAAAA*
*************************************************
```

Figure V-4 The temperature in a fin surrounding a square steam pipe. The temperatures are indicated by the letters A through I, with A designating the coolest temperature.

V.2 PROGRAMMING PROBLEMS

Programming Problem A: Resonant Circuit (Electrical Engineering)

A common problem in electrical engineering is the design of a tuning circuit to select radio waves with a particular frequency from the thousands always present as background noise. This is accomplished by means of a resonant circuit which will only sustain frequencies in a narrow band, $f_0 \pm \Delta f$, on both sides of the desired frequency f_0. Since I suspect that you may be more familiar with mechanical than electrical oscillations, I'll first describe the problem in terms of an analogous mechanical system consisting of a spring, weight, driving force, and damping fluid as illustrated in Figure V-5.

In introductory physics you determined the natural frequency of a spring as

$$(\mathbf{V.2}) \qquad f_0 = \frac{1}{2\pi} \left(\frac{k}{m} \right)^{1/2}$$

The mass, if displaced and released, will oscillate with this frequency. Next, if the top of the spring is forced up and down with a different frequency, $f \neq f_0$, the mass and the driving force will be out of phase. Occasionally, when the mass is moving up, the driving force will be moving down and the result is that there will be no net motion of the mass. However, if the driving frequency is adjusted so that $f = f_0$, the two oscillations are in phase so that the motion of the weight is reinforced with every oscillation, resulting in very large amplitude oscillations. Thus this system will sustain an input frequency $f = f_0$ and no others, and by varying the spring stiffness or the weight, we can tune this system to a variety of frequencies.

Figure V-5 A mechanical system to demonstrate resonant vibrations.

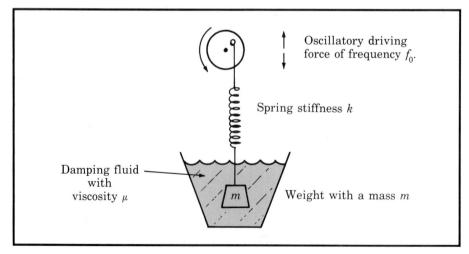

Oscillatory driving force of frequency f_0.

Spring stiffness k

Damping fluid with viscosity μ

m

Weight with a mass m

If the oscillations take place in a fluid such as air, water, oil, or molasses, the analysis is a bit more complicated. The most important new feature is that the system will now sustain frequencies in a band of half-width Δf about f_0. The half-width is given by

$$\Delta f = \frac{1}{4\pi} \frac{\mu}{m}$$ (V.3)

where μ is the viscosity of the fluid. Thus the more viscous the fluid, the less selective the system is in discriminating the tuned frequency.

The mechanical system has a precise analogy with an electrical circuit. The role of the spring is played by a capacitor, a device for storing charge; the viscous force is represented by a resistor which dissipates energy; and the inertia or weight is represented by a coil which resists abrupt changes in the current. The equivalent circuit is shown in Figure V-6.

The correspondence between the mechanical and the electrical components is

Viscosity $\quad\quad \mu \leftrightarrow \quad$ resistance R

Spring stiffness $k \leftrightarrow \quad$ (capacitance)$^{-1}$, $1/C$

Mass $\quad\quad\quad m \leftrightarrow \quad$ inductance L

Thus from Equation (V.2) the natural frequency of the circuit would be

$$f_0 = \frac{1}{2\pi} \left[\frac{1}{LC} \right]^{1/2}$$ (V.4)

and the half-width or selectivity of the circuit is

$$\Delta f = \frac{1}{4\pi} \frac{L}{R}$$ (V.5)

Figure V-6 A resonant electric circuit.

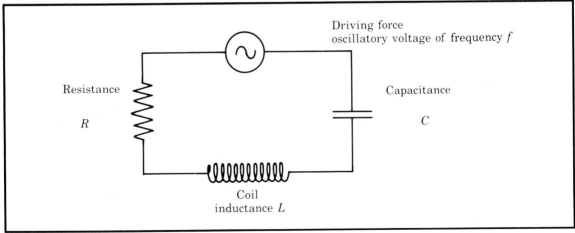

Furthermore, the amplitude of the oscillations (the voltage across the resistor) can be expressed as a function of the driving frequency as

(V.6)
$$V_r(f) = V_0 \left[1 + \frac{(f^2 - f_0^2)^2}{f^2 \, \Delta f^2} \right]^{-1/2}$$

where f_0 and Δf are given by Equations (V.2) and (V.3).

A good tuner will be selective enough (small Δf) so two signals at slightly different frequencies can be distinguished. However, if it is too selective you may have difficulty in finding the station. The actual tuning is done by varying the capacitance.

Problem Specifics

Write a program to

1. Read the inductance L of a coil and the driving voltage V_0. (Use $L = 10^{-5}$, $V_0 = 10$.)
   a. Compute the capacitance C required for a frequency of $f_0 = 710,000$ hertz (Hz).
   b. Compute the resistance necessary for half-widths of

$$\Delta f = \frac{f_0}{5}, \frac{f_0}{20}, \frac{f_0}{80}$$

   c. Print all the above with appropriate labels.
2. Produce three simultaneous printer plots of $V_r(f)$ for the three resistances above. The plot should extend from 500,000 Hz to 1,000,000 Hz along the f axis. Use different symbols for each curve.
3. In addition, the program should determine the actual selectivity of the circuit and compare it with the quantity Δf. To obtain a value for the actual half-width at half-maximum, your program should scan the computed values of V_r and determine the values of f on the left and right of f_0 that minimize $|V_r(f) - \frac{1}{2}V_0|$. The actual half-width is then $\frac{1}{2}[f_{\text{right}} - f_{\text{left}}]$.

Programming Problem B: Compressibility Factors for Real Gases

As I indicated in Programming Problem II-C, the understanding of the properties of real gases in contrast to ideal gases is of great importance to chemical engineers. There have been numerous attempts to formulate equations of state for a substance that would incorporate both liquid and gaseous phases. A rather accurate summary of the experimental properties of imperfect gases is contained in the Beattie-Bridgeman equation, which can be written as

$$P = \frac{RT(1 - \varepsilon)}{v^2}(v + B) - \frac{A}{v^2} \qquad \text{(V.7)}$$

where P = pressure (Pa)

T = temperature (K)

v = molar volume (m^3/mol)

R = ideal gas constant = 8.317 J/mol-K

A = $A_0(1 - a/v)$

B = $B_0(1 - b/v)$

ε = c/vT^3

and A_0, B_0, a, b, and c are the five experimental parameters in the equation that depend upon the specific gas in question.

A quantity of frequent use in chemical engineering problems is the compressibility of a gas defined as

$$Z = \frac{Pv}{RT} \qquad \text{(V.8)}$$

which for an ideal gas would be constant and equal to one. For a real gas this quantity is then a measure of the deviation of the real gas from ideal gas properties and is useful in isolating thermodynamic regions where the ideal gas may or may not be used. That is, when $Z \sim 1$ we would expect minimal error if the ideal gas equation is used in our calculations in place of the more complicated expressions. The purpose of this problem is to generate a plot of Z vs. P which can then be used in this manner.

Problem Specifics

Your program should do the following:

1. Read the Beattie-Bridgeman parameters for a particular gas from a data file containing the data in Table V-1 and print with appropriate labels. Use character variables for the name of the gas.

Gas	v_c (m^3 × 10^{-5})	T_c (K)	A_0 (×10^{-1})	B_0 (×10^{-6})	a (×10^{-6})	b (×10^{-6})	c (×10^0)
Air	80.00	132.5	1.320	46.1	19.30	−10.99	43.35
O_2	37.21	154.4	1.510	46.26	25.60	4.210	47.96
CO_2	95.65	304.3	5.074	104.7	71.36	72.35	660.0
H_2	64.52	33.26	0.200	20.96	−5.06	−43.58	0.507
Ammonia	72.34	405.6	2.424	34.15	170.4	191.1	4769.0

Table V-1 Beattie-Bridgeman parameters for some common gases.

2. Using statement functions for Equations (V.7) and (V.8), fill an array $Z(0{:}60)$ with values of the compressibility factor for 61 equally spaced values of v between $v_c/2$ and $3v_c$ for a temperature $T = T_c/2$. Do the same for an array PRESS(0:60) using Equation (V.7).

3. We next wish to plot Z vs. *pressure* not volume, while our data at this point consists of a table of Z and P vs. volume values. These data must first be rearranged into a table of Z vs. P values. To accomplish this, you will have to execute a pointer-sort on the array PRESS. If the sequencing index is stored in the array INDEX( ), the ordered values of the pressure and the compressibility should be mapped into new arrays, Pseq( ), Zseq( ). For example, if the minimum pressure value in the array PRESS is PRESS(15), then INDEX(1) = 15, and PSEQ(1) = PRESS(INDEX(1)). When this resequencing is completed, the data will consist of a set of $\{Z_i,$ Pseq$_i$, $i = 0.60\}$ values. Note that the pressure values will be in order but *will not* be equally spaced.

4. Since the pressure values are not equally spaced, it is somewhat difficult to produce a printer-plot; instead you will have to make use of the automatic pen-and-ink plotting routines available to you. Unfortunately, the instructions on the use of these routines vary considerably from site to site. Often all that is required, once the two arrays that form the x- and y-axes of the plot are filled, is to insert a CALL to a special plotting machine in the code. For example:

$$\text{CALL PLOTER}(X, Y, N, \langle x\text{-title}\rangle, \langle y\text{-title}\rangle, \langle \text{graph-title}\rangle)$$

where
X, Y = the arrays containing the data to be plotted
N = the number of points in a single plot
$\langle x\text{-title}\rangle, \langle y\text{-title}\rangle$ = character expressions labeling the x- and y-axes
$\langle \text{graph-title}\rangle$ = a character expression labeling the overall graph

Determine the necessary Fortran commands to execute a graph on a special pen-and-ink graphics device.

Repeat the entire calculation for $T = T_c$ and $T = 2T_c$. (Note: Three pen-and-ink plots for each of the five gases in Table V-1 could swamp the plotter if a large number of students are involved. Therefore, the results for only one gas should be plotted.)

Programming Problem C: Density of the Electron Cloud in Hydrogen (Physics/Physical Chemistry)

Understanding the structure of atoms is central to many areas of physics and chemistry. The mathematical description of atomic structure in terms of electrons orbiting an atomic nucleus is called Quantum Mechanics and is certainly one of the greatest achievements of the human mind. Quantum Mechanics states that the precise location of an individual electron is unknowable and is instead replaced by the probability of the electron being at a particular point at a given time. This "probability density" then can be accurately computed and can be used to predict where the electron is likely to be found.

The probability density for an electron in a particular state in hydrogen is

$$P(r, \theta) = \left(\frac{2}{6561\pi}\right) e^{-2r/3}(6r - r^2)^2 \cos^2 \theta \qquad \text{(V.9)}$$

where r is measured in natural atomic units of Bohr radii (one Bohr radius $= 5.28 \times 10^{-11}$ m).[1] And θ is the polar angle, illustrated as follows:

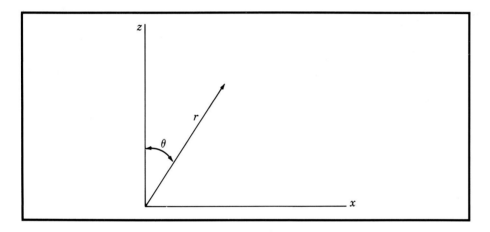

In this problem, we wish to obtain a two-dimensional plot of this function in the x-z plane. Note the relationship between (x, z) and (r, θ) is given by

$$z = r \cos \theta$$
$$x = r \sin \theta \qquad \text{(V.10)}$$

or

$$r = [x^2 + z^2]^{1/2}$$
$$\theta = \tan^{-1}(x/z) \qquad \text{(V.11)}$$

Thus, for a square grid of (x, z) values, we simply compute (r, θ) from Equations (V.11) and then the probability density from Equation (V.9). The values are stored in a square array PROB( ) and a contour plot is finally executed.

----

[1] In spectroscopic notation this is designated the 3p state of hydrogen.

Details

Your program should:

1. Declare an array PAGE(−25:25, −25:25) to be of type CHARACTER*1 and an array PROB(−25:25, −25:25) to be of type REAL.
2. Include statement functions for Equations (V.9) and (V.11).
3. As in the sample problem, assign single alphabetic and blank characters to the elements of the array LETTER(0:27)
4. For all values of x and z from −10. to +10. in steps of 0.4, compute a value of the probability density. [Note: Be careful when $z = 0$. If $z = 0$, then $\theta = \pi/2$. Also, exclude the point $x = z = 0$.]
5. Scan the array PROB( ) to find the maximum and the minimum.
6. Assign symbols to the array PAGE( ) as in the sample problem by

```
IJ  =  (PROB(I,J)  -  PMIN)/(PMAX  -  PMIN)*27
PAGE(I,J)  =  LETTER(IJ)
```

7. Assign the symbol * to the center of the graph ($x = z = 0$); that is, the nucleus. Print the array PAGE.

From the graph, determine the local minimum and maximum of the probability density.

Some other probability density functions you may wish to graph are:

$$\text{(2s state)} \quad P(r, \theta) = \frac{1}{32\pi}(2 - r)^2 e^{-r}$$

$$\text{(2p state)} \quad P(r, \theta) = \frac{1}{32\pi}r^2 e^{-r} \cos^2 \theta$$

$$\text{(3s state)} \quad P(r, \theta) = \frac{1}{19683\pi}(27 - 18r + 2r^2)^2 e^{-2r/3}$$

$$\text{(3d state)} \quad P(r, \theta) = \frac{1}{39366\pi}r^4(3 \cos^2 \theta - 1)^2 e^{-2r/3}$$

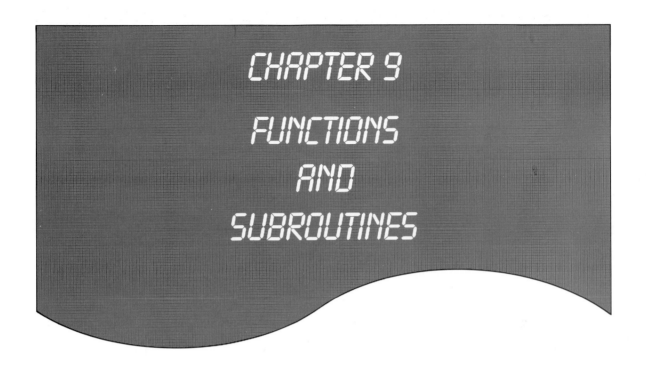

CHAPTER 9
FUNCTIONS
AND
SUBROUTINES

9.1 INTRODUCTION

In the not-too-distant future, in subsequent engineering courses or in your career, you will be solving problems by using the computer, and these problems will very likely be much more complicated and involve much longer Fortran code than we have encountered thus far. As an example, suppose you wanted to

Compute a very complicated function, $y = f(x)$
Obtain a rough sketch of the function on the line printer
Find the zeros of the function by bisection
If all of the above went well, graph the function on an automatic
    pen-and-ink graphics device.

If you do this more than once in a lifetime, it would be a waste of effort and time to recode and retype those elements of the program that remain basically unchanged from your earlier programs. Fortran allows the segmentation of a long program into subprogram blocks that are compiled separately. These blocks are called statement functions, functions, and subroutines.

Statement functions have been discussed in Section 4.2. In this chapter I will discuss primarily Fortran functions and subroutines. Each of these subprogram types may be used as elements of a larger program and later reused in other codes performing similar tasks. For example, we shall shortly construct a subroutine that will graph a function, any function, on the line printer. This code may then be stored on cards, on a disk, or on magnetic tape. Anytime in later programs you wish to have a quick plot of some function, you simply call up this subroutine. Obviously this is an extremely convenient feature of Fortran, and indeed, subprograms are a fundamental element of almost all relatively long Fortran programs. The ability to use previously tested code instead of reprogramming the same ideas also frees us from the frustration and agony of making the same dumb errors twice.

Repeatedly I have stressed the importance of top-down programming; constructing a program in a block pattern with each block or segment being to some extent independent from the others. Subprograms permit a further layer of blocks or modules that will be quite useful in building a program up from parts that can be separately tested.

9.2 REVIEW OF STATEMENT FUNCTIONS

Before we begin the description of subroutines and functions, I will quickly review the properties of statement functions. You will recall that the rules relating to statement functions were concerned with their definition and use.

Definition

1. A statement function definition must precede the first executable statement and must follow all type or dimension declaration statements in a program.
2. All variables in the statement function expression should appear in the argument list of the function. For example,

$$FNC(A,B,T) = A + B*SIN(T)$$

Use

1. The statement function is called up by simply using the function name followed by its argument list in any arithmetic expression or Fortran statement.
2. When using the function, the variables or numbers appearing in the argument list must agree as to type (INTEGER, REAL, CHARACTER) with the variable names in the function definition.
3. The order or sequence of the variables in the argument list is critical.

An example follows:

```
X = 2.
A = 3.
B = 4.
W = A/FNC(A,B,X)            ⟨X not T is used in the
                            reference to the function.⟩
Y = 3./FNC(3.,4.,2.*A - B)  ⟨Y has the same value as W.⟩
```

9.3 SUBROUTINE SUBPROGRAMS

Frequently the operations that we wish the program to execute are too extensive to be represented by a statement function and yet too distracting to the logical flow of the program to include in the body of an otherwise clear algorithm. For example, if in the middle of a program we find that we need the smallest four positive roots of a function $f(x)$, it would be very helpful if we could simply instruct the computer to

Go off to the side, compute the smallest roots of the function $f(x)$, and when they have been found, return and we will continue from there.

The mechanism for doing this in Fortran is the subroutine. A Fortran subroutine is a complete subprogram that is compiled separately from the main program and will appear after the END statement in the main program. As with statement functions, the rules for subroutines relate to their definition and use.

9.3.1 Defining and Referencing a Subroutine

Definition: A Fortran subroutine is a self-contained subprogram that can be initiated in a main program or in other subroutines or functions. The subroutine is compiled separately from the main program and all other subprograms, and the beginning and end of this compilation unit are delimited by a special header line at the beginning and closed with an END statement. The form of the first line of a subroutine is

```
SUBROUTINE    name(argument list)
```

The name of the subroutine may be any valid Fortran variable name; however, since it will designate a procedure, not a variable, it cannot be assigned a value or used within the body of the subroutine. In short, the subroutine *will not* return a value for "name." All computed quantities are passed to and returned from the subroutine via the argument list.

It is not required that a subroutine have an argument list, although most will. The parameters in the argument list of a subroutine definition are called "dummy" arguments, meaning that they will be assigned specific values only when the subroutine has been called by some other program unit. For example, the subroutine LENGTH below computes the length of a vector in the first quadrant. If the vector is not in the first quadrant, the length is assigned a value of -1.0. When the subroutine is compiled the variable names X, Y, L have not yet been assigned values and are dummy arguments.

```
SUBROUTINE LENGTH(X,Y,L)
REAL X,Y,L
IF(X .LT. 0. .OR. Y .LT. 0.)THEN
     L = -1.0
ELSE
     L = SQRT(X * X + Y * Y)
END IF
RETURN
END
```

Subroutines usually contain at least one RETURN statement. When this statement is encountered, the operation of the program transfers back to the line from which the subroutine was called and continues from that point. The END statement is of course used to mark the end of a compilation unit and is not used to terminate the operations of the subroutine.

All variables and statement numbers defined within a subroutine are only locally defined and will not conflict with variable names or statement numbers used in the main program or other subprograms. This is important as it would detract considerably from a subroutine's portability if we had to eliminate such potential conflicts every time a subroutine was used in a large program. The only communication of the calling program with the subroutine is by means of the argument list, which can be thought of as the only "window" in the structure.

Referencing Subroutines

A subroutine is accessed by means of a CALL statement

```
CALL name(argument list)
```

where the "name" agrees with the name in the subroutine definition and the variable names in the argument list agree in number and by type (REAL, INTEGER, CHARACTER) with the dummy arguments in the defining argument list. The number of variable names in the argument list of the subroutine CALL must agree with the number of arguments in the subroutine definition or an execution error will result. Ordinarily, when a subroutine is called, the computation proceeds from the first line of the subroutine until a RETURN or STOP statement is encountered. For example, a program employing the simple subroutine above is

```
PROGRAM VECTOR
REAL RX,RY,R
READ *,RX,RY

CALL LENGTH(RX,RY,R)

IF(R .LT. 0.)THEN
      PRINT *,'VECTOR NOT IN FIRST QUADRANT'
ELSE
      PRINT *,'THE LENGTH OF THE VECTOR IS ',R
END IF
STOP
```

The parameters are transferred to and returned from the subroutine by position in the argument list. Thus, if the values read for RX and RY were 5.0 and 12.0, the values assigned to X and Y within the subroutine would be 5.0 and 12.0 and the subroutine would compute a value of 13.0 for the length L. (See the code below.) When the subroutine returns to the main program, this value is then assigned to the variable R. That is, the input to the subroutine is 5.0, 12.0 (assigned to RX, RY) and the output is 13.0 (assigned to R). I emphasize that the things passed back and forth between the main program and the subroutine are numerical values, not variable names (at least not yet).

```
PROGRAM VECTOR
REAL X,Y,L
X = 5.0              ⟨In the main program, X has
L = 12.0             a value of 5.0 and L is 12.⟩

CALL LENGTH(L,X,Y)   ⟨A value of 13.0 is computed
                     by LENGTH and assigned to Y.⟩

   ...
   ...
END
SUBROUTINE LENGTH(X,Y,L)
REAL X,Y,L
                     ⟨Within subroutine LENGTH,
   ...               X has a value of 12.0, Y a
   ...               value of 5.0, and L is computed
                     to be 13.0.⟩

RETURN
END
```

A subroutine may not directly or indirectly call itself.

9.3.2 Protecting Dummy Arguments

When a subroutine is referenced by a CALL statement, the input variables in the argument list are assumed to have been assigned numerical values, and after the CALL the output variables will also have numerical values. These variables are called *actual* arguments, and the assignments to the dummy arguments in the subroutine definition are made by position in the argument list when the subroutine is entered *and* when it is exited. This can result in unintentional changes in the input variables. For example, a subroutine is given in Figure 9-1 that will compute the root of a function $f(x)$ by using the bisection algorithm of Section 4.6. The function F(X) must be included elsewhere in the code as a separate Fortran FUNCTION subprogram (to be discussed in Section 9.4). The root of the function is known to be in the interval $x_1 \le$ root $\le x_3$. The input to this subroutine is the interval X1, X3, the convergence criterion, EPS, and the maximum number of iterations, IMAX, and the output is the actual number of iterations I, and the answer, ROOT. However, even though X1 and X3 are input to the code, their values change within the subroutine and therefore when a RETURN is executed the actual variables in their position in the CALL will be assigned the final values of X1, X3. Thus if the reference to BISEC in the main program were

```
A = 4.0
B = 6.0
IMAX = 25
EPS = 1.E-6
CALL BISEC(A,B,EPS,IMAX,I,ANS)
PRINT *,A,B,I,ANS
```

the values of A and B would have been altered by the subroutine and the original size of the search interval has been lost. To avoid this, the following safeguard is recommended:

Protect Dummy Input Arguments: Within a subroutine, all dummy input variables should be replaced by locally defined variables. Dummy input variables should never appear to the left of =.
   Thus subroutine BISEC should be amended as

```
SUBROUTINE BISEC(A,B,EPS,IMAX,I,ROOT)
      ...                       ...
      ...                       ...
X1 = A                              The input variables A, B are
X2 = B                              protected and will retain the
      ...                           same values before and after
      ...                           the execution of BISEC.
END
```

Figure 9-1 A subroutine for roots of a function by bisection.

```
      SUBROUTINE BISEC(A,B,EPS,IMAX,I,ROOT)
*--
*--   A ROOT OF THE FUNCTION F(X) IS FOUND BY SUCCESSIVE
*--   INTERVAL HALVING. THE FUNCTION MUST BE A FUNCTION OF A
*--   SINGLE VARIABLE. THE CODE IS SIMILAR TO THAT OF FIGURE 4-13.
*--   IF AN ERROR IS ENCOUNTERED A DIAGNOSTIC IS PRINTED, THE ROOT
*--   IS ASSIGNED THE VALUE 1.E99 AND THE SUBROUTINE RETURNS.
*-------------------------------------------------------------------
* VARIABLES
*
      REAL A,B,EPS,ROOT,X1,X2,X3,F1,F2,F3,D,D0
      INTEGER I,IMAX
*
*              A,B      --  THE ORIGINAL SEARCH INTERVAL
*              X1,X3,X2--  THE LEFT, RIGHT, AND MIDPOINT OF
*                          THE CURRENT INTERVAL SEGMENT
*              F1,F3,F2--  THE FUNCTION EVALUATED AT THESE X'S
*              D        --  THE RATIO OF THE CURRENT INTERVAL
*                          TO THAT OF THE ORIGINAL INTERVAL
*              ROOT     --  THE ROOT OF THE FUNCTION (OUTPUT)
*              I        --  ITERATION COUNTER
*              IMAX     --  MAXIMUM NUMBER OF ITERATIONS
*              EPS      --  CONVERGENCE CRITERION BASED ON D
*-------------------------------------------------------------------
* INITIALIZATION
*
      X1 = A
      X3 = B
      F1 = F(X1)
      F3 = F(X3)
      I  = 0
      D  = 1.0
*
*                          VERIFY THAT THERE IS A ROOT
*                          WITHIN THE ORIGINAL INTERVAL
*
      IF(F1 * F3 .GT. 0.)THEN
         WRITE(*,10)A,B
         ROOT = 1.E99
         RETURN
      END IF
*
*                          THE INTERVAL IS HALVED AND EACH SIDE
*                          IS CHECKED FOR THE POSITION OF ROOT
*
```

Continued

```
*
    1 X2 = (X1 + X3)/2.
      F2 = F(X2)
*
*                                    CONVERGENCE TESTS
*
      IF(D .LT. EPS)THEN
          ROOT = X2
          RETURN
      ELSE IF(I .GT. IMAX)THEN
*
*                                    FAILURE EXCESSIVE ITERATIONS
*
          WRITE(*,11)I,X2,F2
          ROOT = 1.E99
          RETURN
      END IF
*
*                                    CHECK FOR CROSSING ON THE LEFT
*
      IF(F1 * F2 .LT. 0.0)THEN
          D = (X2 - X1)/(B - A)
          F3 = F2
          X3 = X2
*
*                                    OR ON THE RIGHT
*
      ELSE IF(F2 * F3 .LT. 0.0)THEN
          D = (X3 - X2)/(B - A)
          F1 = F2
          X1 = X2
*
*                                    IF CROSSING IS IN NEITHER HALF
*                                    EITHER F2 = 0.0 OR AN ERROR.
*
      ELSE IF(F2 .EQ. 0.0)THEN
          ROOT = X2
          RETURN
      ELSE
          WRITE(*,12)I,X1,X3
          ROOT = 1.E99
          RETURN
      END IF
*
*                                    INCREMENT COUNTER AND REPEAT
*
      I = I + 1
      GO TO 1
```

```
*----------------------------------------------------------------
*FORMATS
*
   10 FORMAT(//,5X,'----ERROR----',/,10X,'FROM BISEC',/,
     +       10X,'THE INTERVAL ',F8.4,' TO ',F8.4,//,
     +       10X,'DOES NOT CONTAIN A ROOT')
*
   11 FORMAT(//,5X,'----ERROR----',/,10X,'FROM BISEC',/,
     +       10X,'EXCESSIVE ITERATIONS, I = ',I5,/,
     +       10X,'LAST VALUE WAS F(',F9.5,') = ',E9.2)
*
   12 FORMAT(//,5X,'----ERROR----',/,10X,'FROM BISEC',/,
     +       10X,'THE SUB-INTERVAL ',F8.4,' TO ',F8.4,/,
     +       10X,'DOES NOT CONTAIN A ROOT')
      END
```

Lest you think this is unduly scrupulous programming, I have one more example; this time not protecting dummy variables will lead to very startling results. You should execute the simple Fortran program in Figure 9-2. The output from this program will probably be

$$1 + 1.0 = 3.0$$

The error in the code is rather subtle. The first value in the CALL statement (1.) is probably intended as input, but in the subroutine this value is changed (to 2.) and upon the return to the main program the value 2. is assigned to the first position in the CALL argument list. The net effect is that in this instance the *constant* 1.0 has been overwritten with the value 2.0. Thereafter, everywhere in the code where we have used 1.0 the computer will use a value 2.0. This could obviously cause considerable confusion. It is, however, easily avoided by following the suggestion for protecting dummy input variables.

```
      PROGRAM GRIEF
      CALL NONO(1.,X)
      PRINT *, '1 + ', X, ' = ', 1. + X
      STOP
      END
      SUBROUTINE NONO(A,B)
      B = 1.
      A = 2.
      RETURN
      END
```

Figure 9-2 An example of a subroutine overwriting a constant.

9.3.3 Arrays as Elements of an Argument List

An entire array can be transferred to or from a subroutine by simply including the name of the array in the calling and defining the subroutine argument list. Of course, the array must be dimensioned in the main (or calling) program. The array must likewise be dimensioned in the subroutine as well. The purpose of this secondary dimensioning is *not* to allot memory space to the array, but merely to inform the subroutine that this variable is an array, already dimensioned elsewhere. For this reason the size of the array as dimensioned in the subroutine may be any value that is less than or equal to the actual dimension size in the main program, or may even be variable. The following examples are correct illustrations of this.

```
PROGRAM XXX
 REAL A(11),B(50),
+ C(-2:7),D(10,10)
```

```
CALL XX(A)                          SUBROUTINE XX(R)
                                     REAL R(10)
```
$$a_1 \longleftrightarrow r_1 \qquad \cdots$$
$$a_{10} \longleftrightarrow r_{10} \qquad \cdots$$
```
                                     END
```

```
CALL YY(B)                          SUBROUTINE YY(S)
                                     REAL S(-20:29)
```
$$b_1 \longleftrightarrow s_{-20} \qquad \cdots$$
$$b_{50} \longleftrightarrow s_{29} \qquad \cdots$$
```
                                     END
```

```
N = 3
CALL ZZ(D,N)                        SUBROUTINE ZZ(T,I)
                                     REAL T(I,I)   ⎛ I must be ⎞
```
$$d_{11} \longleftrightarrow t_{11} \qquad\qquad \text{less than}$$
$$d_{91} \longleftrightarrow t_{33} \qquad \cdots \qquad \text{or equal}$$
```
                                              ⋯      ⎝ to 10. ⎠
                                     END
```

```
CALL WW(A,C,M)                      SUBROUTINE WW(U,V,J)
                                     REAL V(1)   ⎛ legal but ⎞
                                                 ⎜ not recom- ⎟
                                                 ⎝ mended ⎠
                                     REAL U(-J:J) ⎛ J must be ⎞
```
$$a_1 \longleftrightarrow u_{-j} \qquad\qquad \text{less than}$$
```
                                              ⋯      or equal
                                              ⋯    ⎝ to 5. ⎠
 END                                 END
```

In this program

The element	in the subroutine	has the same value as the	element	in the program.
R(9)	XX		A(9)	
S(-20)	YY		B(1)	
S(0)	YY		B(21)	
T(1,1)	ZZ		D(1,1)	
T(3,3)	ZZ		D(9,1)	
V(5)	WW		C(2)	
U(5)	WW		A(11)	

Of course any dummy variables in the defining argument list that refer to the dimension size of an array must be integers.

When the array being transferred has two or more subscripts, special care must be used in the secondary dimensioning. For example, if the array A were dimensioned in the main program as

```
PROGRAM MANE
REAL A(3,3)
CALL ALPHA(A,2)
```

the elements are stored in the sequence (see Section 7.3)

$$a_{11} \quad a_{21} \quad a_{31} \quad a_{12} \quad a_{22} \quad a_{32} \quad a_{13} \quad a_{23} \quad a_{33}$$

If the array elements are transferred to a subroutine with a dummy variable for the size of the array, as

```
SUBROUTINE ALPHA(X,N)
REAL X(N,N)
```

the elements of the dummy variable X would be (for $N = 2$)

$$x_{11} \quad x_{21} \quad x_{12} \quad x_{22}$$

and the correspondence would be

$$x_{11} \leftrightarrow a_{11}$$
$$x_{21} \leftrightarrow a_{21}$$
$$x_{12} \leftrightarrow a_{31}$$
$$x_{22} \leftrightarrow a_{12}$$

Since it is usually the case that the indices are desired to match, the arrays should be dimensioned as the *same* size in both the main program and the subroutine. This could be accomplished in the subroutine with an additional index as

```
SUBROUTINE ALPHA(X,NA,NL)
REAL X(NA,NA)
DO 55 I = 1,NL
DO 54 J = 1,NL
      X(I,J) =   ...
```

where NA is the actual dimension size of the array as specified in the main program and NL is the local size of the array as used in the subroutine.

In the following statements

```
PROGRAM ALPHA
REAL A(25)
CALL BETA(X,Y,A(25))
   ...
   ...
END
```

only the twenty-fifth element of the array A is transferred to the subroutine and not the entire contents of the array as may have been intended. (Note: A reference like A(25) would not be permitted within the argument list of the subroutine definition.)

9.4   FUNCTION SUBPROGRAMS

Another form of Fortran subprogram is the external function, which combines the features of the statement function and the subroutine. Like a subroutine, it is a complete subprogram, separately compiled from the main program and other subprograms. The rules for passing information to and from the function via the argument list are the same as for subroutines. Like a statement function, it is referenced by simply using the function name followed by its argument list in any Fortran arithmetic expression or statement. Fortran functions are used most often when a single value is to be computed and returned to the calling program. The principal differences between functions and subroutines are a consequence of the fact that a function returns a value associated with its name and the subroutine does not. The name of a function is important, the name of a subroutine is usually not. The rules relating to the definition and use of a Fortran function are

1. The header line for a function is one of the following

```
FUNCTION name(argument list)
```

```
REAL FUNCTION name(argument list)
INTEGER FUNCTION name(argument list)
CHARACTER FUNCTION name(argument list)
```

The data type of the value returned through the function name depends on the data type of the name. I would recommend that you use the default typing (I through N for integers, rest real) in selecting function names.

```
FUNCTION KOUNT(A,B,C)     Returns an integer   KOUNT
FUNCTION BGGST(X,N)       Returns a real       BGGST
```

2. Since the function is expected to return a value associated with the name, it is crucial that before any RETURN is encountered in the body of the function subprogram, an assignment statement of the form

```
            NAME = ...
```

be present.

3. Whereas a subroutine is not required to have an argument list, a function subprogram must have an argument list (which may be empty). That is, the compiler has been programmed to recognize that a name followed by a left parenthesis represents something other than a simple variable.

Referencing a Function

A Fortran function is accessed by its name, not with CALL, in precisely the same manner as statement functions.

As a simple example, consider the program in Figure 9-3, which reads a set of numbers from a data file and scales the numbers by dividing each number by the largest number in the set. A function BGGST is used to determine the maximum element. In this program you will note that in the function BGGST a variable BIG is used to keep track of the maximum element and only after the entire search is completed is the assignment BGGST = BIG made. The reason for this is that you must be very cautious when using the function name while inside the function. As a general rule, the function name should never appear to the right of the assignment operator (=) inside the function. A statement like . . . = BGGST could be interpreted as the function calling itself which is not allowed in Fortran.

Also, every cycle of the DO loop in the main program recomputes the maximum element of the array (and in addition each time changes the set X). Even if this would give the correct result, it is a serious error. Needlessly recomputing a function can waste considerable computer time. The loop should be rewritten as

```
      XMAX = BGGST(X,N)
      DO 2 I = 1,N
          X(I) = X(I)/XMAX
    2 CONTINUE
```

Figure 9-3 Use
of a function to
incorrectly scale
a list of numbers.

```
PROGRAM SCALES
REAL X(100)
OPEN(37,FILE='XDATA')
REWIND(37)
READ(37,'(F9.5)',END=1)(X(I),I=1,100)
1 N = I - 1
DO 2 I = 1,N
    X(I) = X(I)/BGGST(X,N)
2 CONTINUE
WRITE(*,'(1X,5F9.6)')(X(I),I=1,N)
STOP
END
```

```
FUNCTION BGGST(A,M)
REAL A(M),BIG
BIG = A(1)
DO 1 I = 2,M
    IF(A(I) .GT. BIG)BIG = A(I)
1 CONTINUE
BGGST = BIG
RETURN
END
```

Even though functions may return values in addition to their name by means of the argument list in precisely the same manner as subroutines, this should be avoided. If the subprogram is to return one value, use a function; if it is to return more than one value, use a subroutine. Using a function structure when a subroutine would be more appropriate results in a very confusing logical flow to a program.

Finally, none of the dummy arguments in the argument list of a function should be altered within the body of the function. This suggestion is most easily adhered to if you agree to never use variables in the argument list as output from a function. Also, just as with the arguments in a subroutine argument list, function variable names should always be protected by introducing local replacements.

9.5 EXAMPLE PROGRAMS USING FUNCTIONS AND SUBROUTINES

In addition to the example programs in this section, essentially all of the procedures discussed to this point would profit by recasting in a modular form using functions and subroutines. The algorithms for roots of a function by bisection, sorting an array, computing the minimum/maximum and average of an array, etc. should be rewritten in a structured style. Furthermore, you may find it useful when constructing a program to employ subroutines for reading extensive data lists and for printing elaborate tables. In the future your main programs may consist of little else than CALLs to subroutines to read data, print output, and execute the computations.

9.5.1 Fractions

You certainly remember, perhaps with some pain, the rules for adding fractions and reducing them to simplest form by dividing both numerator and denominator by common factors. To add a large number of fractions exactly, i.e., to express the answer as a fraction rather than a decimal, usually involves a considerable amount of tedious arithmetic. Thus the evaluation of

$$1 + \frac{1}{2} + \frac{1}{3} + \frac{1}{4} + \ldots + \frac{1}{25}$$

is easily done on a pocket calculator if the numbers are expressed as decimals. (The answer is 3.815958177. .). If you attempt to add the fractions exactly you will see that the problem quickly becomes hopelessly complicated. But hopelessly complicated and terribly tedious arithmetic is the computer's strong suit. To code this problem we will have to construct the following main and subprograms:

Function IGCF(ITOP, IBOT): An integer function that determines the greatest common factor contained in both the numerator, ITOP, and the denominator, IBOT, of a given fraction

Subroutine ADD(IA, IB, IC): A subroutine to add two fractions, IA, IB, and return the result as a fraction, IC. The fractions are represented as two-element arrays, the numerator of IA is IA(1) and the denominator is IA(2). Thus

$$\frac{a_1}{a_2} + \frac{b_1}{b_2} = \frac{c_1}{c_2}$$

where $c_1 = a_1 b_2 + a_2 b_1$ and $c_2 = a_2 b_2$

The fraction is then reduced to lowest terms by using the function IGCF

Program series: A program that will sum the series by repeatedly calling the subroutine ADD

The main program is quite similar to an ordinary summation and is given in Figure 9-4. The subroutine to add two fractions and reduce the result to lowest terms is fairly simple and is given in Figure 9-5.

The algorithm to determine the greatest common factor in two integers was developed by Euclid and consists of dividing the first integer (the larger, P) by the second (the smaller, Q) and determining the remainder R. If the remainder is not zero, the pair (P, Q) is replaced by the pair (Q, R) and the

Figure 9-4 The main program to add a series of fractions.

```
          PROGRAM SERIES
          INTEGER SUM(2),TERM(2),K(2),LIMIT
*
*                         ALL THE ARRAYS REPRESENT FRACTIONS,
*                         THE FIRST POSITION IS THE NUMERATOR
*                         THE SECOND IS THE DENOMINATOR
*                         THE SUM IS FROM 1 TO 1/LIMIT.
*
*         READ *,LIMIT
*                         INITIALIZE THE SUM = 0/1
          SUM(1) = 0
          SUM(2) = 1
*
          DO 1 I = 1,LIMIT
*
*                         ASSIGN A VALUE TO TERM
*
              TERM(1) = 1
              TERM(2) = I
*
*                         ADD SUM + TERM AND CONVERT THE RESULT
*                         TO LOWEST TERMS. THE RESULT IS IN K
*
              CALL ADD(SUM,TERM,K)
*
*                         REPLACEMENT, SUM = SUM + TERM
*
              SUM(1) = K(1)
              SUM(2) = K(2)
    1     CONTINUE
          WRITE(*,2)LIMIT,SUM
          STOP
*
    2     FORMAT(5X,'THE SUM OF FRACTIONS 1/N FOR N = 1 TO ',I3,//,
         +        20X,'IS',//,
         +        15X,I12,/,15X,12('-'),/15X,I12)
          END
```

Figure 9-5
A subroutine to
add two fractions.

```
SUBROUTINE ADD(A,B,C)
INTEGER A(2),B(2),C(2),ID
C(1) = A(1) * B(2) + A(2) * B(1)
C(2) = A(2) * B(2)
*
*                   The greatest common factor in
*                   C(1), C(2) is obtained from the
*                   function IGCF.
*
ID = IGCF(C(1),C(2))
C(1) = C(1)/ID
C(2) = C(2)/ID
RETURN
END
```

process repeated. If the remainder is zero, the greatest common factor (GCF) is the last value of Q. For example, starting with $P = 221$, $Q = 91$, Euclid's algorithm yields

P	Q	R
221	91	39
91	39	13
39	13	0

So the GCF is 13. The function to execute this algorithm is then shown in Figure 9-6. By the way, the result of the problem with LIMIT = 25 is

$$\frac{34052522467}{8923714800}$$

Of course, problems involving integer arithmetic are not often very useful in engineering applications; however, the structure of this program illustrates several features of functions and subroutines and should be thoroughly understood before you attempt to code your own subprograms. There are several series involving fractions in the problem section of this chapter and you are invited to attempt to compute answers to these problems as fractions in lowest terms.

9.5.2 The Function ROUNDR

In Chapter 8 I outlined the procedures that are followed when graphing a function. One crucial step was omitted. If the range of the x values that are to be plotted is, say, -13.63 to $+8.77$, you would naturally set the x axis from

Figure 9-6
A function to
compute the
greatest common
factor.

```
        FUNCTION IGCF(ITOP,IBOT)
*
*                              In order not to destroy the
*                              dummy variables ITOP, IBOT,
*                              we introduce local variables.
*
        IA = ITOP
        IB = IBOT
*
*                              IR is the remainder, note the
*                              use of integer arithmetic.
*
    1   IR = IA - IA/IB * IB
        IF(IR .EQ. 0)THEN
                IGCF = IB
                RETURN
        ELSE
                IA = IB
                IB = IR
        END IF
        GO TO 1
        END
```

−15 to +10. That is, the minimum and maximum values are rounded up in magnitude. The function that accomplishes this is given in Figure 9-7. It is somewhat tricky, but you should be able to follow the algorithm. The function ROUNDR is used primarily in constructing axes for a printer plot, as in the next example.

9.5.3   A Subroutine for Printer Plots

We are now in a position to assemble a modular program to execute a printer plot of an arbitrary function $f(x)$. The code will consist of several parts:

Main Program: The main program will compute a list of values, $(x_i, y_i, i = 0, 40)$ to be graphed by calling a function subprogram F(X). For this example it is assumed that the x values are equally spaced between zero and XMAX. The main program then calls the subroutine PLOT, which will graph the tabulated values.

SUBROUTINE PLOT (X,Y,N):   The subroutine will then:

1. Determine the range of y values (y_{min}, y_{max}) by calling a subroutine MIN-MAX.

Figure 9-7 A Fortran function to round an arbitrary number.

```
        FUNCTION ROUNDER(Q)
*
* Roundr will round the real number Q up in magnitude so that
* the result will have the first two digits divisible by 5, e.g.
* 771.3 becomes 800, and -0.08341 becomes -0.085.
*
        A = Q
*                              The input dummy variable Q
*                              is protected. If Q < 0 use -Q.
*
        IF(A .LT. 0.)A = -A
*
        B = LOG10(A)
        IB = B
        B = B - IB
*
*                              If B is pos. it is the mantissa, and
*                              IB is the characteristic; e.g. if
*                              Q = 15, log(15) = 1.1761, IB = 1, B = 0.1761.
*                              If B is neg, we express the log as
*                              0.XXXX-1.0, So
*
        IF(B .LT. 0.)THEN
           B = B + 1.
           IB = IB - 1
        END IF
*
*                              If Q = 0.8, IB = 0, B = -0.0969 are
*                              replaced by B = 0.9031, IB = -1,
*                              i.e. log(0.8) = 0.9031 - 1.
*
        C = 10.**B
*
*                              C is the number Q without its sign
*                              or exponent.
*
        IC = 2.*C + 1.
*                              This line does the actual rounding;
*                              next reattach the correct power of 10.
*
        R = IC/2. * 10.**IB
        IF(Q .LT. 0.)R = -R
        ROUNDR = R
        RETURN
        END
```

2. Round the minimum and maximum values of y to produce a y axis with whole numbers at the beginning and end.
3. Print the y axis horizontally with y values and tic marks.
4. Step through the x values, one by one, and print a horizontal line of all blanks except a * in the corresponding column for $y(x_i)$.
5. The code will also print an x value for every tenth line and the graph will have a border on all four sides.

FUNCTION F(X): A function subprogram for the function we wish to graph. The function may require more information in its argument list.

SUBROUTINE MINMAX: A subroutine to scan a single subscripted array A of size N and return the minimum AMIN and the maximum AMAX of the values in the array.

FUNCTION ROUNDR: A function to round values up to magnitude and used to establish the limits of the axes. (See Figure 9-7.)

The code to handle all this is shown in Figure 9-8.

The function chosen for the graph represents damped harmonic motion, such as the gradual shrinking of the amplitude of a pendulum (small damping) or the removal of almost all "bounce" on a bumpy road in a car with good shock absorbers (large damping). The mathematical function describing such motion is the product of a decaying exponential and a periodic function such as a cosine.

(9.1)
$$y(t) = e^{-\alpha t} \cos(\beta t)$$

where $y(t)$ represents the amplitude of the oscillations at time t, α is the decay constant, and $\beta/(2\pi)$ is the oscillation frequency. The parameters α, β are related to physical quantities by

(9.2)
$$\alpha = \frac{\gamma}{2m}$$

(9.3)
$$\beta = \left(\frac{k}{m} - \frac{\gamma^2}{4m^2} \right)^{1/2}$$

where m = mass (kg)

k = restoring spring force constant (N/m)

γ = damping or viscosity coefficient (N-sec/m)

The output from this program is shown in Figure 9-9.

Figure 9-8 A Fortran program to printer plot a function.

```
          PROGRAM MAINE
*--
*--    THE MAIN PROGRAM COMPUTES A TABLE OF X(), Y() VALUES BY
*--    CALLING THE FUNCTION F(X). AFTER THE TABLE IS COMPLETED,
*--    THE SUBROUTINE PLOT IS USED TO OBTAIN A LINE PRINTER
*--    SKETCH OF THE COMPUTED VALUES.
*-------------------------------------------------------------------
* VARIABLES
*
       REAL X(0:50),Y(0:50)
*
*--              COMPUTE THE TABLE
*--
       X(0) = -10.
       DX = 0.4
       Y(0) = F(X(0))
       DO 1 I = 1,50
          X(I) = X(I - 1) + DX
          Y(I) = F(X(I))
     1 CONTINUE
*--
*--              CALL PLOT TO EXECUTE A PRINTER PLOT
*--
       CALL PLOT(X,Y,50)
*--
*--
       STOP
       END
*-------------------------------------------------------------------
       FUNCTION F(X)
*--
       F = ...
*--
       RETURN
       END
*-------------------------------------------------------------------
       SUBROUTINE PLOT(X,Y,N)
*--
*--    THIS SUBROUTINE WILL SKETCH THE TABULATED VALUES IN THE
*--    ARRAYS X(),Y(). IT IS ASSUMED THAT THE X VALUES ARE
*--    EQUALLY SPACED FROM A = X(0) TO B = X(N). THE RESULTS WILL
*--    BE DISPLAYED ON THE TERMINAL SCREEN OR ON THE LINE
*--    PRINTER. THE GRAPH WILL BE N LINES LONG (X) AND 60
*--    COLUMNS WIDE (Y).
```

Continued

```
*----------------------------------------------------------
* VARIABLES
*
      REAL A,B,X(0:N),Y(0:N),YMAX,YMIN,XSTEP,RNGY
      INTEGER N,ISTEP,STRCOL
      CHARACTER*1 LINE(0:60),BLANK,STAR
*
*                 A,B        --   THE LIMITS OF THE X AXIS(INPUT)
*                 YMIN,      --   THE MIN/MAX VALUES OF THE
*                   YMAX          COMPUTED Y VALUES
*                 XSTEP      --   THE STEP SIZE ALONG THE X AXIS
*                 N          --   THE NUMBER OF STEPS ALONG X
*                 X(),Y()    --   ARRAYS FOR STORING THE COMPUTED
*                                 VALUES FOR X AND Y
*                 LINE()     --   THE ARRAY CONTAINING THE SYMBOLS TO
*                                 BE PRINTED FOR ONE VALUE OF X
*                 ISTEP      --   THE CURRENT STEP IN THE PLOT
*                 STRCOL     --   THE COLUMN NUMBER TO POSITION THE
*                                 STAR FOR THIS LINE
*                 RNGY       --   THE RANGE OF THE ROUNDED Y VALUES
*                 BLANK      --   CONTAINS ONE BLANK CHARACTER
*                 STAR       --   CONTAINS THE SYMBOL *
*----------------------------------------------------------
* INITIALIZATION
*
      BLANK = ' '
      STAR = '*'
      DO 1 I = 0,60
         LINE(I) = BLANK
    1 CONTINUE
*
      A = X(0)
      B = X(N)
      XSTEP = (B - A)/N
*----------------------------------------------------------
*
*                 DETERMINE THE MIN/MAX VALUES OF Y()
*
      CALL MINMAX(Y,N + 1,YMIN,YMAX)
*
*                 NEXT SCALE THE AXES BY ROUNDING THE VALUES
*                 (IF BOTH YMIN AND YMAX ARE THE SAME SIGN,
*                 ONLY ROUND THE LARGER.)
*
      IF(YMAX * YMIN .LT. 0.)THEN
         YMAX = ROUNDR(YMAX)
```

```
            YMIN = ROUNDR(YMIN)
        ELSE IF(YMAX .GT. 0.)THEN
            YMAX = ROUNDR(YMAX)
            YMIN = 0.0
        ELSE
            YMIN = ROUNDR(YMIN)
            YMAX = 0.0
        END IF
*----------------------------------------------------------------
*  PRINT THE GRAPH
*
*                        FIRST PRINT THE INPUT DATA ALONG THE Y AXIS
*                        INCLUDING TIC MARKS
*
        RNGY = YMAX - YMIN
        WRITE(*,10)(YMIN + RNGY/3.*I,I = 0,3)
*
*                        NEXT STEP ALONG THE X AXIS PRINTING THE CON-
*                        TENTS OF LINE() AT EACH STEP. POSITION THE
*                        STAR AT THE PROPER COLUMN EACH STEP AND EVERY
*                        TENTH STEP PRINT AN X VALUE AS WELL.
*
        DO 3 ISTEP = 0,N
            STRCOL = (Y(ISTEP) - YMIN)/RNGY * 60.
            LINE(STRCOL) = STAR
            IF(ISTEP/10 * 10 .EQ. ISTEP)THEN
                WRITE(*,11)X(ISTEP),(LINE(K),K=0,60)
            ELSE
                WRITE(*,12)(LINE(K),K=0,60)
            END IF
*
*                        AFTER LINE() IS PRINTED, IT IS AGAIN SET
*                        EQUAL TO ALL BLANKS
*
            LINE(STRCOL) = BLANK
*
    3 CONTINUE
*
*                        PRINT A BOTTOM BORDER ON THE GRAPH
*
        WRITE(*,13)
        RETURN
*----------------------------------------------------------------
*  FORMATS
```

Continued

```
*
   10 FORMAT(////,20X,'A PLOT OF Y VS. X',//,25X,'Y(X)',/,
     +          3X,F7.3,3(13X,F7.3),/,
     +          7X,'+',3(19X,'+'),/,
     +          6X,63('+'))
*
   11 FORMAT(F7.3,61A1,'*')
   12 FORMAT(6X,'*',61A1,'*')
   13 FORMAT(6X,63('*'),/,7X,'+',3(19X,'+'))
*
      END
*---------------------------------------------------------------
      SUBROUTINE MINMAX(A,N,AMIN,AMAX)
*--
*-- MINMAX DETERMINES THE MINIMUM (AMIN) AND THE MAXIMUM (AMAX)
*-- OF THE ARRAY A() WHICH CONTAINS N ELEMENTS.
*---------------------------------------------------------------
*
      REAL A(N),AMIN,AMAX
*
      AMIN = A(1)
      AMAX = A(1)
      DO 1 I = 2,N
        IF(A(I) .LT. AMIN)AMIN = A(I)
        IF(A(I) .GT. AMAX)AMAX = A(I)
    1 CONTINUE
      RETURN
      END
*---------------------------------------------------------------
      FUNCTION ROUNDR(Q)
*--
*--   See Figure 9-7
*--
      END
*---------------------------------------------------------------
*
```

Figure 9-9 A printer plot of damped harmonic motion.

```
INPUT X LIMITS OF PLOT (A,B) 0.0001, 10.0
ENTER SPRING STIFFNESS, MASS, VISCOSITY 25.0, 10.0, 5.0


            A PLOT OF DAMPED HARMONIC MOTION

    SPRING STIFFNESS    =   25.0000 N/M
```

```
     OBJECTS MASS       =   10.0000 KG
     VISCOSITY COEF.    =    5.0000 N-SEC/M

COMPUTED QUANTITIES
     DECAY CNST.(ALPHA) =    .250E+00 /SEC
     FREQUENCY (BETA)   =    .156E+01 /SEC


       -.650              -.100              .450              1.000
         +                  +                  +                  +
        *******************************************************************
  .000   *                                                        *  *
         *                                                   *    *
         *                                              *         *
         *                                       *                *
         *                                  *                     *
         *                            *                           *
         *                     *                                   *
         *                *                                        *
         *  *          *                                           *
         * *                                                       *
         *   *                                                     *
 2.500   *          *                                              *
         *                 *                                       *
         *                      *                                  *
         *                          *                              *
         *                             *                           *
         *                                *                        *
         *                                  *                      *
         *                                *                        *
         *                             *                           *
         *                        *                                *
         *                    *                                    *
 5.000   *                *                                        *
         *             *                                           *
         *           *                                             *
         *        *                                                *
         *        *                                                *
         *        *                                                *
         *           *                                             *
         *                *                                        *
         *                 *                                       *
         *                   *                                     *
 7.500   *                     *                                   *
         *                      *                                  *
         *                      *                                  *
         *                      *                                  *
         *                    *                                    *
         *                 *                                       *
         *                *                                        *
         *             *                                           *
         *            *                                            *
         *          *                                              *
10.000   *          *                                              *
        *******************************************************************
         +                  +                  +                  +
        STOP
```

9.6 CONSTRUCTING MODULAR PROGRAMS

9.6.1 The EXTERNAL and INTRINSIC Statements

The subroutine in the previous section could be restructured to be somewhat more modular by having the subroutine PLOT graph an *arbitrary* function from x_1 to x_2. Thus, if we wanted to graph $\sin(x)$ from 0 to 2π we could simply insert the following lines anywhere we wish:

```
X1 = 0.0
X2 = 2. * 3.1415926
CALL PLOT(X1,X2,SIN)
```

Or to graph a more complicated function called, say, BESSEL(X), where BESSEL is a function subprogram included elsewhere in the complete code, the command would be

```
CALL PLOT(2.5,12.3,BESSEL)
```

However, you should have noticed that we are using the argument list of a subroutine in a manner significantly different from all previous examples. Up to this point we have transferred either ordinary variables or dimensioned arrays by means of the argument list. These are things with numerical values. The transfer considered here is the transfer of a *name*. In the above call to PLOT the function names SIN or BESSEL appear, while in the body of the subroutine a dummy name like FNC is used.

```
SUBROUTINE PLOT(A,B,FNC)
REAL X(50), Y(50)
CHARACTER*1 LINE(66)
    ...              ...
F1 = FNC(A)
F3 = FNC(B)
    ...              ...
    ...              ...
RETURN
END
```

There is no function called FNC and in all the statements in PLOT that refer to this function, the computer must be instructed to replace the name FNC with SIN or BESSEL or some other existing function subprogram.

This is accomplished in Fortran by means of the EXTERNAL and INTRINSIC statements, which have a form

```
EXTERNAL name₁, name₂,...

INTRINSIC name₃, name₄,...
```

where $name_1$, $name_2$, . . . are names of functions that are not called directly, but which appear in the argument list of referenced subroutines or functions. The EXTERNAL statement is used to identify a name as that of a user-written function subprogram that appears elsewhere in the code, while the INTRINSIC statement identifies a name as a library function.

The EXTERNAL and INTRINSIC statements are required only in the program unit that makes indirect reference to the function through the argument list of a referenced subprogram. It is *not* required in the subprogram that makes a *direct* reference to the function. Thus, in the above example, the statements

```
EXTERNAL BESSEL
INTRINSIC SIN
```

would be required in the main program that calls PLOT (which in turn calls SIN and BESSEL), but would not be needed in PLOT itself. The point is this: the compiler must be instructed that in the call to PLOT in the main program, the name BESSEL *is not* a variable but the name of an entire function. Since the function BESSEL is defined "externally" to the main program, the compiler has no way of determining this, unless we specify that BESSEL is a function name defined elsewhere. Naturally then, this does not apply to statement functions which are defined internally. Statement function names can never be included in an EXTERNAL or INTRINSIC statement. The EXTERNAL and INTRINSIC statements are nonexecutable and must appear before the first executable statement. Subroutine names may also be declared EXTERNAL.

A sketch of the complete code to graph the functions SIN and BESSEL is given in Figure 9-10.

9.6.2 An Example Program Illustrating EXTERNAL Statements

The theory of probability is concerned with describing the results of measurements that contain a degree of randomness and has applications in a variety of fields. For example, suppose that you wish to determine whether investing in a gas station on the new interstate is a good idea. The most important factor in the decision is of course the volume of traffic passing the station. The traffic flow can be characterized by λ, the average number of cars that pass per minute in the daylight hours. To aid in your decision, you set up some expensive electronic equipment to measure how many cars pass in each minute during the day. When you return in a few days, you find a

Figure 9-10
Outline of the
code to execute
printer plots.

```
PROGRAM XXX
REAL A,B,C,D                    plotting intervals, a − b, c − d
EXTERNAL BESSEL
INTRINSIC SIN

READ *,A,B,C,D
CALL PLOT(A,B,SIN)
CALL PLOT(C,D,BESSEL)
    ...          ...
    ...          ...
END
```

```
SUBROUTINE PLOT(X1,X2,FNC)
REAL X(50),Y(50)                              PLOT will printer-
CHARACTER*1 LINE(66),BLANK,STAR               plot an arbitrary
                                              function which has
DX = (X2 - X1)/49.                            the dummy name
DO 37 I = 1,50                                FNC. When it is
    X(I) = (I - 1.) * DX                      called, the replace-
    Y(I) = FNC(X(I))                          ment FNC → SIN or
37 CONTINUE                                    BESSEL is made.
CALL MINMAX(Y,50,YMIN,YMAX)
TOP = ROUNDR(YMAX)
    ...          ...
    ...          ...
RETURN
END
```

```
FUNCTION BESSEL(S)
BESSEL = ...
RETURN
END
```

```
SUBROUTINE MINMAX(A,N,AMIN,AMAX)
    ...          ...
RETURN
END
```

```
FUNCTION ROUNDR(Q)
    ...          ...
ROUNDR = ...
RETURN
END
```

problem. The electronics could not keep up with the counting when the flow was extremely heavy. Whenever 60 or more cars pass per minute, the counting apparatus malfunctions and effectively registers "tilt." This is found to occur in 20 percent of the measurements. Is all lost? No, but to unravel the results may take a bit of work and will require a few assumptions.

First of all, probability theory says that if λ is the average number of events (i.e., cars passing) in an interval, the probability of k events in a given interval is given approximately by the Poisson function:

$$P_k(\lambda) = \frac{\lambda^k e^{-\lambda}}{k!} \tag{9.4}$$

Where $P_k \sim 1$ implies near certainty that k cars will pass in any given minute.

Now, your faulty measurements tell you that the probability of 60 *or more* cars per minute is 20 percent, in other words

$$0.2 = \sum_{k=60}^{\infty} P_k(\lambda) \tag{9.5}$$

This equation must be solved for the average λ.

To evaluate the infinite sum, we once again determine the ratio of successive terms

$$\text{Ratio} = \frac{P_{k+1}(\lambda)}{P_k(\lambda)} = \frac{\lambda^{k+1} e^{-\lambda}}{(k+1)!} \frac{k!}{\lambda^k e^{-\lambda}} = \frac{\lambda}{k+1} \tag{9.6}$$

Also, since the summation begins at $k = 60$, the value of 60! must be evaluated first. We will express Equation (9.5) as a function called TILT(x), i.e.,

$$\text{Tilt}(x) = \sum_{k=60}^{\infty} P_k(x) - 0.2 \tag{9.7}$$

The Fortran code for this function is given in Figure 9-11.

The program to solve the problem must determine the root of Equation (9.7) and is given in Figure 9-12. The setup for this program would have the main program first, followed by the subprograms in any order. I would suggest that they be ordered in relation to the sequence that they are called. That is PLOT calls MINMAX and should precede it in the listing.

The result of the calculation is

```
AVERAGE NUMBER OF CARS/MINUTE = 53.403
PROBABILITY OF THIS MANY CARS/MIN = 0.05463
PROBABILITY OF ZERO CARS/MIN = 0.64E-23
```

9.6.3 The SAVE Statement

Occasionally, when a subprogram is called more than once, we may wish to make use of a *locally* defined variable that was computed in an earlier call.

Figure 9-11
The Fortran
code for the func-
tion in Equation
(9.7).

```
          FUNCTION TILT(X)
*
*                      Be careful with X = 0. Since
*                      0.**K is zero, set TILT(0.) = -0.2.
*
          IF(X .EQ. 0.)THEN
               TILT = -0.2
               RETURN
          END IF
*
          K = 60
*
*                      The sum begins at K = 60, we need
*                      to compute 60-factorial to start.
*
          FACT60 = 1.
          DO 1 I = 1,K
             FACT60 = FACT60 * I
       1  CONTINUE
*
*                      Sum from K = 60 until terms are
*                      smaller than 1.E-6.
*
          TERM = X**60/FACT60
          SUM = 0.
          DO 2 I = 1,100
             SUM = SUM + TERM
             IF(ABS(TERM) .LT. 1.E-6)THEN
                  TILT = EXP(-X) * SUM - 0.2
                  RETURN
             ELSE
                  RATIO = X/(I + 1.)
                  TERM = TERM * RATIO
             END IF
       2  CONTINUE
*
*         PRINT *,'SUM NOT CONVERGING, X = ',X
          STOP
          END
```

Ordinarily, all values associated with variable names within a subprogram that are not also in the argument list are lost after leaving the subprogram. To preserve the value associated with a variable from one use of a subprogram to the next, the SAVE statement is inserted before the first executable statement in the subprogram. The form of the SAVE statement is

$$\text{SAVE name}_1, \text{name}_2, \ldots$$

```
       PROGRAM CARS
*--
*--    BASED ON THE INFORMATION THAT THE PROBABILITY OF 60 OR MORE
*--    CARS PASSING PER MINUTE IS 20 PERCENT, THE AVERAGE NUMBER OF
*--    CARS PASSING PER MINUTE IS ESTIMATED BY ASSUMING A POISSON
*--    DISTRIBUTION AND SOLVING THE EQUATION
*--
*--                 SUM( P(K,AVG) ) = 0.20
*--
*--    WHERE THE SUM IS FROM 60 TO INFINITY AND AVE IS THE ROOT OF
*--    THE EQUATION. TO AVOID THE INFINITE SUM, THE EQUATION IS RE-
*--    WRITTEN AS
*--
*--        TILT(AVG) = 0.20 - SUM( P(K,AVG) )
*--
*--    WHERE THE SUM IS NOW FROM 0 TO 59.
*-------------------------------------------------------------------
* VARIABLES
*
       REAL A,B,EPS,AVG
       INTEGER IMAX,IAVG
*
*                  A,B      --  LEFT AND RIGHT ENDS OF THE SEARCH
*                               INTERVAL.
*                  EPS      --  CONVERGENCE CRITERION FOR BISEC
*                  IMAX     --  MAXIMUM NO. OF ITERATIONS IN BISEC
*                  AVG      --  THE ROOT OF THE EQ. RETURNED
*                               BY BISEC
*                  IAVG     --  THE ROOT TRUNCATED TO AN INTEGER
*-------------------------------------------------------------------
* DECLARE THE FUNCTION TILT AS EXTERNAL
*
       EXTERNAL TILT
*-------------------------------------------------------------------
* INITIALIZATION
*
       A = 25.
       B = 60.
       IMAX = 50
       EPS = 1.E-6
*-------------------------------------------------------------------
* COMPUTATION
*                  THE ONLY FUNCTION OF THE MAIN PROGRAM IS TO
*                  CALL BISEC TO FIND THE ROOT OF TILT.
```

Continued

```
*
      CALL BISEC(A,B,EPS,IMAX,I,AVG,TILT)
*-----------------------------------------------------------------------
* OUTPUT
*                    AND PRINT THE RESULTS
*
      IAVG = AVG
      WRITE(*,10)AVG
      WRITE(*,11)P(IAVG,AVG)
      WRITE(*,12)P(0,AVG)
*-----------------------------------------------------------------------
* FORMATS
*
   10 FORMAT(10X,'AVERAGE NUMBER OF CARS/MINUTE   = ',F7.3)
   11 FORMAT(10X,'PROBABILITY OF THIS MANY CARS/MIN = ',F7.5)
   12 FORMAT(10X,'PROBABILITY OF ZERO CARS/MINUTE  = ',1E9.2)
      END
*-----------------------------------------------------------------------
      FUNCTION P(K,X)
*--
*--   P(K,X) IS THE POISSON DISTRIBUTION WHICH GIVES THE APPROX.
*--   PROBABILITY THAT K EVENTS OCCUR IN A TIME INTERVAL IF THE
*--   AVERAGE IN THAT INTERVAL IS KNOWN TO BE X.
*-----------------------------------------------------------------------
* VARIABLES
*
      REAL X,FACT,P
      INTEGER K,I
*               X      --   THE AVERAGE NUMBER OF EVENTS/INTERVAL
*               K      --   EVENTS PER TIME INTERVAL
*               FACT   --   K - FACTORIAL
*               I      --   A COUNTER
*-----------------------------------------------------------------------
*
*               THE CASE K = 0 MUST BE HANDLED SEPARATELY
*
      IF(K .EQ. 0)THEN
         P = EXP(-X)
      RETURN
      END IF
*-----------------------------------------------------------------------
      FACT = 1.
      DO 1 I = K,1,-1
         FACT = FACT * I
    1 CONTINUE
      P = X**K * EXP(-X)/FACT
      RETURN
      END
```

```
        SUBROUTINE BISEC(A,B,EPS,IMAX,I,ROOT,F)
              ...            ...
        SEE FIGURE 9-1
        NOTE: A DUMMY FUNCTION NAME HAS BEEN ADDED TO THE
              ASSIGNMENT LIST.
              ...            ...
    END
```

```
    FUNCTION TILT(X)
          ...            ...
        SEE FIGURE 9-11
          ...            ...
    END
```

For example, in the previous section, the function TILT computes 60! every time the function is referenced, which may be hundreds of times. This could be remedied by inserting a flag in the function which is undefined the first time the function is called and is defined and saved after the initial call.

```
    FUNCTION TILT(X)
    SAVE FLAG,FACT60
*                         The variable FLAG has not yet
*                         been defined and it is very
*                         unlikely that it has the value
*                         of 1357.1.
    IF(FLAG .NE. 1357.1)THEN
                          Compute FACT60 = 60!
                          Assign FLAG = 1357.1
    END IF
```

Note that there is never a need to explicitly SAVE variables in the argument list of a subprogram, and attempts to do so will result in compilation time errors. SAVEd variables may, however, appear as actual arguments in the argument list of a subprogram that is called *from* the current subprogram.

If the SAVE statement appears without a list of variables, then *all* the locally defined variables in the subprogram are saved from one call to the next.

9.6.4 A Note on Floating-Point Overflow

If you attempt to execute the program CARS of Figure 9-12, there is a good chance, depending on the word length characteristic of your machine, that the program will fail. The Fortran code itself is perfectly valid; however, the function TILT requires the calculation of very large factorials, (e.g.,

$60! = 8.32 \times 10^{81}$) and these may easily exceed the capacity of a typical computer. The maximum real number on a computer, of course, depends on the word length of the machine and a typical maximum real number might be 9.0×10^{99}. If a computed number exceeds this maximum, the result is an execution time error called *floating-point overflow* and the correction of the problem can often be extremely difficult. Often the only recourse is to rewrite the entire code so as to avoid the problem if possible. This usually requires considerable ingenuity. In the present problem, however, the correction is rather easy. Recall that

$$\text{Tilt}(x) = \sum_{k=60}^{\infty} P_k(x) - 0.2$$

and note that

(9.8)
$$\sum_{k=0}^{\infty} P_k(x) = 1$$

Equation (9.8) is a statement that the sum of the probabilities of all possibilities must be one. (It is a certainty that either zero cars or some cars pass in 1 minute.) Combining these two equations, we obtain

(9.9)
$$\text{Tilt}(x) = \left[\sum_{k=0}^{\infty} P_k(x) - \sum_{k=0}^{59} P_k(x) \right] - 0.2$$

$$= 0.8 - \sum_{k=0}^{59} P_k(x)$$

and thus the function Tilt may be rewritten without any infinite summations. Once again, a careful analysis of the problem before a Fortran code is attempted can frequently reduce enormously the time spent in later patching up a poorly thought out and casually constructed program.

9.6.5 The COMMON Statement

Up to now, the only communication that subroutines and functions have had with each other and with the main program is via the variables that appear in their argument lists. There is one additional mechanism for transferring information, called COMMON blocks. The idea is to reserve special blocks of memory that may be accessed by one or more program units. The form of the statement that assigns variables to these blocks is

```
COMMON/blockname/variable₁, variable₂,...
```

or

```
COMMON variable₁, variable₂,...
```

In the first example, called *labeled* COMMON, a block of memory is assigned a name, "blockname," which is any valid Fortran name and is set off by slashes. The list of variables contained in this block then follows, separated by commas. The variables may be a mix of integers, reals, and arrays, but the list may not contain any function names. Additionally, if any of the variable names in the block is of type CHARACTER, *all* the variable names in the block must be of type CHARACTER.

The second example is called *blank* COMMON and the only difference is that the reserved block of memory has been left unnamed. Examples of COMMON statements follow:

```
COMMON/ABLOCK/X,Y,J,K
COMMON/BBLK/A(50),C,D/CBLC/F(35)    ⟨BBLK is 52 memory words,
                                     followed by block CBLC,
                                     35 memory words.⟩
COMMON/W/W                          ⟨Block W is one word and
                                     contains the variable W.⟩
```

In this text I will use labeled COMMON blocks exclusively.

The rules pertaining to COMMON statements are

1. A pair of slashes is used to separate the name given to a group of variables, all stored together in a block.
2. A COMMON statement is nonexecutable and must appear before the first executable statement.
3. A similar Fortran line, with the *same* block name, must appear in both the program units that share the use of some or all the variables in the block.
4. The assignment of values to variables in the block proceeds in the same manner as for argument lists: i.e., values are assigned by position in the list, not by variable name.
5. If a variable is in a COMMON block, it *cannot* simultaneously be in an argument list. This is because such an arrangement would require two distinct memory addresses for the same variable. For example,

```
        SUBROUTINE PROD(X,Y,N)
        COMMON/AAA/X                    ⟨Incorrect⟩
```

6. A DIMENSION declaration may be combined with COMMON statements, for example,

```
        DIMENSION X(50),Y(25)
        COMMON/CCCC/X,Y
```

has the same effect as

COMMON/CCCC/X(50),Y(25)

You will have no difficulty in understanding COMMON statements if you recognize that they are simply a replacement for an argument list. Thus,

```
PROGRAM MANE                            PROGRAM MANE
READ *,D,E,F                            COMMON/COEF/D,E,F
Z = 2,                                  READ *,D,E,F
T = F(D,E,F,Z)                          Z = 2,
                                        T = F(Z)
STOP                                    STOP
END                                     END

FUNCTION F(A,B,C,X)         and         FUNCTION F(X)
F = A * X**2 + B * X + C                 COMMON/COEF/A,B,C
RETURN                                  F = A * X**2 + B * X + C
END                                     RETURN
                                        END
```

are essentially interchangeable.

COMMON blocks are most often employed in two very frequently occurring situations in Fortran:

Eliminating Long Argument Lists: If a subprogram is referenced many times and if the argument list of the subprogram is quite long, it is very easy to make errors. The order of the variables or their type may be entered incorrectly, or some accidentally omitted. In these cases it is suggested that most or all of the variables be passed through a COMMON block. This also makes the program somewhat easier to read.

Matching the Argument List to a Dummy Function: Frequently a subprogram module will make reference to a function of a *single* variable, F(X), while the particular function in question requires several parameters in addition to X. The only recourse is to pass the additional parameters through a COMMON block. Thus if we wished to PLOT the function F above, the first representation would not be suitable.

PROBLEMS

1. Write a short program segment that reads two positive numbers corresponding to the x, y coordinates of a point in the first quadrant and by using subroutine LENGTH of Section 9.3.1 determines whether the point lies between the two circles of radii 1. and 3.

2. Properties of subroutines:
   a. Write a subroutine that does something useful and has no argument list at all.
   b. A subroutine will compile without a RETURN statement. Construct a subroutine that again does something useful and does not have a RETURN statement.
   c. Is it possible to do the following from a subroutine?
      i. Stop
      ii. Call a different subroutine
      iii. Call a different subroutine and from the second subroutine return directly to the main program
      iv. Print output
      v. Read data
3. Redo the above problem applied to function subprograms.
4. Write a subroutine that takes two real variables (A, B) and returns the values interchanged—i.e. (B, A).
5. Write the code for a subroutine that evaluates the "product" of two arrays A(N,N) and B(N,N) defined by the equation

$$C_{ij} = \sum_{k=1}^{n} A_{ik} B_{kj} \qquad \text{for all } i, j = 1 \text{ to } n$$

The arrays have been dimensioned in the main program to be N by N, but the subroutine should accommodate any square array of size $M \leq N$.
6. Rewrite the code for a pointer sort found in Chapter 8 as a subroutine which is to take a list of N items stored in an array A(N) and return the sequencing index array INDX(N).
7. University records for all students are stored on a file called STUDAT as follows:

	Format	
Last name (begins in col. 1)		(A10)
First and middle initial		(A2)
Student ID		(I9)
Class (1-freshman, 4-senior)		(I1)
Sex (M or F)		(A1)
SAT(Verbal)		(I3)
SAT(Math)		(I3)
College(A+S = 1, Engr = 2, Bus = 3,		
Educ = 4)		(I1)
Current GPA		(F4.2)
Last semester GPA		(F4.2)

   a. Write a subroutine called INPUT which will read the file, store and return the information in appropriately named arrays, and prints the total number of students. The name of the file should be passed to the subroutine through its argument list. You can assume that there are less than 8000 students in the university.

b. Write a main program that then:

    i. Counts the number of students in each college.

    ii. Counts the number of females in each college with SAT(Math) $\geq$ 600 and prints the result.

    iii. Produces an alphabetical list of all students and prints the list including all the related information. You can assume you have available a subroutine to execute a pointer sort that returns the array INDX.

    iv. Produce an alphabetical list of males in A+S college who expect to graduate next year but who are currently on probation (current GPA $\leq$ 1.8). The quickest way to proceed is to find all such students, copy the list into a separate array (including a separate indexing array) and then call for a pointer sort.

8. A rather famous infinite series for early computations of π is

$$\frac{\pi}{6} = \frac{1}{2} + \frac{1}{2} \cdot \frac{1}{3 \cdot 2^3} + \frac{1 \cdot 3}{2 \cdot 4} \cdot \frac{1}{5 \cdot 2^5} + \frac{1 \cdot 3 \cdot 5}{2 \cdot 4 \cdot 6} \cdot \frac{1}{7} \cdot \frac{1}{2^7} + \cdots$$

This series converges quite rapidly. Determine an expression for the general term in the series and for the ratio of successive terms. Write a program to add the first P terms as *fractions* reduced to lowest terms to obtain an accurate fractional approximation to π.

9. The two linear equations in two unknowns x, y

$$ax + by = e$$
$$cx + dy = f$$

are easily solved by writing the first as

$$x = \frac{e - by}{a}$$

and substituting this into the second equation and solving for y. The result is

$$x = \frac{ed - fb}{ad - bc} \qquad y = \frac{af - ec}{ad - bc}$$

Write a program that will read *fractions* for the constants a, b, c, d, e, f (i.e., IA(1), IA(2), the numerator and denominator of a, etc.) and will obtain the solutions for x and y expressed as a fraction in lowest terms. Use FUNCTION IGCF and write two subroutines MINUS and MULTP to subtract and multiply two arbitrary fractions and return the result as a fraction.

10. Write a subroutine CHANGE that will determine the appropriate change to be returned when an amount PAY is submitted for an item that has a price COST. Assume that PAY is less than $100 and that there are no $2 bills. The subroutine should return

$$(I10D,15D,I1D,I25C,I10C,I5C,I1C)$$

where I10D is the number of $10 bills, I25C is the number of $0.25 coins, etc. to be returned. The subroutine should, of course, minimize the amount of small change returned.

11. Rewrite function ROUNDR to round an arbitrary real number *down*. Thus if $Q = 0.16731$, then ROUNDR(Q) = 0.15.

12. Rewrite the PLOT program (Figure 9-8) as a subroutine that will graph an arbitrary function FNC(X) in 50 steps from X = A to X = B.

13. Find any errors in the following:

    a. SUBROUTINE AB(X,I + 1,EPS,ANSWER)
    b. CALL CD(X,I + 1,EPS,ANS)
    c. FUNCTION EF(X,A(12),I)
    d. Z = GH(Y,A(3),K)
    e. FUNCTION SUM(X,Y)
      Z = X + Y
      RETURN
      END
    f. SUBROUTINE DIFF(X,Y)
      DIFF = X - Y
      RETURN
      END

14. All of the student data is contained on the file described in Problem 9.7. Also, the pointer array INDX(I) has been determined that arranges all the names in alphabetical order (i.e., NAME(INDX(1)) is the first name in the alphabetized list). Write a subroutine that (a) reads a complete set of information for a new student and adds it to the end of the data file; (you will have to rewrite the entire file to a new data file); (b) puts the new student's name in the correct alphabetical position. (You only need change the array INDX.)

15. Polynomial evaluation
    a. Write a function to evaluate a polynomial POLY(X,N,C) for a particular value of x. The polynomial is of degree n and the coefficients are supplied in c_n. Thus

    $$\text{Poly}(x) = c_n x^n + c_{n-1} x^{n-1} + \cdots + c_2 x^2 + c_1 x^1 + c_0$$

    $$= \sum_{i=0}^{n} c_i x^i$$

    b. Rewrite the function POLY so that *no exponentiation* is employed. For example,

$$c_2 x^2 + c_1 x + c_0 = [(c_2 x + c_1)x + c_0]$$

For n large, this method of evaluating a polynomial is far more efficient.

16. The combinatorial function is

$$C(n, p) = \frac{n!}{(n - p)! \, p!}$$

Write a function subprogram to compute all the $C(n, i)$'s for $i = 1$ to p. Note: Your function should *not* compute any factorials. Instead, express $C(n,p)$ as an extended product, where

$$C(n, i + 1) = (\text{Ratio}) \times [C(n, i)] \qquad \text{for } i = 1, p$$

Start with $C(n, 1) = n$, compute $C(n, 2)$, etc. up to $C(n, p)$. Surprisingly it is about as efficient to compute all $C(n, i)$ in this manner as it is to compute a single combinatorial using factorials.

17. Write a function subprogram that returns the cube root of a real number x. (**Note:** x can be negative.)

18. A vector $\mathbf{a} = [a_x, a_y, a_z]$ can be represented by an array A(I), I = 1,3; i.e., $[a_1, a_2, a_3]$. The dot product of two vectors is a number defined by

$$\mathbf{a} \cdot \mathbf{b} = a_1 b_1 + a_2 b_2 + a_3 b_3$$

And the cross product is a *vector* whose components are defined by

$$\mathbf{c} = \mathbf{a} \times \mathbf{b}$$
$$c_1 = a_2 b_3 - a_3 b_2$$
$$c_2 = a_3 b_1 - a_1 b_3$$
$$c_3 = a_1 b_2 - a_2 b_1$$

Write a function for the dot product and a subroutine for the cross product of two vectors. Write a main program that tests whether the identity

$$(\mathbf{a} \times \mathbf{b}) \cdot (\mathbf{c} \times \mathbf{d}) = (\mathbf{a} \cdot \mathbf{c})(\mathbf{b} \cdot \mathbf{d}) - (\mathbf{a} \cdot \mathbf{d})(\mathbf{b} \cdot \mathbf{c})$$

is correct for values of $\mathbf{a}, \mathbf{b}, \mathbf{c}, \mathbf{d}$ that are read in.

19. A point on a sphere can be characterized by two angles, the polar angle α, and the azimuthal angle β as shown on Figure 9-13. These angles are related to the longitude and latitude as follows:

$$\alpha = 90° - \text{latitude (if north)}$$
$$\alpha = 90° + \text{latitude (if south)}$$
$$\beta = \text{longitude (if west)}$$
$$\beta = 360° - \text{longitude (if east)}$$

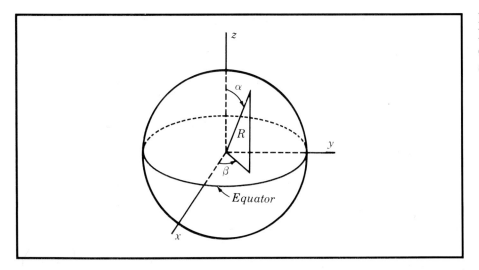

Figure 9-13
Polar coordinates
of a point on a
sphere.

Furthermore, if the angular separation between two points on the sphere is θ_{12}, then the expression for the cosine of this angle is

$$\cos(\theta_{12}) = \cos(\alpha_1)\cos(\alpha_2) + \sin(\alpha_1)\sin(\alpha_2)\cos(\beta_2 - \beta_1)$$

and once the angle θ_{12} is determined, the surface distance between the two points is given by $d = \theta_{12}R$, where R is the radius of the sphere.

a. Write a subroutine subprogram that will convert the coordinates of a point on the earth given in terms of longitude and latitude in degrees into coordinates in terms of the angles (α, β) in radians. The input angles should be in the form:

$$\text{Latitude:} \quad \text{if north (e.g., XX°XX'N) use} \rightarrow +\text{XX.XX}$$
$$\text{south (e.g., XX°XX'S) use} \rightarrow -\text{XX.XX}$$
$$\text{Longitude:} \quad \text{if west (e.g., XX°XX'W) use} \rightarrow +\text{XX.XX}$$
$$\text{east (e.g., XX°XX'E) use} \rightarrow -\text{XX.XX}$$

b. Write a function subprogram that will return the surface distance between two points on the earth whose coordinates are given in terms of the pairs of angles (α_1, β_1), (α_2, β_2). The radius of the earth is 3958.89 miles.

c. Use these subprograms to determine which two cities in the following list have the largest surface separation.

	Latitude	Longitude
Chicago	41°49'N	87°37'W
Los Angeles	35°12'N	118°02'W
Montreal	45°30'N	73°35'W

	Latitude	Longitude
London	51°30′N	0°07′W
Rio de Janeiro	22°50′S	43°20′W
Melbourne	35°52′S	145°08′E
Vladivostok	43°06′N	131°47′E
Johannesburg	26°08′S	27°54′E

20. True or false? Explain!

 a. The name of a COMMON block must not be the same as that of any variable in the list.

 b. To use a COMMON block in a subroutine, the block must also appear in the main program.

 c. Two COMMON blocks can appear on the same line, i.e., COMMON/AA/X,Y,Z/GG/D,E,F

 d. The variables X, Y in the two COMMON blocks

```
COMMON/SS/W(3:12),T,II,Y,R(7)
COMMON/SS/U(-1:5),MM(6),X,SS,B,C,D,E,F
```

have the same value.

 e. All variables in a COMMON block must be of the same type.

21. You have at your disposal a subroutine

```
SUBROUTINE ROOT(A,ANSWER,F)
```

that will find a root of a function of a single variable $f(x)$ when an initial guess (A) is given for the root. The root is returned as ANSWER. You also have a function subprogram,

```
FUNCTION XINTGL(A,B,FNC)
```

which computes the definite integral of fnc(t) from $t = A$ to $t = B$; i.e.,

$$XINTGL(A,B,FNC) = \int_a^b fnc(t) \, dt$$

(Don't panic—You do not need to understand integration to do this problem.) Write a program that will find the root of the equation

$$g(x) = 5e^{-2x^2} - \sin\left(\frac{\pi x}{2}\right) + \int_0^x (4t^2 - 5)e^{-t^2} \, dt$$

starting with an initial guess of $x_1 = 1.0$. (Notice that the variable is x and it appears in the limits of the integral.)

22. Consider the following program:

```
PROGRAM XXX
COMMON/B/B(9,9),D,E,F
READ *,B                    Reads 81 zeros
READ *,D,E,F                Reads 0.0, 4.0, 6.0
E = E/2, - 1,
D = -G(E)
F = B(9,9)
PRINT *,D,E,F
STOP
END
FUNCTION G(B)
COMMON/B/D(8,8),E(20)
E(17) = 9,
F = B - E(20)
G = 7,*F
B = -7,
RETURN
END
```

What values are printed?

PROGRAMMING ASSIGNMENT VI

VI.1   SAMPLE PROGRAM

Aeronautical/Aerospace Engineering

Among the youngest of the engineering disciplines, aeronautical/aerospace engineering is concerned with all aspects of the design, production, testing, and utilizing of vehicles or devices that fly in air (aeronautical) or in space (aerospace); from hang gliders to the space shuttle. Since the science and engineering principles involved are so broad-based, the aeroengineer will usually specialize in a subarea which may overlap with other engineering fields such as mechanical, metallurgical/materials, chemical, civil, or electrical engineering. Such subareas are

Aerodynamics: The study of the flight characteristics of various structures or configurations. Typical considerations are the drag and lift associated with airplane design, or the onset of turbulent flow. A knowledge of fluid dynamics is essential. Additionally, the modeling and testing of all forms of aircraft is part of this discipline.

Structural Design: The design, production, and testing of aircraft and spacecraft to withstand the wide range of inflight demands expected of these vehicles. In addition, similar problems involving other types of vehicles, such as underwater vessels, are in the province of the structural engineer.

Propulsion Systems: The design of internal combustion, jet, and liquid- and solid-fuel rocket engines and their coordination in the overall design of the vehicle. Rocket engines, especially, require innovative engineering to accommodate the extreme temperatures of storing, mixing, and burning fuels such as liquid oxygen.

336

Instrumentation and Guidance: The aerospace industry has been a leader in developing and utilizing solid-state electronics in the form of microprocessors to monitor and adjust the operations of hundreds of air- and spacecraft functions. This field makes use of the expertise of both electrical and aeroengineers.

Navigation: The computation of orbits within and outside the atmosphere, and the determination of the orientation of a vehicle with respect to points on the earth or in space.

Sample Program: The Range of a Rocket Trajectory

The trajectory that maximizes the range of a rocket is obviously a topic of importance to the aerospace engineer. In general the problem is quite complex, but, as always, by making some simplifying assumptions, we can reduce the problem to one that is amenable to a computer solution and use the results to suggest the desirable characteristics of rocket engines.

The major assumptions we will make are (1) to neglect air resistance, (2) to assume that the earth is flat (or equivalently that the trajectory is of short range), and (3) that the inclination of the thrust of the rocket is held fixed with respect to the horizontal. Furthermore, we wish to compare two types of rocket engines:

Constant Thrust: Characterized by rapid burn rates, usually solid fuel, not suitable for manned flight, as the acceleration on board increases with time without limit. It is called constant thrust as the output of the engine is constant until the fuel is expended.

Constant Acceleration: Characterized by slower burn rates and use of liquid fuels. The ratio of the thrust output of the engine to the current total mass (spacecraft plus fuel) is held constant. As the rocket burns fuel and becomes lighter, the thrust is correspondingly reduced.

The force diagram for the rocket in flight is shown Figure VI-1. In the figure,

T = thrust of the engine (N)

θ = fixed angle of inclination of T with respect to the horizontal

m = current mass of the spacecraft plus unexpended fuel

g = gravitational acceleration (m/sec^2) assumed constant.

Newton's equations ($F = ma$) can be written down and solved for this problem, and for strictly vertical flight ($\theta = \pi/2$), this problem is often

---

Figure VI-1
Force diagram for
a rocket in flight.

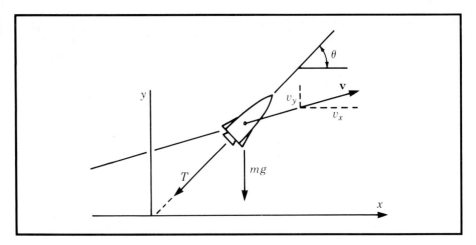

analyzed in introductory physics. For motion in both the x and y directions, the analysis is more complicated but straightforward.[1]

The results depend on the following parameters:

v_e = exhaust velocity of the propellant relative to the spaceship

g = gravitational acceleration (value at the earth's surface; i.e., 9.8 m/sec^2)

M_0 = original mass (including fuel) of the rocket

Ω = ratio of the mass of the fuel to the total mass

β = burn rate of the rocket engines (kg/sec)

b = ratio of the burn rate to the original mass (β/M_0)

θ = angle the thrust makes with the horizontal

t_c = burn time for the engines, time to use up all the fuel

t_{hit} = total flight time, the time that the rocket hits the earth's surface

$\dfrac{x(t)}{y(t)}$ = the x and y positions of the rocket as a function of time

x_c, y_c = the x and y positions at the time of cutoff of the engines.

The properties of the trajectories associated with each type of rocket engine are given in Table VI-1.

In this problem you will compute a table of values of $x(t)$, $y(t)$, for $t = 0$ to t_{hit} for each of the two rocket characteristics given in Table VI-2. If you have available to you a pen-and-ink graphics device, you should also plot $y(t)$

---

[1] See Angelo Miele, *Flight Mechanics,* vol. 1, Addison-Wesley, Reading, Mass., 1962.

Constant thrust	Constant acceleration
Before Cutoff	

$$v_x(t) = [v_e \cos(\theta)][-\ln(1 - bt)] \qquad\qquad = [v_e \cos(\theta)]bt$$
$$v_y(t) = [v_e \sin(\theta)][-\ln(1 - bt)] - gt \qquad = [v_e \sin(\theta)]bt - gt$$
$$x(t) = [v_e \cos(\theta)]f(b, t) \qquad\qquad\qquad = [v_e \cos(\theta)](bt^2)/2$$
$$y(t) = [v_e \sin(\theta)]f(b, t) - gt^2/2 \qquad\quad = [v_e \sin(\theta)]bt^2/2 - gt^2/2$$

where

$$f(b, t) = t + \frac{1}{b}(1 - bt)\ln(1 - bt)$$

The cutoff occurs at a time t_c defined by

$$bt_c = \Omega \qquad\qquad\qquad\qquad = -\ln(1 - \Omega)$$

After Cutoff

$$x(t) = x_c + v_x(t_c)t \qquad\qquad = x_c + v_x(t_c)t$$
$$y(t) = y_c + v_y(t_c)t - gt^2/2 \qquad = y_c + v_y(t_c)t - gt^2/2$$

where x_c, y_c, $v_x(t_c)$, $v_y(t_c)$ are the position and velocity at cutoff.
The optimum angle θ is determined from the root of

$$Q\sigma^3 - (\sigma^2 - \tfrac{1}{2}) = 0$$

where $\sigma = \sin(\theta)$, and

$$Q = [-\Omega + \ln(1 - \Omega)[\ln(1 - \Omega)]^2]\frac{g}{bv_e} \qquad = \frac{1}{2}\frac{g}{bv_e}$$

The rocket hits the ground at $t = t_{\text{hit}}$, the positive root of

$$y(t = t_{\text{hit}}) = 0$$

Table VI-1
Trajectory characteristics for two rocket types.

vs. $x(t)$ for both rockets simultaneously and indicate the point of engine cutoff for each engine on the graph.

Problem Specifics

Your program should

1. Read the rocket engine parameters from Table VI-2.
2. Determine the value of Q for each engine.
3. Solve the cubic equation for $\sigma = \sin(\theta)$ to obtain the optimum angle for maximum range. Do not use the bisection algorithm for this simple equa-

	v_e [exhaust velocity (m/sec²)]	M_0 [original total mass (kg)]	Ω [fuel to total mass ratio]	β [burn rate (kg/sec)]
Constant thrust	5000	30,000	0.6	550
Constant acceleration	3000	65,000	0.5	900

Table VI-2
Characteristic parameters for two types of rocket engines.

tion. Instead, write a function subprogram that starts with a guess for σ ($\frac{1}{2} < \sigma < 1$.) and employs successive substitutions to obtain the root. That is, first write the cubic as

$$\sigma = (Q\sigma^3 - \tfrac{1}{2})^{1/2}$$

Then use your guess for σ in the expression on the right and compute a new value for σ. Repeat this until successive values change by less than 1.E−5%. The function should contain appropriate error diagnostics.

4. Compute the cutoff time for each engine.
5. Compute the positions and velocities of the rockets at the time of cutoff.
6. Solve the quadratic equation for $y(t)$ to determine the value of t_{hit}. Also use this value to determine the range (i.e., $x(t = t_{hit})$ for each engine. Use a subroutine.
7. Using the larger of the t_{hit} values, divide the time interval into fifty steps and compute a table of $x(t)$, $y(t)$ values. (Note: If the computed value of y is negative, replace with zero.)
8. Print all the results in a neat table.
9. If you have access to an automatic graphics plotting device, graph both trajectories on the same graph and indicate on the graph the position at time of cutoff for each rocket.

Your program should have the following subprograms:

FUNCTION CUBRT(Q,X)	*Solves the cubic equation by successive substitution. The maximum number of iterations should be 50.*
SUBROUTINE QUAD(THIT)	*Returns the positive root of the equation $y(t) = 0$. The rest of the parameters should be passed through labeled COMMON.*
FUNCTION X(T,TC,TYPE)	*Computes $x(t)$ for $t \le t_{cutoff}$, or for $t >$ t_{cutoff} and for the two types of engines [TYPE = 'CT' (constant thrust), or 'CA' (constant acceleration)]. Also similar functions for $y(t)$, and the velocity components.*
FUNCTION F(BT)	*This function will be used by $x(t < t_c)$ and $y(t < t_c)$ for type 'CT'.*

Whenever possible, use the argument list for independent variables and common blocks for parameters.

The solution for this problem is given in Figures VI-2 and VI-3.

Figure VI-2 The Fortran code for the rocket trajectory program.

```fortran
      PROGRAM ROCKET
*--
*--   THIS PROGRAM WILL COMPUTE THE FEATURES OF THE TRAJECTORIES
*--   OF TWO TYPES OF ROCKET ENGINES: 1-CONSTANT THRUST,2-CONSTANT
*--   ACCELERATION. THE ROCKET CUTOFF TIME AND POSITION ARE COM-
*--   PUTED AND THE OVERALL RANGE OF EACH ROCKET IS DETERMINED.
*--   ADDITIONALLY THE COMPLETE TRAJECTORIES OF BOTH ROCKETS ARE
*--   DETERMINED AND PLOTTED ON AN AUTOMATIC PEN AND INK PLOTTING
*--   DEVICE. THE MAJOR ASSUMPTIONS MADE WERE THAT THE INCLINATION
*--   ANGLE OF THE ROCKET TO THE HORIZONTAL WAS FIXED AND THAT AIR
*--   RESISTANCE COULD BE NEGLECTED.
*----------------------------------------------------------------
* VARIABLES
*
      REAL VEX,MZERO,FRATIO,BURNRT,OPANGL(2),TCUT,
     +     Q,SIGMA,LN,OMEGA,VXCUT,VYCUT,XCUT,YCUT,
     +     G,THIT(2),TLIMIT,B,RANGE(2),VCUT
      REAL X(0:50,2),Y(0:50,2),X1(51),X2(51)
      INTEGER TYPE
      CHARACTER NAME*12
      COMMON/PARAM/B(2),VEX(2),VCUT(2),VCUT(2),XCUT(2),YCUT(2),TCUT(2),G,
     +     VXCUT(2),VYCUT(2)
*
* INPUT PARAMETERS
*
*        VEX()    --   VELOCITY OF EXHAUST FOR EACH
*                      ROCKET TYPE
*        MZERO    --   ORIGINAL ROCKET MASS INCL. FUEL
*        FRATIO   --   RATIO OF FUEL TO TOTAL MASS
*                      FOR EACH TYPE ENGINE
*        BURNRT   --   BURN RATE OF EACH ROCKET ENGINE
*        NAME     --   IDENTIFYING NAME OF ENGINE TYPE
*
* INTERMEDIATE COMPUTED QUANTITIES
*
*        B()      --   RATIO OF BURN RATE TO MZERO
*        Q        --   PARAMETER IN EQ FOR OPTIMUM
*                      ANGLE
*        SIGMA    --   SINE OF OPTIMUM ANGLE
*        LN       --   LOG(1 - FRATIO)
*        FACTR    --   TEMPORARY CONSTANTS USED
*        VFACTR        IN EQUATIONS
*        XFACTR
*        DISCR
*        T, DT    --   TIME AND TIME STEP
*        TYPE     --   EQUALS 1 OR 2 FOR TWO TYPES OF
*                      ROCKET ENGINES
*
* COMPUTED RESULTS
*
*                      COMPUTE POSITION AND VELOCITY AT CUTOFF
*
      VXCUT(TYPE) = VEX(TYPE)*COS(OPANGL(TYPE))*VFACTR
      VYCUT(TYPE) = VEX(TYPE)*SIN(OPANGL(TYPE))*VFACTR -
     +              G*TCUT(TYPE)
      VCUT(TYPE) = SQRT(VXCUT(TYPE)**2 + VYCUT(TYPE)**2)
      XCUT(TYPE) = VEX(TYPE)*COS(OPANGL(TYPE))*XFACTR
      YCUT(TYPE) = VEX(TYPE)*SIN(OPANGL(TYPE))*XFACTR -
     +             G*(TCUT(TYPE))**2/2.
*
*                      COMPUTE TIME OF HIT AND RANGE
*
      DISCR = VYCUT(TYPE)**2 + 2.*G*YCUT(TYPE)
      THIT(TYPE) = (VYCUT(TYPE) + SQRT(DISCR))/G
      RANGE(TYPE) = DIS(TYPE,THIT(TYPE),1,OPANGL(TYPE))
    1 CONTINUE
*
*                      PRINT THE RESULTS FOR BOTH ROCKET ENGINES
*                      (THE OPTIMUM ANGLE IS PRINTED IN DEGREES)
*
      WRITE(*,13)(OPANGL(K)*360./2./PI,TCUT(K),XCUT(K),
     +   YCUT(K),VCUT(K),THIT(K),RANGE(K),K = 1,2)
*
*                      COMPUTE THE TWO ROCKET TRAJECTORIES IN 50
*                      STEPS FROM T = 0. TO T = TLIMIT (THE LARGER
*                      OF THE TWO FLIGHT TIMES)
*
      IF(THIT(1) .GT. THIT(2))THEN
         TLIMIT = THIT(1)
      ELSE
         TLIMIT = THIT(2)
      END IF
      DT = TLIMIT/50.
*
*                      THE POSITIONS ARE COMPUTED IN THE FUNCTION
*                      DIS(), 1 FOR X, 2 FOR Y, IF A COMPUTED
*                      VALUE FOR Y IS NEGATIVE, SET Y = 0.0.
*
      T = 0.0
*
      DO 3 I = 0,50
         DO 2 TYPE = 1,2
            X(I,TYPE) = DIS(TYPE,T,1,OPANGL(TYPE))
            Y(I,TYPE) = DIS(TYPE,T,2,OPANGL(TYPE))
            IF(Y(I,TYPE) .LT. 0.)THEN
               Y(I,TYPE) = 0.0
            END IF
```

Continued

```fortran
*         OPANGL( ) --    ANGLE THE ENGINE MAKES WITH THE
*                         HORIZONTAL (COMPUTED FOR MAXIMUM
*                         RANGE)
*         VXCUT( ) --     POSITIONS AND VELOCITIES AT THE
*         VYCUT( ) --     TIME OF CUTOFF
*         VCUT( )
*         XCUT( )
*         YCUT( )
*         TCUT( )  --     TIME OF CUTOFF
*         THIT( )  --     TIME ROCKET STRIKES THE EARTH
*         TLIMIT   --     THE LARGER OF THE TWO THIT'S
*         RANGE( ) --     THE RANGE OF EACH TRAJECTORY
*         X( ) Y( ) --    THE COORDINATES OF EACH
*                         TRAJECTORY
*-----------------------------------------------------
* INITIALIZATION AND COMPUTATION
*
      G = 9.8
      PI = ACOS(-1.)
      OPEN(UNIT = 1,FILE = 'ROKDAT')
      REWIND 1
      WRITE(*,11)
      DO 1 TYPE = 1,2
      READ(1,10)VEX(TYPE),MZERO,FRATIO,BURNRT,NAME
      WRITE(*,12)NAME,VEX(TYPE),MZERO,FRATIO,BURNRT
      LN = LOG(1. - FRATIO)
      B(TYPE) = BURNRT/MZERO
*
*         COMPUTE THE OPTIMUM THRUST ANGLE
*
      IF(TYPE .EQ. 1)THEN
         FACTOR = -(FRATIO + LN)/LN**2
      ELSE
         FACTOR = 0.5
      END IF
*
      Q = FACTOR * G/B(TYPE)/VEX(TYPE)
      SIGMA = CUBRT(Q,.75)
      OPANGL(TYPE) = ASIN(SIGMA)
*
*         COMPUTE THE TIME OF CUTOFF
*
      IF(TYPE .EQ. 1)THEN
         TCUT(TYPE) = FRATIO/B(TYPE)
         VFACTR = -LOG(1. - B(TYPE)*TCUT(TYPE))
         XFACTR = F(B(TYPE),TCUT(TYPE))
      ELSE
         TCUT(TYPE) = -LN/B(TYPE)
         VFACTR = B(TYPE) * TCUT(TYPE)
         XFACTR = 0.5 * B(TYPE) * (TCUT(TYPE)**2)
      END IF


  2   CONTINUE
      T = T + DT
  3   CONTINUE
*
*         PRINT THE COORDINATES OF BOTH TRAJECTORIES
*
      WRITE(*,14)
      DO 6 K = 0,50
         WRITE(*,15)(X(K,TYPE),Y(K,TYPE),TYPE = 1,2)
  6   CONTINUE
*
* ***************************************************
* INSERT HERE THE CODE NECESSARY TO GRAPH THE
* TRAJECTORIES OF AN AUTOMATIC PEN AND INK PLOTTER.
* ***************************************************
*
      STOP
* FORMATS
*
 10   FORMAT(2F6.0,F3.0,F4.0,A12)
*
 11   FORMAT(10X, 'A COMPARISON OF THE PERFORMANCES',//,
     +       10X,'OF TWO ROCKET ENGINES',///,
     +        5X,'INPUT PARAMETERS',//,
     +       32X,'ORIGINAL   FUEL TO',//,
     +       22X,'EXHAUST    TOTAL     TOTAL     BURN',/,
     +       21X,'VELOCITY    MASS       MASS      RATE',//,
     +       23X,'(M/S)      (KG)      RATIO   (KG/S)')
*
 12   FORMAT(7X,'CONSTANT',//,7X,A12,T23,F8.2,F8.2,
     +       4X,F4.1,3X,F8.2)
*
 13   FORMAT(///,10X,'RESULTS OF THE CALCULATION',////,
     +   T29,'X',T39,'Y',//,
     +   4X,'OPTIMUM    TIME      POSITION   POSITION   VELOCITY',
     +   ' TIME         RANGE',//,
     +   4X,' THRUST       OF              AT         AT        AT     ',
     +   '       OF       OF',/,
     +   4X,' ANGLE     CUTOFF             CUTOFF     CUTOFF    CUTOFF  ',
     +   '     FLIGHT    FLIGHT',//,
     +   4X,' (DEG)     (SEC)             (M)        (M)       (M/S)   ',
     +   '     (S)       (M)',//,
     +   1X,'CT',4X,F6.2,6(1X,E9.2),//,
     +   1X,'CA',4X,F6.2,6(1X,E9.2))
*
 14   FORMAT(//,5X,'THE TRAJECTORIES OF THE TWO ROCKETS',//,
     +   T10,'CONSTANT',T45,'CONSTANT',/,
     +   T10,'THRUST',T45,'ACCELERATION',//,
     +   T9,'X',T20,'Y',T49,'X',T60,'Y',//)
*
 15   FORMATS(T5,2(1X,E9.2,1X),T45,2(1X,E9.2,1X))
      END
```

```
*-----------------------------------------------------------------
      FUNCTION DIS(I,T,IXY,THETA)
*-----------------------------------------------------------------
*--  THIS FUNCTION WILL COMPUTE THE X (IXY = 1) OR Y (IXY = 2)
*--  COORDINATES OF THE ROCKET'S TRAJECTORY FOR THE CASES OF
*--          T   <    T-CUTOFF
*--          T   >    T-CUTOFF
*--          I   =    ROCKET TYPE 1 OR 2
*-----------------------------------------------------------------
      COMMON/PARAM/B(2),VEX(2),VCUT(2),XCUT(2),YCUT(2),
     +    TCUT(2),G,VXCUT(2),VYCUT(2)

      IF(T .LT. TCUT(I))THEN

      IF(I .EQ. 1)THEN
         FACTR = F(B(I),T)
      ELSE
         FACTR = B(I) * T**2/2.
      END IF

      FACTR = FACTR*VEX(I)
      IF(IXY .EQ. 1)THEN
         DIS = FACTR*COS(THETA)
         RETURN
      ELSE
         DIS = FACTR*SIN(THETA) - G*T**2/2.
         RETURN
      END IF

      ELSE
      IF(IXY .EQ. 1)THEN
         DIS = VXCUT(I)*T + XCUT(I)
         RETURN
      ELSE
         DIS = VYCUT(I)*T + YCUT(I) - G*T**2/2.
         RETURN
      END IF

      END IF
      END IF
*
      END
*-----------------------------------------------------------------
      FUNCTION F(B,T)
*-----------------------------------------------------------------
*--  THIS FUNCTION IS USED IN COMPUTING THE TRAJECTORIES OF A
*--  CONSTANT THRUST ROCKET.
*-----------------------------------------------------------------

      F = T + (1. - B * T) * LOG(1. - B*T)/B
      RETURN
      END

*-----------------------------------------------------------------
      FUNCTION CUBRT(Q,X)
*-----------------------------------------------------------------
*--  CUBRT SOLVES FOR THE ROOT OF THE EQUATION
*--
*--             X = SQRT(Q * X**3 + .5)
*--
*--  BY SUCCESSIVE SUBSTITIONS
*-----------------------------------------------------------------

      I = 0
    1 TEST = SQRT(Q * X**3 + 0.5)
      I = I + 1
      IF(I .GT. 50)THEN
         PRINT *,'EXCESSIVE ITERATIONS IN CUBRT (MORE THAN 50)'
         PRINT *,'RUN TERMINATED'
         STOP
      ELSE IF(ABS(X - TEST) .GT. 1.E-5)THEN
         X = TEST
         GO TO 1
      ELSE
         CUBRT = TEST
         RETURN
      END IF
      END
```

Figure VI-3
The output from
the rocket
trajectory
program.

A COMPARISON OF THE PERFORMANCES
OF TWO ROCKET ENGINES

INPUT PARAMETERS

	EXHAUST VELOCITY (M/S)	ORIGINAL TOTAL MASS (KG)	FUEL TO TOTAL MASS RATIO	BURN RATE (KG/S)
CONSTANT THRUST	5000.00	30000.00	.6	950.00
CONSTANT ACCELERATION	3500.00	65000.00	.5	500.00

RESULTS OF THE CALCULATION

	OPTIMUM THRUST ANGLE (DEG)	TIME OF CUTOFF (SEC)	X POSITION AT CUTOFF (M)	Y POSITION AT CUTOFF (M)	VELOCITY AT CUTOFF (M/S)	TIME OF FLIGHT (S)	RANGE OF FLIGHT (M)
CT	45.48	.19E+02	.26E+05	.25E+05	.45E+04	.64E+03	.21E+07
CA	49.63	.90E+02	.71E+05	.43E+05	.18E+04	.23E+03	.44E+06
CA OPTIMUM		.90E+02	POSITION	POSITION	VELOCITY	.23E+03	.44E+06

THE TRAJECTORIES OF THE TWO ROCKETS

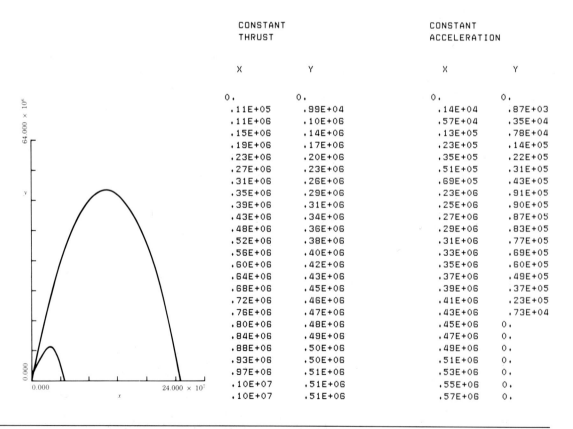

CONSTANT THRUST		CONSTANT ACCELERATION	
X	Y	X	Y
0.	0.	0.	0.
.11E+05	.99E+04	.14E+04	.87E+03
.11E+06	.10E+06	.57E+04	.35E+04
.15E+06	.14E+06	.13E+05	.78E+04
.19E+06	.17E+06	.23E+05	.14E+05
.23E+06	.20E+06	.35E+05	.22E+05
.27E+06	.23E+06	.51E+05	.31E+05
.31E+06	.26E+06	.69E+05	.43E+05
.35E+06	.29E+06	.23E+06	.91E+05
.39E+06	.31E+06	.25E+06	.90E+05
.43E+06	.34E+06	.27E+06	.87E+05
.48E+06	.36E+06	.29E+06	.83E+05
.52E+06	.38E+06	.31E+06	.77E+05
.56E+06	.40E+06	.33E+06	.69E+05
.60E+06	.42E+06	.35E+06	.60E+05
.64E+06	.43E+06	.37E+06	.49E+05
.68E+06	.45E+06	.39E+06	.37E+05
.72E+06	.46E+06	.41E+06	.23E+05
.76E+06	.47E+06	.43E+06	.73E+04
.80E+06	.48E+06	.45E+06	0.
.84E+06	.49E+06	.47E+06	0.
.88E+06	.50E+06	.49E+06	0.
.93E+06	.50E+06	.51E+06	0.
.97E+06	.51E+06	.53E+06	0.
.10E+07	.51E+06	.55E+06	0.
.10E+07	.51E+06	.57E+06	0.

CONSTANT THRUST		CONSTANT ACCELERATION	
.11E+07	.51E+06	.59E+06	0.
.11E+07	.50E+06	.61E+06	0.
.12E+07	.50E+06	.63E+06	0.
.12E+07	.49E+06	.65E+06	0.
.13E+07	.49E+06	.67E+06	0.
.13E+07	.48E+06	.69E+06	0
.13E+07	.47E+06	.71E+06	0.
.14E+07	.45E+06	.73E+06	0.
.14E+07	.44E+06	.75E+06	0.
.15E+07	.42E+06	.77E+06	0.
.15E+07	.41E+06	.79E+06	0.
.15E+07	.39E+06	.81E+06	0.
.16E+07	.37E+06	.83E+06	0.
.16E+07	.35E+06	.85E+06	0.
.17E+07	.32E+06	.87E+06	0.
.17E+07	.30E+06	.89E+06	0.
.17E+07	.27E+06	.91E+06	0.
.18E+07	.24E+06	.93E+06	0.
.18E+07	.21E+06	.95E+06	0.
.19E+07	.18E+06	.97E+06	0.
.19E+07	.15E+06	.99E+06	0.
.19E+07	.11E+06	.10E+07	0.
.20E+07	.77E+05	.10E+07	0.
.20E+07	.39E+05	.11E+07	0.
.21E+07	0.	.11E+07	0.

VI.2   PROGRAMMING PROBLEMS

Programming Problem A: The Determination of the Diffusion Constant (Chemistry/Chemical Engineering)

The diffusion of one fluid into another is a phenomenon that is of great importance in several areas of engineering, especially chemistry and chemical engineering. The rate at which diffusion takes place is governed by the diffusion constant D, which represents the average rate of unit displacement by a fluid particle. One of the most common methods of measuring the diffusion constant is to fill a small capillary tube with one liquid, immerse the tube in a second liquid, and measure the concentration of molecules of one fluid as it diffuses into the other as a function of time. This may be done by labeling the intruder fluid with radioactive isotopes or by simply observing the color change of, for example, ink diffusing into water.

The relationship between the concentration at a later time, $C(t)$, and the initial concentration at time $t = 0$, C_0 is given by the equation

(VI.1)
$$\frac{C(t)}{C_0} = \frac{8}{\pi^2} \sum_{k=0}^{\infty} \frac{1}{(2k+1)^2} \exp\left[-\frac{(2k+1)^2 \pi^2 Dt}{4L^2}\right]$$

where L = length of the capillary tube (m)

$\quad\quad$ t = time (sec)

Once the concentrations $C(t)$, C_0, are measured, this equation is then solved for the diffusion constant D. We will use Newton's method (see Section 4.7) to find the root of the equation

(VI.2)
$$\frac{C(t)}{C_0} - \text{SUM}(D) = 0$$

Problem Specifics

Write and execute a computer program that will

1. Read the parameters required by the root-solving routine, namely, EPS, the convergence criterion; IMAX, the maximum number of iterations; SMALL, criterion for terminating the sum in Equation (VI.1); NMAX, the maximum number of terms that are to be included in a sum.
2. Read the measured ratio of concentrations $C(t)/C_0$, at a time t, the time, the length of the tube L, and the name of the intruder fluid. Echo print with appropriate labels.
3. Code Equation (VI.1) as a function of a single variable DIFF(D). The other parameters should be passed to the function via a labeled COMMON statement. The function should sum the series until the absolute value of a term is less than SMALL.
4. Code a second function DDIF(D), which is the first derivative of the function DIFF(D). Recall

$$\frac{d(e^{ax})}{dx} = ae^{ax}$$

so that

(VI.3)
$$\text{DDIF}(D) = \frac{d[\text{DIFF}(D)]}{dD} = \frac{2t}{L^2} \sum_{k=0}^{\infty} \exp\left[\frac{-(2k+1)^2 \pi^2 Dt}{4L^2}\right]$$

Again you will need a COMMON statement and to terminate the sum when terms are less than SMALL.

5. Code a subroutine NEWTON that implements Newton's root solving procedure. The convergence criterion should be on the size of the x increment—that is, Δx. The subroutine should have adequate safeguards.

6. Solve for the root of Equation (VI.1) using the Newton subroutine. Print the root, the number of iterations required, and the value of the function at the root.

Suggested Input Data

$$\text{Ratio} = C(t)/C_0 = 0.5$$

$$\text{Time} = t = 43{,}200 \text{ sec}$$

$$L = 0.1 \text{ m}$$

$$I_{max} = 40$$

$$\text{EPS} = 10^{-13}$$

$$\text{SMALL} = 10^{-7}$$

$$\text{NMAX} = 50$$

$$\Delta x_{max} = 10^{-6}$$

Initial
search
interval $(x_1 = 10^{-8})$ to $(x_3 = 10 \times 10^{-8})$

Initial
guess for
root $x_0 = 3 \times 10^{-8}$

Name of
liquid   = cadmium sulfide

Programming Problem B: Pouring Concrete in Cold Weather (Civil Engineering)

Pouring a concrete foundation for structures in winter can be somewhat risky. If the water in the concrete mix is permitted to freeze before the concrete has a chance to set, the result could easily be a crumbling wall. On the other hand, waiting until spring is usually unacceptable. It is therefore important to the engineer to obtain a reasonable estimate of the time scales involved in the competing processes of setting vs. freezing of cement. We will assume that experience has supplied the engineer with a fairly good estimate of the time required for a given mass of concrete to set and will concentrate on the freezing problem.

To simplify the problem, consider the situation of forming a wall by pouring concrete into a trench dug out of the soil. The wall is many times taller than it is thick. The important parameters are illustrated in Figure VI-4.

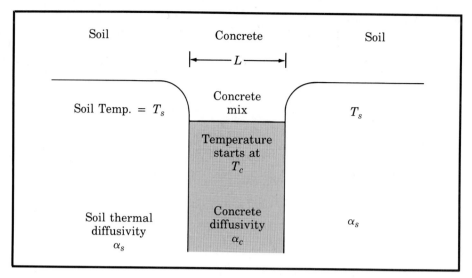

The temperature of the concrete mix is originally T_c and it is poured at time $t = 0$ into the soil, which is at a temperature of T_s ($T_s < 0$ °C). The temperature of the mix at later times is sketched as follows:

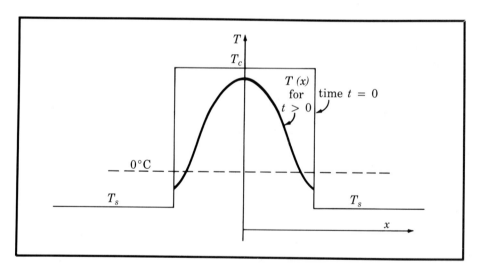

Thus the surface will quickly cool to the temperature of the soil. The important question is, how long will it take for the temperature to drop below 0° C to a significant depth within the wall? To solve this problem, we must first solve the heat-flow equations in one dimension for the temperature $T(x)$ at a position x within the wall. This equation will be introduced and solved in later courses in differential equations. The rate of temperature drop depends on a single physical parameter, the thermal diffusivity, α. The thermal diffusivity of soil and concrete are about equal. Under these conditions the equation for $T(x)$ is found to be

$$\frac{T(x) - T_s}{T_c - T_s} = \frac{1}{2}[\text{Erf}(\gamma(L - x)) + \text{Erf}(\gamma x)] \tag{VI.4}$$

where

$$\text{Erf}(y) = \frac{2}{\sqrt{\pi}} \int_0^y e^{-t^2} \, dt \tag{VI.5}$$

is the so-called error function and was briefly described in programming problem II-B. The parameter γ is defined as

$$\gamma^2 = \frac{1}{4\alpha t} \tag{VI.6}$$

Thus, if we wish to find the *time* required for the temperature to fall to a temperature of $0°$ C at a depth of 5 cm ($L = 50$ cm), we must first solve the equation

$$\frac{0 - T_c}{T_c - T_s} = \frac{1}{2}[\text{Erf}(45\gamma) + \text{Erf}(5\gamma)] \tag{VI.7}$$

for the value of γ. The time is then obtained from Equation (VI.6).

Details

1. Code a Fortran function subprogram

```
FUNCTION SIMPSN(A,B,N,F)
```

   that uses the Simpson rule approximation (See Section 4.5) to evaluate the integral of an arbitrary function $F(\tau)$ from $\tau = A$ to $\tau = B$ using N panels (N is even).
2. Code a Fortran function Erf(τ) based on Equation (VI.5) that employs the function SIMPSN.
3. Code a Fortran function FNC

```
FNC(TS,TC,X,ERF)
```

   that is based on Equation (VI.7) and is zero for the desired value of γ, once T_c, T_s, and x have been specified. That is,

$$\text{FNC}() = \frac{1}{2}[\text{Erf}(\gamma(L - x)) + \text{Erf}(\gamma x)] + \frac{T_s}{T_c - T_s}$$

4. Code a Fortran subroutine subprogram BISEC,

```
SUBROUTINE BISEC(X1,X3,IMAX,EPS,G,ROOT)
```

that will find a root of the arbitrary function $G(x)$ if a root is known to be in the interval $x_1 \leftrightarrow x_3$. (See Section 4.6.)

5. Write and execute the Fortran program employing the above subprograms to find the root γ of Equation (VI.7). Once γ is obtained, determine the time t. Print your results neatly.

Finally, after the root γ_r has been found, use this value to determine the temperature at the center of the concrete mix at the critical value of time, t. That is, from Equation (VI.4),

(VI.8)
$$\frac{T_{center} - T_s}{T_c - T_s} = \text{Erf}(\gamma_r L/2)$$

The values for the parameters used in this problem are given in Table VI-3.

Table VI-3
Values for the parameters in the concrete wall problem.

T_s	$= -5\,°C$	—soil temperature
T_c	$= +9\,°C$	—concrete mix temperature
x	$= 0.05\ m$	—critical wall depth
L	$= 0.50\ m$	—wall thickness
α	$= 5.8 \times 10^{-7} (\text{sec/m}^2)$	—thermal diffusivity of concrete and soil
γ_1, γ_3	$= 0.5, 1.5$	—initial search interval for root of Equation (VI.7)
IMAX $= 20$		—maximum number of iterations used by bisection code
EPS	$= 1 \times 10^{-5}$	—convergence criterion used in bisection code

Programming Problem C: Functions Describing Diffusion (General Engineering)

Diffusion is a process that is encountered in most areas of science and engineering. It describes the movement of neutrons through a nuclear reactor, the transport of heat through materials, the intrusion of one fluid into an adjacent fluid, the propagation of a disturbance in a line of freeway traffic, and numerous other phenomena. The solution for each case depends on the specifics of the problem, but the mathematical function describing diffusion is always a function that decays exponentially in time and the dependence on the coordinates is a function of the geometry of the problem. A particularly simple geometry is that of a sphere.

Consider the problem of a sphere of radius a which is at an initial temperature of T_0. If the material is characterized by a diffusivity α^2 and it is placed in 0 °C air, the sphere will slowly cool by convection and radiation.[2]

---

[2] The diffusivity of a material is defined as

$$\alpha^2 = \frac{\text{thermal conductivity}}{(\text{specific heat})(\text{density})} = \frac{\lambda}{C\rho}$$

If the rate of temperature decrease at the surface is characterized by an experimental constant h, the solution for the temperature at the center of the sphere is

$$T_c = T_0 \left(\frac{ah}{\sigma}\right) \sum_{n=1}^{\infty} C_n e^{(-\beta_n^2 t)} \qquad \text{(VI.9)}$$

where

$$\beta_n^2 = \theta_n \frac{\alpha^2}{a^2} \qquad \text{(VI.10)}$$

$$C_n = \frac{4 \sin(\theta_n)}{2\theta_n - \sin(2\theta_n)} \qquad \text{(VI.11)}$$

and θ_n is the nth root of the equation

$$\tan(\theta_n) = \frac{1}{1 - (ah/\sigma)} \theta_n \qquad \text{(VI.12)}$$

The values of the physical parameters to use in this problem are

$$T_0 = 250 \ °C$$

$$\alpha^2 = 1.2 \times 10^{-5} \ m^2/sec$$

$$h = 23.0 \ W/m^2\text{-}°C$$

$$\sigma = 46.0 \ W/m\text{-}°C$$

$$a = 0.1 \ m$$

and the problem is to determine and tabulate the first 10 roots of Equation (VI.12) and then to use Equation (VI.9) to compute the temperature at the center of the sphere for values of time $t = 0$ to $t = 3600$ sec (1 hour) in steps of 60 sec.

Problem Specifics

1. Code a Fortran function subprogram NEWTON,

        NEWTON(XO,IMAX,EPS,ROOT,F,DFDX)

   that will return the root (ROOT) of the arbitrary function $f(x)$. The subprogram will start with an initial guess x_0, and also requires the derivative of $f(x)$—that is, $df(x)/dx$. The maximum number of iterations is I_{max} and the convergence criterion based on Δx is EPS (see Section 4.7).
2. Since Newton's method will be used to find the roots, you will have to code function subprograms for a function based on Equation (VI.12) and a second Fortran function for the derivative. Note that the derivative of the tangent is

$$\frac{d\,(\tan(x))}{dx} = \frac{1}{\cos^2(x)}$$

3. The main program should start with $\theta_n = 0.1$, step in values of 0.1 until Equation (VI.12) changes sign, then CALL NEWTON to find the first root. The first root is easily found to be near 0.4. You must be extremely careful in searching for the remaining nine roots. If you continue to step in θ, you will find that the function changes sign as θ crosses $\pi/2$ but does *NOT* go through zero. That is $\tan(\pi/2) = \infty$, and the function is not continuous. From a sketch of both sides of Equation (VI.12) you can see that the subsequent roots are slightly to the left of multiples of $\pi/2$. For all roots after the first you should begin with the initial guess:

(VI.13)
$$\theta_n \simeq \left(n - \frac{1}{2}\right)\pi\left[1 - \frac{1}{(n - \frac{1}{2})^2\pi^2 k^2}\right] \qquad n = 2, 3, \ldots, 9$$

where

$$k = 1 - \left(\frac{ah}{\sigma}\right)$$

4. Once the 10 roots have been computed and neatly printed, the sum in Equation (VI.9) is to be evaluated for each value of time and the value printed and stored in the array TCENTR.

5. Finally, execute a printer-plot of TCENTR vs. time.

CHAPTER 10

ADDITIONAL FORTRAN FEATURES

There are several advanced features in Fortran that have not been discussed to this point, some very specialized that we can, for now, do without; others that are more frequently encountered and which you may find useful. These are described in this chapter. For a more complete compendium, see the Appendix.

10.1 ADDITIONAL DATA TYPES AVAILABLE IN FORTRAN

The majority of programs only require the three data types that we have encountered thus far: integer, real, and character. In special situations these may not be suitable or sufficient for the solution of the problem at hand. Three additional data types are available in Fortran for use in more advanced programming problems. They are *double-precision, complex,* and *logical* variable types.

10.1.1 Type DOUBLE PRECISION

As you are aware, the real number computations done by a computer are only approximate arithmetic due to the finite length of a computer word. A computer will carry anywhere from 8 to 14 significant digits for each real variable and therefore every arithmetic operation then involves some round-off error, a loss of significant digits. Most of the time this does not cause problems, but there are situations where round-off error can invalidate a calculation. If the number you want is $C = A - B$ and A and B are very nearly equal, say $A = 1.0000032$, $B = 1.0000031$, both with eight significant figures. Then $C = 1.\text{E}-7$ with only *one* significant figure. In cases like this, Fortan allows you to double the number of significant digits allotted to a particular variable by reserving two computer words to that variable. This is then a new type of variable in addition to the three we have seen: real, integer, character. The form of the type statement for these variables is

$$\text{DOUBLE PRECISION } name_1, \ name_2, \dots$$

which like other type statements is nonexecutable and must appear before the first executable statement.

Double-precision constants with an exponent use the letter D to designate the exponent part of the number in place of the letter E that is used for this purpose when writing real numbers. That is, the number 3.5D+06 is type DOUBLE PRECISION, while 3.5E+6 is type REAL or single precision.

Similarly, when printing or reading double-precision numbers, the E format should be replaced by a D format (i.e., E7.1 → D7.1). The D format instructs the computer to expect a number that occupies two words of memory. In all other respects the D format is the same as the E format. Double-precision numbers may be read or written using F or E formats without execution time errors but a loss of accuracy may result. Arithmetic expressions involving reals and double-precision numbers or integers and double-precision numbers are now a new form of mixed-mode arithmetic and special care must be exercised when you use them. Double-precision variables are defined to have "dominance" over reals, which in turn have dominance over integers. (See Section 2.2.2.)

```
PROGRAM MIXMOD
INTEGER I
REAL     R
DOUBLE PRECISION DP1,DP2,DP3

I = 5
R = 4.2
DP1 = 6.3
DP2 = 2.D+2        ⟨D replaces E in the exponent on double
                     precision constants.⟩
```

```
DP3 = DP1/R    ⟨R is converted to double precision before
                 dividing, the result is DP3 = 0.15D+1⟩

I = DP2/DP1    ⟨The result of DP2/DP1, .3125D+2 is
                 truncated to I = 31⟩
```

Double-precision variables should only be invoked when absolutely necessary since the computer executes double-precision arithmetic at a much slower rate than ordinary real arithmetic. (It is more than a factor of 2 slower.)

Function subprograms may also be "typed" as double precision:

```
DOUBLE PRECISION FUNCTION A(X)
```

In addition, all of the commonly used library functions like SIN(X), EXP(X), LOG(X) that accept real values for input will accept double precision values as input. The value returned by these functions will then be double-precision. This is *not* the case with a user-written subprogram. If a double-precision value is used in the argument of a type REAL function, the value returned to the referencing program will be real, not double-precision, unless the function has been explicitly typed as a double-precision function.

10.1.2 Type COMPLEX

In algebra you learned that a generalization of the ordinary set of real numbers is the set of complex numbers written in the form

$$a + ib$$

where i is used to designate the square root of -1 ($i = \sqrt{-1}$), a is called the *real* part and b the *imaginary* part of the number. Note that both a and b are themselves ordinary real numbers. Complex numbers occur frequently in scientific and engineering applications and the common arithmetic operations of addition, subtraction, and multiplication can be generalized to include complex numbers.

$$
\begin{array}{cc}
(a_1 + ib_1) & (a_1 + ib_1) \\
+ (a_2 + ib_2) & - (a_2 + ib_2) \\
\hline
(a_1 + a_2) + i(b_1 + b_2) & (a_1 - a_2) + i(b_1 - b_2)
\end{array}
$$

$$(a_1 + ib_1)(a_2 + ib_2) = (a_1 a_2 + i^2 b_1 b_2) + i(a_1 b_2 + a_2 b_1)$$
$$= (a_1 a_2 - b_1 b_2) + i(a_1 b_2 + a_2 b_1)$$

Arithmetic operations can also be effected between complex numbers in

Fortran. First, variable names that will be used to store complex numbers are declared as complex with a type statement of the form

COMPLEX   ⟨list of variable and/or array names⟩

Next, complex constants are written in Fortran in terms of two real numbers, the real and imaginary parts *enclosed in parentheses* and separated by a comma. Some examples are

```
COMPLEX Z,ROOT,S,T
Z = (3.0,4.0)
S = (0.0,2.0)
T = S * Z
ROOT = SQRT((-4.,0.))
```

i.e., $z = 3 + 4i$

$s = 2i$

$t = 2i(3 + 4i) = -8 + 6i$

Notice, if the argument of SQRT is complex, the result is likewise,
$root = 2i$, *i.e.* $(0., 2.)$

When a complex number is read using a list-directed READ statement, the form entered must include the enclosing parentheses and the comma separating the real and imaginary parts. Similarly, when a complex number is printed using a list-directed output statement, the form displayed will include parentheses and a separating comma.

```
COMPLEX A,B
READ *,A,B
PRINT *,'A*A = ',A*A
```

Enter (4., 5.), (1., 0.)

Output is $A*A = (-9.0, 40.0)$

Formatted READ and WRITE statements may also be used to input and output complex numbers provided *two* real-number fields are provided for each complex number.

Mixed-mode arithmetic operations combining complex numbers and real- or integer-type numbers is permitted with the expression evaluated by assigning complex numbers dominance over reals. That is, the result of 2. * (−2., 3.) + 2 is a complex number whose value is (−2., 6.). However, complex and double-precision numbers may never appear in the same arithmetic operation. Attempts to do so will result in a compilation time error.

It is quite interesting that many of the common numerical techniques developed with real-number arithmetic in mind remain valid when the number set is generalized to include complex numbers. For example, Newton's method for finding square roots may be generalized to obtain complex results by simply typing the relevant variables as complex.

10.1.3   Type LOGICAL

The result of a logical relation of the form (2.**2 .EQ. 4.) must be a value of either ⟨true⟩ or ⟨false⟩. The Fortran data type that is used to store these values is the type LOGICAL. A variable name may be declared to be of type LOGICAL with a type statement of the form

LOGICAL   ⟨*list of variables and/or arrays*⟩

Variable names declared in this manner are permitted to contain *only* LOGICAL constants. There are only two logical constants,

.TRUE.   .FALSE.

The periods at the beginning and end of the word are part of the constant.

```
LOGICAL TEST
TEST = .TRUE.
IF(2**2 .EQ. 4 .AND. TEST)THEN     ⟨true⟩ .and. ⟨true⟩ → ⟨true⟩
```

In addition to the logical combinatorial operators .AND. and .OR., there are two more operators that you may find useful. These are

.EQV.	*Equivalent*
.NEQV.	*Not equivalent*

which may be used to compare two logical variables, constants, or expressions to determine whether or not they have the same value. Thus if *both* sides of .EQV. are the same, either ⟨true⟩ or ⟨false⟩, the entire expression is ⟨true⟩, otherwise it is ⟨false⟩.

((4 .LT. 0) .EQV. (1 .GE. 5))   →   ⟨*true*⟩

To my mind the use of LOGICAL variables and constants is rather artificial and unnecessary in most programs and I recommend that you avoid employing them.

10.1.4   IMPLICIT Type Statements

Explicit typing of variable names is achieved by means of the six type statements: REAL, INTEGER, CHARACTER, DOUBLE PRECISION, COMPLEX, and LOGICAL. In addition Fortran automatically will implicitly type variable names, unless instructed otherwise, as

Names that begin with the letters	
I through N	→ type INTEGER
All others	→ type REAL

The Fortran IMPLICIT statement allows you to alter this. The form of the statement is

$$\text{IMPLICIT } \langle\text{type}\rangle \ (a_1\text{-}a_2)$$

where a_1, a_2 are single letters, and $\langle\text{type}\rangle$ can be any of the six Fortran number types. This statement then forces all variable names that begin with the letters a_1 through a_2 to be of the specified data type. Thus,

```
IMPLICIT REAL (A-Z)
```

will cause all variables, unless otherwise explicitly typed, to be real. The IMPLICIT statement is, of course, nonexecutable and must precede all executable statements. More significantly, it affects only those statements that come after it and so should be the very first line after the PROGRAM line.

A very common programming error is simply misspelling or mistyping a variable name. Usually, then, the misspelled variable will not have been assigned a value, and when it is used in an arithmetic expression an execution time error will result. These errors are sometimes very difficult to trace. Using the IMPLICIT statement you can construct a trick to detect misspellings immediately at compilation time. The idea is to implicitly type all variables as CHARACTER*1 at the start of the code, and explicitly type all variables that appear later. If any variables are found in the program that have not been explicitly typed (i.e., misspellings) they will be of type CHARACTER and therefore arithmetic expressions involving these variables will be illegal.

```
PROGRAM SPELL
IMPLICIT CHARACTER*1 (A-Z)
REAL X(50),Y(50)
INTEGER LOW,HI,I
X(1) = 0.0
DO 1 I = 1,50
    X(I) = I*(HI - LO)/49.
    Y(I) = F(X(I))
1 CONTINUE
```

The variable LOW is misspelled as LO in the DO loop. Since LO is implicitly typed as character, this statement will result in a compilation time error.

10.2   INITIALIZING VARIABLES AT COMPILATION TIME

Up to this point, values are stored in memory locations assigned to variable names only during execution of the program by either an assignment statement (X = ...) or by a READ statement (READ *,X). There are, however, situations when it is wasteful or distracting to use valuable execution time to assign values. Setting a large array to all zeros or initializing PI = 3.1415926 are but two examples of occasions of when it would be useful to have values already stored in memory *before* the program begins execution. Most programs require the initialization of numerous constants before the computation can commence. These assignments differ in nature from the initialization of the variables of the problem, which will be changed from one run to the next by reading input data. Each Fortran assignment or READ statement is, of course, an executable statement and will add to the execution time of the program. If the assignments occur in a subprogram that is referenced hundreds of times, this can accumulate to a significant cost. Fortran provides two mechanisms for assigning values to variable names *during compilation.* These are the DATA and the PARAMETER statements.

10.2.1 The DATA Statement

The form of a DATA statement is

$$\text{DATA} \langle \text{namelist} \rangle / \langle \text{valuelist} \rangle /$$

where ⟨namelist⟩ is a list of Fortran variables to be initially assigned a corresponding value in the ⟨valuelist⟩. Of course, the values in the ⟨valuelist⟩ must agree by type to the corresponding name.

```
DATA PI,ILOW,IHI/3.1415926,0,50/
```

Note that the names and the values are separated by commas and that the ⟨valuelist⟩ is enclosed by slashes.

Also, the constants in the ⟨valuelist⟩ can be repeated by including an unsigned nonzero integer as a replication factor. This is very useful in "zeroing" an array.

```
REAL A(10,10)
DATA A/100*0./
```

In addition, implied DO loops may be used in the ⟨namelist⟩ to specify portions of an array.

```
REAL B(50)
DATA (B(I),I = 1,49,2),(B(J),J = 2,50,2)/25*0.,25*1./
```

Thus, the elements of B are 0., 1., 0., 1., . . . If you had wanted to assign the odd elements of B the value −1., you could try 25*(−1.) in the ⟨valuelist⟩; however, standard Fortran 77 does not permit parentheses in a ⟨valuelist⟩ and, depending on the local version of Fortran installed on your machine, it may or may not work. A more contrived version that makes use of the PARAMETER statement is described in the next section.

A variable may not appear twice in the same or different DATA statements, and function names may not appear at all. Also, any variables in *blank* COMMON may not be initialized by a DATA statement. (Variables in labeled COMMON are, however, permitted.) Finally, the number of variables in the ⟨namelist⟩ must match the number of values in the ⟨valuelist⟩.

The rules pertaining to the initialization of character variables are similar to those for ordinary assignment statements:

> If the length of the character variable in the ⟨namelist⟩ is longer than the corresponding character string in the ⟨valuelist⟩, the right end of the character variable is filled with blanks.
> If the length of the character variable in the ⟨namelist⟩ is shorter than the corresponding character string in the ⟨valuelist⟩, the additional symbols to the right are ignored.

```
CHARACTER WORD1*6,WORD2*3
DATA WORD1,WORD2/'BIG','BIGGER'/
```

WORD1	*contains*	⎸B⎸I⎸G⎸ ⎸ ⎸ ⎸
WORD2	*contains*	⎸B⎸I⎸G⎸

10.2.2 The PARAMETER Statement

Occasionally a Fortran variable is to be assigned a value and it is intended that this value *never* be altered. An example is PI = 3.1415926. All variables defined in Fortran, whether they are initialized by an assignment statement, a READ statement, or a DATA statement can be reassigned a different value later in the code. All, that is, except variables initialized by means of the PARAMETER statement. The use of the PARAMETER statement is rather like that of the DATA statement, except for two important distinctions: First, once a variable name has been assigned a value in a PARAMETER statement, the compiler will not permit the value of that variable to be altered thereafter in the program. Second, variable names initialized in a PARAMETER statement are formally called *named constants*, meaning that they may then be used in place of ordinary numerical constants in all subsequent Fortran statements except format edit specifications or as designations for statement numbers. The form of the PARAMETER statement is

```
PARAMETER (name₁=value₁,name₂=value₂,....)
```

where $name_1$, $name_2$, . . . are Fortran variable names and $value_1$, $value_2$, . . . are the values they are to be permanently assigned. These values may be constants, constant expressions, or character strings. For example:

```
INTEGER SIZE,EXIT,INFILE
PARAMETER (PI=3.1415926, INFILE=12, SIZE=20, EXIT=99)

REAL A(SIZE,SIZE)              〈This is now a valid dimension
                                 statement.〉

OPEN(INFILE,FILE='DATA3')
REWIND INFILE
READ(INFILE,*,END=EXIT)        〈Error, the use of EXIT is
                                 incorrect. You cannot use a
                                 name in place of a statement
                                 number.〉
```

Using a PARAMETER statement to assign a value to the various array sizes at the beginning of a program is one of several useful applications. If the array sizes need to be changed in subsequent runs of the program, only a single line of the code need be changed.

More exotic applications can be constructed by using character strings as elements of a PARAMETER statement:

```
CHARACTER LIST*11
REAL X(50)
PARAMETER (LIST = '(1X,10F6.2)')
READLIST,X                     〈Note: This is equivalent
                                 to READ '(1X,10F6.2)',X〉
```

In addition to permanently fixing the value of Fortran variable names, an important feature of PARAMETER constants is that they may then be used in subsequent DATA statements. Thus, the attempt to assign $-1.$ to elements of the array B(I) in the previous section may now be accomplished as follows:

```
REAL XX,B(50)
INTEGER TOP
PARAMETER (XX = -1.0,TOP = 25)
DATA (B(I),I = 1,2*TOP-1,2),(B(I),J = 2,2*TOP,2)
+      /TOP*XX , TOP*1.0/
```

The imaginative use of PARAMETER and DATA statements is a distinctive feature of modern programming style.

10.3   THE ORDER OF FORTRAN STATEMENTS

Numerous new Fortran statements have been introduced in this chapter and you may be uncertain regarding the relative ordering of the various types of executable and nonexecutable statements. The correct arrangement is indicated in Table 10-1. Within each group, the various statements of the same classification may appear in any order, but the groups must be arranged as shown. Statements that can appear anywhere within more than one group are indicated in vertical columns that overlap two or more groups. For example, FORMAT statements may appear anywhere after the PROGRAM line.

Table 10-1 The ordering of the various types of Fortran statements.

Fortran statements				
PROGRAM/SUBROUTINE/FUNCTION				Comments
IMPLICIT		PARAMETER	FORMAT	
INTEGER REAL CHARACTER DOUBLE PRECISION	*Type specifications*			
DIMENSION COMMON EXTERNAL INTRINSIC LOGICAL	*Specification statements*			
STATEMENT FUNCTIONS				

Assignment Statements
DO
CONTINUE
IF
ELSE
ELSE IF
END IF
GO TO
CALL
RETURN
STOP
OPEN
CLOSE
REWIND
READ
WRITE
PRINT
	DATA			

END				

The following points regarding statement ordering are worth repeating:

Comment lines can appear anywhere in a program. Those that are placed after an END statement will be listed with the next program unit.

FORMAT statements can appear anywhere within a program or subprogram.

The END statement is the last statement of a program unit.

Generally, specification statements precede executable statements. The arrangement of statements within the specification grouping is:

PARAMETER statements can appear anywhere in the group, but must be placed before any reference to variables they define.

Variable names should be type-declared prior to being assigned values in a PARAMETER statement.

IMPLICIT statements *must* precede *all* other specification statements (except PARAMETER).

Statement functions must appear after the specification statements and before the first executable statement. When there is more than one statement function, the ordering must be such that each function references only those placed above it.

DATA statements can be placed anywhere after the specification statements and before, after, or among the statement functions. The recommended placement is after the specifications and before all statement functions.

10.4   THE DoWhile STRUCTURE IN EXTENDED FORTRAN

In Section 3.4 the construction of Fortran loops was described in terms of a DoWhile-EndDo structure. At that time you were advised that present versions of Standard Fortran 77 do not support explicit DO WHILE and END DO statements. However, a great many local installations of Fortran 77 do permit these statements and you should determine whether your computing center is one of these. A version of Fortran 77 that allows for DO WHILE statements is usually called *extended Fortran*. Any program that employs one or more loops will benefit from a rewrite using the new loop control statements.

The form of the Extended Fortran DO WHILE statement is

```
DO [statement number] WHILE(logical expression)
```

The statement number is enclosed in brackets to indicate that it is optional. The brackets do not appear in the actual Fortran statement. If a statement number is included, it specifies the loop-terminating statement, which may be any of the allowed executable statements used to terminate a normal DO loop, or it may be an END DO statement. (See below.) The logical expression within the parentheses is tested at the beginning of each cycle of the loop, including the first. If the expression is evaluated as ⟨true⟩ the body of the loop is executed, otherwise control is transferred to the statement following the loop terminator.

The preferred terminus of a DO WHILE structure is the END DO statement, which has the form

```
END DO
```

An END DO statement *must* be used to terminate a DO or a DO WHILE if the optional terminal statement number has been *omitted*. An END DO may have a statement number. A few examples are

$$I_{sum} = \sum_{i=1}^{100} i$$

```
I = 0
ISUM = 0
DO WHILE (I .LT. 100)
    ISUM = ISUM + 1
    I = I + 1
END DO
```

$$e^x = \sum_{n=1}^{\infty} \frac{1}{n!} x^n$$

```
READ *,X
N = 1
SUM = 0.
TERM = X
DO WHILE (ABS(TERM) .LT. 1.E-6)
    SUM = SUM + TERM
    TERM = TERM * X/(N + 1.)
    I = I + 1
    IF(I .GT. 100)THEN
        PRINT *,'NOT CONVERGING'
        STOP
    END IF
END DO
```

The DO WHILE structure can be used with or without an END DO statement:

```
      READ *,C
      X = 0.5 * (C + 1.)
      I = 0
      DO 99 WHILE (ABS(X * X - C)) .GT. 1.E-6)    〈Newton's
          X = 0.5 * (X + C/X)                        algorithm for
          I = I + 1                                  square roots〉
          IF(I .GE. 50)THEN
              PRINT *,'NOT CONVERGING'
              STOP
          END IF
   99 CONTINUE                                    〈or 99 END DO〉
```

Note that most of the structured features of Fortran 77 reduce the need for statement number labels and as a result make possible the writing of code that is to a large extent readable in a continuous path from beginning to end. Generally, associating statement numbers with Fortran statements is done to provide alternative paths through the program which in turn can make the program difficult to decipher. Except for FORMAT statements, with the introduction of the DO WHILE and END DO statements into the language there is little compelling reason for using statement numbers *at all*.

10.5   ADDITIONAL FORTRAN INTRINSIC FUNCTIONS

The most commonly used intrinsic functions in scientific and engineering applications were introduced in Section 2.7. There are numerous other intrinsic function available in Fortran and several of these are listed in Tables 10-2 and 10-3.

Most Fortran mathematical intrinsic functions come in a variety of forms depending upon the data type of the argument. For example, DSQRT(D) can be used if the argument is double-precision and the computed result will then also be double-precision. Similarly, CSQRT(C) can be used if the argument is a complex number and the result will be of type COMPLEX. However, a very convenient alternative to choosing the function to fit the type of its argument is to use the *generic* function names that are provided in Fortran. For example, the generic name for computing a square root is SQRT(X). If the argument is real, the result returned is likewise real; if the argument is double-precision, the result returned is double-precision, etc. The specific function names (DSQRT, CSQRT, etc.) have been retained in Fortran to provide compatibility with earlier versions, and I suggest that your programs employ generic function names exclusively, when possible. In Tables 10-2 and 10-3 the symbol I is used to denote an integer argument, X for real, D for double-precision, C for complex, CH for character, and L for logical.

Table 10-2 Fortran mathematical intrinsic functions.

Function name	Description	Argument	Result
SQRT(X)	$\sqrt{x}$ square root, generic	Real Double precision Complex	Real Double precision Complex
EXP(X)	e^x exponential, generic	Real Double precision Complex	Real Double precision Complex
LOG(X)	$\ln(x)$ natural logarithm, generic	Real Double precision Complex	Real Double precision Complex
LOG10(X)	$\log_{10}(x)$ base-10 logarithm, generic	Real Double precision	Real Double precision
ABS(X)	$\|x\|$ absolute value, generic; for complex argument returns $(x^2 + y^2)^{1/2}$	Integer Real Double precision Complex	Integer Real Double precision Real
SIN(X) COS(X) TAN(X)	Trigonometric functions sine, cosine, tangent argument is radians	Real Double precision Complex (sin, cos, only)	Real Double precision Complex (sin, cos, only)
ASIN(X) ACOS(X) ATAN(X)	Inverse trigonometric functions: $\sin^{-1}(x)$, $\cos^{-1}(x)$ $\tan^{-1}(x)$ result in radians; if $x = \tan(\theta)$ then $\theta = \tan^{-1}(x)$	Real Double precision	Real Double precision
SINH(X) COSH(X) TANH(X)	Hyperbolic functions: $\sinh(x) = (e^x - e^{-x})/2$; $\cosh(x) = (e^x + e^{-x})/2$; $\tanh(x) = \sinh(x)/\cosh(x)$	Real Double precision	Real Double precision
MOD(X,Y)	Remainder of division of x by y	Real Double precision	Real Double precision
MAX(X$_1$,X$_2$, ...)	Maximum element of the list x_1, x_2, ...	Real Double precision	Real Double precision
MIN(X$_1$,X$_2$, ...)	Minimum element of the list x_1, x_2, ...	Real Double precision	Real Double precision

10.6 CONCLUSION

We have now covered all of the elements of Fortran 77 grammar necessary to construct programs to solve almost any problem that is amenable to solution by a computer. There are a few additional Fortran statements that have not been discussed and which are described in the Appendix. Also there are thousands of subtle points that have been glossed over or even purposely

Table 10-3 Fortran intrinsic functions for converting data types.

Function name	Description	Argument	Result
REAL(X)	Converts argument to real	Real	Real
		Integer	Real
		Double Precision	Real
		Complex	Real
CMPLX(X)	Converts argument to complex	Real	Complex
		Integer	Complex
		Double Precision	Complex
		Complex	Complex
INT(X)	Converts argument to integer by truncation	Real	Integer
		Integer	Integer
		Double precision	Integer
		Complex	Integer
NINT(X)	Round x to nearest integer	Real	Integer
		Double precision	Integer
LEN(CH)	Return length of character string CH	Character	Integer
INDEX(CH$_1$,CH$_2$)	Position of substring CH$_2$ within string CH$_1$	Character	Integer
ICHAR(CH)	Position of the single character CH in the system established sequence	Character	Integer
CHAR(I)	The single character in the ith position of the collating sequence for character symbols	Integer	Character

disguised or ignored. You do not learn programming or anything else by first memorizing myriad details. These will come naturally with time and experience. You must first develop confidence in your ability to solve complicated problems with what you already know. The analogy with languages is very apt. The best way to become fluent in a language is not to spend great effort in learning a vocabulary by memorization, but rather to begin communicating as soon as you can with the small vocabulary you currently have.

An important responsibility of an instructor is to see to it that the student does not fall into a variety of bad habits that may not appear serious now but will be difficult to break later. Structured programming was devised to aid in the logical construction of programs, and you should try to segment and "layer" even the simplest of codes. In addition, your programs should contain every manner of internal checks for potential errors that you can devise. By now, you are well aware that some execution time errors are extremely difficult to track down and in such cases helpful clues supplied by the programmer would be greatly appreciated.

The problem now is to use the Fortran developed to this point and to attempt to dovetail it with the mathematics, science, and engineering that you have already seen or will soon see. Once that is accomplished, you will be quite fluent in the mathematical analysis of engineering and science problems.

PROBLEMS

1. True or False? Explain!
   a. Any PARAMETER statement must precede all type statements.
   b. Variables typed as real may also be typed as double precision in the same program unit.
   c. Variables implicitly typed as integer may also be typed as real in the same program unit.
   d. Variables initialized in a DATA statement may not be altered in the same program unit.
   e. Character variables may not be initialized via a DATA statement, a PARAMETER statement must be used instead.
   f. Complex variables may not appear in a DATA statement.
2. Use a DATA statement to initialize a square 10 by 10 array, A(I, J), in the following manner.

$$a_{ij} = +1 \quad \text{for } j > i \quad \text{above the main diagonal}$$
$$a_{ii} = 0 \quad\quad\quad\quad\quad\quad\; \text{along the main diagonal}$$
$$a_{ij} = -1 \quad \text{for } j < i \quad \text{below the main diagonal}$$

You will also need a PARAMETER statement (for the -1's), and several nested implied DO loops in the DATA statement.

3. Rewrite the Fortran code in Section 3.4.3 for Newton's method for finding the square roots of numbers to find the square root of a complex number. Test the program by evaluating the square roots of a negative real number and of an arbitrary complex number. Verify your answer by comparing the square of the result with the test number. (Note: For a test for "smallness" of a complex number c use

```
IF((ABS(C)) .LT. 1.E-6)THEN
```

That is, the sum of the squares of the real plus imaginary parts determines the "size" of the complex number.)

4. Write a Fortran program to use the method of successive substitutions (see Problem 3.11) to find a complex root of a function. Test the program by finding a root of

$$f(x) = x^3 - 4x^2 + 6x - 4 = 0 \qquad \text{Exact roots} = 1 + i, 1 - i, +2$$

Hint: Write the equation as $x = \frac{1}{2}[x^3 + 6x - 4]^{1/2}$ and start with $x_0 = (1.0, 0.5)$.

5. Determine the output of the following program

```
PROGRAM ANDOR
LOGICAL A,B,C,D,E
A = .FALSE.
C = .TRUE.
```

```
      DO 1 I = -1,1
         B = I .GT. 0
         D = B .AND. (A .OR. C)
         E = B .AND. A .OR. C
         IF(D .NEQV. E)THEN
                PRINT *,'THE ORDER OF AND/OR MAKES'
                PRINT *,'A DIFFERENCE FOR '
                PRINT *,B,'.AND.',A,'.OR.',C
         END IF
    1    CONTINUE
```

6. Division of complex numbers can be defined in terms of the multiplicative inverse of the number, which can be obtained in the following manner:

$$c = a + ib \qquad (a, b \text{ are known})$$

$$c^{-1} = \alpha + i\beta \qquad (\alpha, \beta \text{ are to be determined})$$

$$c(c^{-1}) = 1 = (a + ib)(\alpha + i\beta)$$

$$= (a\alpha - b\beta) + i(a\beta + b\alpha) = 1 + 0i$$

So

$$b\alpha + a\beta = 0$$

$$a\alpha - b\beta = 1$$

a. Solve these equations for (α, β) in terms of (a, b) to obtain:

$$\alpha = \frac{a}{a^2 + b^2} \qquad \beta = \frac{-b}{a^2 + b^2}$$

b. Test these equations directly on the computer by executing a program to read two complex numbers, X, Y and evaluate and print X/X, X/Y, Y/X, (X/Y)*(Y/X).

7. Determine any compilation errors in the following. If none write OK.

   a. ```
   PARAMETER (PI = ACOS(-1.))
   ```
   b. ```
   PARAMETER (K = 2)
   OPEN (K,FILE = 'DATA')
   ```
   c. ```
   PARAMETER (N = 44)
   GO TO N
   ```
   d. ```
   PARAMETER (N = 4)
   REAL A(N)
   DATA (A(I),I = 1,N)/N*3./
   ```
   e. ```
   PARAMETER (N = 4)
   INTEGER A(N)
   DATA (A(I),I = 1,N)/N*N/
   ```
   f. ```
   CHARACTER NAME*3,STUDENT*12
   PARAMETER (NAME = '(A)')
   READNAME,STUDENT
   ```

PROGRAMMING ASSIGNMENT VII

This set of problems is intended to illustrate the use of COMPLEX and DOUBLE PRECISION variables, and DATA and PARAMETER statements.

VII.1 SAMPLE PROGRAM

Complex Numbers

A Hypothetical Extension of Fermat's Last Theorem

Pierre Fermat (1601–1655), a lawyer and brilliant mathematician, had a habit of scribbling his versions of proofs for mathematical theorems in the margins of his mathematics books. While reading a section of the ancient text *The Arithmetic of Diophantus* dealing with integer solutions of the equation

$$a^2 + b^2 = c^2$$

he entered the following in the margin:

> "On the other hand it is impossible to separate a cube into two cubes, or generally any power except a square into two powers of the same exponent. I have discovered a truly marvelous proof of this, which however, the margin is not large enough to contain."

This is the same as saying that *integer* values cannot be found for a, b, c (a, b, and c all different and all greater than or equal to one) such that

(VII.1)
$$a^n + b^n = c^n \qquad \text{where } n > 2$$

Mathematicians have been trying ever since Fermat's death to either rediscover his proof or to prove him wrong. Although the apparent truth of the

370

conjecture has been demonstrated by testing all integers a, b, c, up to a very large integer N for exponents $n = 3, 4, 5, \ldots$, it has never been formally proved either way. The conjecture is, however, generally believed to be true.

It is amusing to consider a possible generalization of Fermat's theorem; namely:

> There are no solutions of Equation (VII.1) with a, b, c being *complex* numbers with both their real and imaginary parts being positive integers.

For example:

$$a = 1 + 2i$$
$$b = 1 + 2i$$
$$a^2 + b^2 = -6 + 8i = c^2$$

so

$$c = \sqrt{2} + \sqrt{8}i$$

and this is not a solution of the equation with integer real and imaginary parts.

We will construct a program to test the hypothesis for $n = 2, 3, 4, 5$, and for all complex numbers with real or imaginary parts ≤ 20. The Fortran version of the program is given in Figure VII-1.

```
      PROGRAM FERMAT
*--
      COMPLEX A,B,C
      REAL    RC,IC
      INTEGER RA,IA, RB,IB, N, L, COUNT
      PARAMETER (L = 20)
*--
      WRITE(*,10)
      DO 2 N = 2,5
         WRITE(*,11)N
*--
*--            N is the exponent in a**n + b**n = c**n,
*--            COUNT will record the total number of
*--            potentially successful sets of numbers
*--            satisfying the hypothesis.
*--
         COUNT = 0
         DO 1 RA = 1,L
            DO 1 RB = RA,L
               DO 1 IA = 1,L
                  DO 1 IB = IA,L
*--
*--               To create a COMPLEX A we use the
*--               intrinsic function CMPLX(,)
```

Figure VII-1 A Fortran program to test a complex generalization of Fermat's theorem.

Continued

```
*--
                        A = CMPLX(RA,IA)
                        B = CMPLX(RB,IB)
                        C = A**N + B**N
                        C = C**(1./N)
*--
*--         Now, to test whether C has integer or
*--         near integer real and imaginary parts,
*--         the real/imaginary parts are identified
*--         by using the following intrinsic functions.
*--
                        RC = REAL(C)
                        IC = AIMAG(C)
*--
*--         The intrinsic function NINT(X) returns the
*--         nearest integer to the real quantity X.
*--
                        DELTA = ABS(RC-NINT(RC)) + ABS(IC-NINT(IC))
*--
*--         If DELTA is small, the number C may have
*--         integer parts.
*--
                        IF(DELTA .LT. 1.E-6)THEN
                            COUNT = COUNT + 1
                            WRITE(*,12)COUNT,A,B,C,DELTA
                        END IF
      1         CONTINUE
                IF(COUNT .EQ. 0)THEN
                    WRITE(*,13)L,N
                END IF
      2     CONTINUE
            STOP
     10   FORMAT(/////,T5,'A TEST OF A COMPLEX GENERALIZATION OF',/,
          +        T5,'FERMATS LAST THEOREM',//)
     11   FORMAT(T10,'THE RESULTS FOR AN EXPONENT OF N = ',I2,/,
          +        T15,'I',T25,'A',T35,'B',T51,'C',T67,'DELTA')
     12   FORMAT(T14,I2,T20,2(1X,F3.0,2X,F3.0,1X),2F11.8,2X,E8.2)
     13   FORMAT(///,T10,'THERE ARE NO SETS OF NUMBERS WITH INTEGER',/,
          +        T10,'REAL AND IMAGINARY PARTS LESS THAN ',/,
          +        T10,'L = ',I2,' THAT SATISFY THE EQUATION WITH',/,
          +        T10,'AN EXPONENT N = ',I2,/)
          END
```

There are five deeply nested DO loops in this program. It is not difficult to count the total number of cycles that must be executed for the program to run to completion. The result is

$$\text{Number of cycles} = L^2(L + 1)^2$$

For a limit chosen as $L = 20$, this amounts to 176,400 cycles. If we had tried $1 = 50$, the number of cycles would have been greater than $6\frac{1}{2}$ million.

```
A TEST OF A COMPLEX GENERALIZATION OF
FERMAT'S LAST THEOREM


        THE RESULTS FOR AN EXPONENT OF N = 3
            I         A         B              C              DELTA



        THERE ARE NO SETS OF NUMBERS WITH INTEGER
        REAL AND IMAGINARY PARTS LESS THAN
        L = 20 THAT SATISFY THE EQUATION WITH
        AN EXPONENT N = 3.

        THE RESULTS FOR AN EXPONENT OF N = 4
            I         A         B              C              DELTA



        THERE ARE NO SETS OF NUMBERS WITH INTEGER
        REAL AND IMAGINARY PARTS LESS THAN
        L = 20 THAT SATISFY THE EQUATION WITH
        AN EXPONENT N = 4.

        THE RESULTS FOR AN EXPONENT OF N = 5
            I         A         B              C              DELTA



        THERE ARE NO SETS OF NUMBERS WITH INTEGER
        REAL AND IMAGINARY PARTS LESS THAN
        L = 20 THAT SATISFY THE EQUATION WITH
        AN EXPONENT N = 5.
```

Figure VII-2
Output of the complex Fermat's theorem program.

Obviously this program can very easily run up a huge bill, which is why I've included it as a sample program and not as an assignment.

From the results of the program, given in Figure VII-2, it appears, at least from this rather small sampling, that the complex generalization of Fermat's theorem may be true.

VII.2 PROGRAMMING PROBLEMS

Programming Problem A: A Double Precision Root Solver

A common operation in computational algorithms is the addition of a small correction term to a not so small base term. Often a great deal of effort has gone into the evaluation of the correction term, yet, when it is added to the base term most of the significant figures in the correction term are lost. This is, of course, a consequence of the finite word length of the computer. For example, on a machine with an eight-digit word length, if the base term is $x_0 = 1.0000000$ and the correction term is evaluated as $\Delta x = 1.2345678\text{E-}6$, then the sum $x_0 + \Delta x$ becomes:

$$
\begin{array}{r}
1.0000000 \\
+\ 0.0000012345678 \\
\hline
=\ 1.0000012
\end{array}
$$

and six of the eight significant figures of Δx have been lost. If extreme accuracy is important, all parameters in the calculation could be declared DOUBLE PRECISION, however a price is paid in significantly increased execution times. For algorithms of this type there is a better way. Since extreme accuracy is desired in the base and sum only and is not needed in Δx, the trick is to declare x_0 and SUM as DOUBLE PRECISION and Δx as single precision (i.e., REAL). Returning to our eight digit machine, this means the code

```
DOUBLE PRECISION ONE, SUM      ⟨This is mixed mode arithmetic.
REAL DX                         The result is double precision.⟩
ONE = 1.D+00
DX = 1.2345678E-6
SUM = ONE + DX
```

is numerically equivalent to

$$
\begin{array}{r}
1.000000000000000 \\
+\ 0.0000012345678 \\
\hline
=\ 1.000001234567800
\end{array}
$$

Thus, this simple alteration to the program, which does not measurably increase execution times, nonetheless retains all the significant figures of the correction term.

Problem: Radar Speed Traps (Electrical Engineering)

A common highway-patrol radar speed-detection apparatus emits a beam of microwaves at frequency f_0. The beam is reflected off an approaching car and the reflected beam is picked up and analyzed by the apparatus. The frequency of the reflected beam is slightly shifted from f_0 to f due to the motion of the car. The relationship between the speed of the car and the two microwave frequencies is

(VII.2)
$$
v = v(f) = \frac{c}{n}\left(\frac{f^2 - f_0^2}{f^2 + f_0^2}\right)
$$

where c is the speed of light (and of microwaves) and n is the index of refraction of microwaves in air. The emitted waves have a frequency of $f_0 = 2 \times 10^{10}$ sec^{-1}, and the other constants have values

$$c = 2.99792458 \times 10^{-8} \text{ m/sec}$$

$$n = 1.00031$$

A zealous patrolman has been ticketing motorists whose speeds are measured as $v > 55.01$ mph. The problem is to convince a judge that the apparatus is not sufficiently accurate to discriminate between 55.00 and 55.01 mph. The argument will hinge on the selectivity of the electronic equipment. The radar device actually measures the difference Δf between the received frequency f and the emitted frequency f_0 to an accuracy of 1 part in 10^4 (that is, 0.01%).

Problem Specifics

After converting the speeds $v_1 = 55.00$ mph and $v_2 = 55.01$ mph to m/sec, the major task of the program is to solve Equation (VII.2) for the received frequencies f_1, f_2 associated with these speeds. This can be accomplished algebraically without much effort, but here we will use the computer. Once these frequencies are obtained, the corresponding frequency shifts $\Delta f_1 = |f_1 - f_0|$, $\Delta f_2 = |f_2 - f_0|$ are computed. If they are the same to four significant figures, the judge will be convinced, otherwise you pay the fine.

Details

Your program should include the following:

1. A Fortran subroutine subprogram

```
SUBROUTINE DNEWTN(XO,EPS,IMAX,ROOT,F,DFDX)
```

that will return a DOUBLE PRECISION root (ROOT) of the function F(X) starting from the DOUBLE PRECISION initial guess X0. The function for $f(x)$ and df/dx should be single precision, however. Also, the convergence criterion should be based on the fractional size of the improvement term, as in

$$\left| \frac{\Delta x}{x_0} \right| < \text{EPS} = 1 \times 10^{-12}$$

Use the value 30 for IMAX.
2. Fortran functions for the function FNC(F) based on Equation (VII.2),

$$\text{FNC}(f) = v - \frac{c}{n}\left(\frac{f^2 - f_0^2}{f^2 + f_0^2} \right) \tag{VII.3}$$

and on the derivative of this function with respect to f; that is,

(VII.4) $$\text{DFDX}(f) = -4\frac{c}{n}\left[\frac{ff_0^2}{(f^2 + f_0^2)^2}\right]$$

The main program should compute the root of Equation (VII.3) to as many significant figures as needed to compare Δf_1, Δf_2.

Programming Problem B: Complex Numbers and AC Circuits (Electrical Engineering)

Introduction

The fundamental equations governing the effects of the basic elements of electric circuits relate the voltage drop across the element to the current flowing through it. These relations are:

(VII.5) Resistors $\quad$ $I \longrightarrow$ $\quad$ $V_R = RI$
$\quad R$

(VII.6) Inductors (coils) $\quad$ L $\quad$ $V_L = L\dfrac{dI}{dt}$

(VII.7) Capacitors $\quad$ C $\quad$ $\dfrac{dV}{dt} = \dfrac{1}{C}I$

where $\quad V$ = voltage drop (volts)
$\quad\quad\quad I$ = current $\quad\quad$ (amperes)
$\quad\quad\quad R$ = resistance $\quad$ (Ohms)
$\quad\quad\quad L$ = inductance $\quad$ (henries)
$\quad\quad\quad C$ = capacitance $\quad$ (farads)

If each of these elements is connected in series to an oscillating voltage supply,

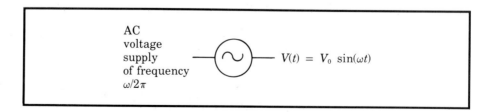

AC voltage supply of frequency $\omega/2\pi$ $\quad$ $V(t) = V_0\sin(\omega t)$

as illustrated in the following figure,

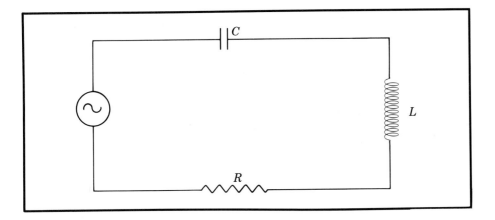

the equation for the circuit simply states that the voltage from the voltage supply must equal the sum of the voltage drops across the three elements.

$$V(t) = V_C + V_L + V_R \qquad\qquad \text{(VII.8)}$$

This equation is then solved for the current I as a function of time. This is not as trivial as it seems, since Equations (VII.5), (VII.6), and (VII.7) are actually differential equations. However, the solution can be made to appear trivial by using complex numbers.

First, recall an important identity from algebra (DeMoivre's theorem)

$$e^{i\theta} = \cos\theta + i\sin\theta$$

Next, since the current is alternating, we expect a solution of the form

$$I(t) = I_0 \cos(\omega t)$$

But if, instead, we make the replacement

$$I(t) \rightarrow I_0 e^{i\omega t} \qquad \langle = I_0 \cos(\omega t) + iI_0 \sin(\omega t)\rangle \qquad \text{(VII.9)}$$

the *actual* current is simply the *real* part of this expression. Inserting this expression into Equations (VII.5), (VII.6), and (VII.7) we obtain

$$V_R = RI_0 e^{i\omega t} = RI(t) \qquad\qquad \text{(VII.10)}$$

$$V_L = LI_0 e^{i\omega t}(i\omega) = i\omega LI(t) \qquad\qquad \text{(VII.11)}$$

$$V_C = \frac{1}{C}I_0 \frac{e^{i\omega t}}{i\omega} = \frac{1}{i\omega}\left[\frac{1}{C}I(t)\right] \qquad\qquad \text{(VII.12)}$$

where I have used the calculus relations

$$\frac{d}{dt}(e^{i\omega t}) = i\omega e^{i\omega t}$$

$$\int e^{i\omega t}\, dt = \frac{1}{i\omega} e^{i\omega t}$$

Inserting these results into Equation (VII.8) results in

$$V(t) = V_R + V_L + V_C$$

(VII.13)
$$= \left[R + i\left(\omega L - \frac{1}{\omega C} \right) \right] I(t)$$

$$= ZI(t)$$

where Z is called the complex impedance of the AC circuit. It is quite analogous to ordinary resistance in DC circuits. The actual current is then

$$I_{\text{actual}}(t) = \text{Real part}\left[\frac{V(t)}{Z} \right]$$

(VII.14)
$$= V_0 \cos(\omega t)\left[\text{Real part}\left(\frac{1}{Z} \right) \right]$$

The measured AC current is an average of $|I_{\text{actual}}(t)|$ over one cycle. The average of the $|\text{cosine}|$ is $1/\sqrt{2}$, so that

(VII.15)
$$\langle I_{\text{actual}} \rangle = \frac{1}{\sqrt{2}} V_0 \, \text{Re}\left(\frac{1}{Z} \right).$$

where Re ( ) means "real part of ( )" and $\langle \ldots \rangle$ designates an average over one cycle.[1] The measured AC voltage, say, across the resistor is then

(VII.16)
$$\langle V_r \rangle = R \langle I_{\text{actual}} \rangle$$

Problem

Use the values

$$V_0 = 10.0 \text{ volts}$$
$$C = 3.35 \times 10^{-7} \text{ farads}$$
$$L = 1.5 \times 10^{-7} \text{ henries}$$

---

[1] More precisely, $[\langle \cos^2(\omega t) \rangle_{\text{one cycle}}] = 1/2$.

for the supply voltage, capacitance, and the inductance. Then, for each of the values of resistance

$$R = 0.0167,\ 0.066,\ 0.250\ \text{ohms}$$

compute 51 values of the complex impedance Z defined in Equation (VII.13) for values of ω from $\omega = 2\pi \times 5 \times 10^5$ to $2\pi \times 10 \times 10^5$ in steps of $2\pi \times 1 \times 10^4$. For each complex value of Z, compute the voltage across the resistor using Equation (VII.16). Plot separate curves for each of the three values of the resistance. The results may surprise you. You should compare with the results of Programming Problem V-A.

Details

In your program, the computed values of the complex impedance should be stored in the COMPLEX array $Z(3,0:50)$. If the real part of a complex number C is A (a real number) and the imaginary part is B (also a real number)—that is, $C = A + iB$, then the Fortran assignment for C is

$$C\ =\ \text{CMPLX}(A,B) \qquad \langle \textit{CMPLX}(\ )\ \textit{is an intrinsic-}$$
$$\textit{type conversion function} \rangle$$

Similarly, the real part of a complex number is obtained in Fortran by using the intrinsic function REAL( ), as

$$A\ =\ \text{REAL}(C)$$

and the imaginary part by using the intrinsic function AIMAG( ),

$$B\ =\ \text{AIMAG}(C)$$

Thus

$$\text{IACTUL}\ =\ (V0/\text{SQRT}(2.))*\text{REAL}(1./Z)$$

APPENDIX SUMMARY OF FORTRAN STATEMENTS AND GRAMMAR RULES

The information contained in this appendix is intended for quick reference. A detailed account of most of the items listed can be found in the text as indicated. For the sake of completeness, several Fortran statements are included here that are not described in the text. These are denoted by asterisks.

In this Appendix I have used the following notation:

[. . .]   indicates an optional part of a statement
⟨ . . . ⟩   a list or numerical value that is to be supplied by the programmer.
CAPITALS   Fortran statements

Fortran statements are entered in columns 7 through 72. Blanks are ignored. A line beginning with a "*" or a "C" in column 1 is ignored by the compiler and is treated as a comment. Any character other than "blank" or zero in column 6 indicates that this line is a continuation of the previous Fortran line. Some Fortran statements may be given an identifying statement number of 1 to 5 digits in columns 1 through 5.

A.1 PROCEDURE STATEMENTS

Statement	Comment
PROGRAM ⟨name⟩ [p.43]	Defines the program name that is used as an entry point for the program execution. If omitted, the compiler assigns a name.
SUBROUTINE ⟨name⟩ [arg. list] [pp.295–297]	The symbolic name defines the main entry point. The subroutine name is not to be assigned a value. The optional argument list contains dummy variable names that can be variables, arrays, dummy procedure names, or alternate return addresses of the form *⟨stmt. no.⟩. Multiple ENTRYs and alternate RETURNs are permitted.

FUNCTION ⟨name⟩ ([arg. list])
[p.304]

The symbolic name defines the main entry point. If the type of the function is specified (e.g., REAL FUNCTION ⟨name⟩), the name must not appear in a type statement. The name must be assigned a value before any RETURN. Control is back to the referencing program unit when a RETURN or an END statement is encountered. Multiple ENTRY points are permitted, but alternate RETURNs are not.

Statement Functions
[pp.108–111, 294]

A user-defined, single statement computation that is valid only within the program unit containing the definition. It is a nonexecutable statement. The argument list may not contain an array or a function name. The statement function name must not appear in an INTRINSIC or EXTERNAL statement.

(*)BLOCK DATA [name]

A nonexecutable subprogram unit that is used to initialize variables in COMMON blocks by means of one or more DATA statements. It may contain specification statements but may not contain any executable Fortran statements.

END

All Fortran procedures must have END as their last line. No procedure may reference itself either directly or indirectly.

Argument Lists

Statement	Comment
Actual Arguments: [pp.111, 298]	Arguments that appear in the "call" to the subprogram. They can be variable names, expressions, array names or elements. They cannot contain a statement function name.
Dummy Arguments: [pp.111, 296, 298]	Arguments that appear in the definition of a subprogram. They are associated with the actual arguments when the subprogram is referenced and when it returns. Dummy arguments that refer

to arrays must be dimensioned within the subprogram to a size less than or equal to the actual dimensioning. The association with actual arguments is by position in the list, and both the total number and types of variables in both lists must agree. If the dummy argument list contains a procedure name, it must be available at the time of the call. The subprogram must not redefine a dummy argument that is a constant, a name of a function, an expression using operators, or any expression enclosed in parentheses.

A.2 SPECIFICATION STATEMENTS

Type Declaration Statements

Statement	Comment
INTEGER ⟨name list⟩ [p.34]	Used to define the names of variables, arrays, functions, or dummy procedures to be of type INTEGER. The names in the name list must be separated by commas.
REAL ⟨name list⟩ [p.34]	Similar to INTEGER, used to define names to be of type REAL.
CHARACTER[*s], ⟨name[*s_1], ... ⟩ [p.35]	Defines a variable, array, function, or dummy procedure to be of type CHARACTER. If the optional [*s] is present, each of the elements of the name list is of length s, where s is a positive (unsigned) constant or an asterisk enclosed in parentheses. The latter is used in subprograms for the assigning of dummy argument names to have whatever length the associated actual argument has at the time of the call. Alternately, each name in the name list can be assigned different lengths by using the form ⟨name⟩*s_1, where s_1 is the length of the string and satisfies the

same rules as does *s*. The string lengths (s, s_1) may be variable names only if the variable has been initialized in a previous PARAMETER statement and is enclosed in parentheses.

DOUBLE PRECISION ⟨name list⟩
[pp.354–355]

Similar to INTEGER and REAL. Used to define names to be of type DOUBLE PRECISION.

COMPLEX ⟨name list⟩
[pp.355–356]

Similar to INTEGER and REAL. Used to define names to be of type COMPLEX. COMPLEX numbers consist of two real numbers corresponding to the real and imaginary parts. Thus if C is complex, an assignment statement would be

```
C = (3.0,4.0)
```

and SQRT(C) would return the complex result

```
SQRT(C) → (2.0,1.0)
```

LOGICAL ⟨name list⟩
[pp.356–357]

Similar to INTEGER and REAL. Used to define names to be of type LOGICAL. LOGICAL variables can have only two values, namely,

```
.TRUE.     .FALSE.
```

(The periods are part of the value.) Thus if X is of type LOGICAL, the expression

```
X = 4 .GT. 7
```

assigns a value of .FALSE. to X.

IMPLICIT ⟨type⟩ ($a_1 - a_2$)
[pp.357–358]

Used to override or augment the default typing. ⟨type⟩ is any of the above six variable types and a_1, a_2, are single letters. The IMPLICIT statement must precede all statement specifications except PARAMETER. Explicit typing overrides an IMPLICIT specification.

Other Specification Statements

Statement	Comment
PARAMETER (⟨name⟩=⟨exp⟩, . . .) [pp.360–361]	Assigns a symbolic name to a constant. ⟨exp⟩ is a constant expression (can be of type CHARACTER). Any variable name in the expression must have been previously defined in a PARAMETER statement. The parentheses may contain more than one assignment. Variables in one PARAMETER statement may not be redefined in another. Variables initialized in PARAMETER statements may be used in a DATA statement but not in a FORMAT statement.
DATA ⟨name list⟩/ ⟨value list⟩/ [,⟨name list⟩/⟨value list⟩/ . . .] [pp.359–360]	Used to initialize variables, arrays, array elements, and substrings at compilation time. DATA statements are nonexecutable and must appear after other specification statements and should appear before the first executable statement. The same name should not appear in two DATA name lists. The values in the value list are assigned, one-to-one, to the elements of the name list, which should agree in type and must agree in total number of elements. The name list may contain an array name and an implied DO loop (or nested loops) of the form: (A(I),I=ILO,IHI,ISTEP). The values in the limits of the implied DO and in the value list must be constants or named constants defined in a PARAMETER statement. The values in the value list may be repeated by preceding by a positive (unsigned) integer or named integer constant (defined in a PARAMETER statement). Variables in blank COMMON cannot be assigned values with a DATA statement.
DIMENSION ⟨array name⟩(n_d [,n_d, . . .]) [pp.219–221]	Designates a name as an array name and defines the subscript bounds. More than one array can be declared in a sin-

gle DIMENSION statement. The form of the subscript bound definition, n_d, is either one of the forms

i_{top} limits are 1, 2, . . . , i_{top}; $i_{top} > 0$

$i_{bot} : i_{top}$ limits are i_{bot}, $i_{bot} + 1$, . . . , i_{top}

If only the upper bound (i_{top}) is given, the default value for i_{bot} is 1. The bounds must be integers or integer expressions. In the initial dimensioning, the bounds must be integer constants; while in subsequent dimensioning (in subprograms) the bounds can be dummy integer variables or expressions.

EXTERNAL ⟨proc. name⟩
[pp.318–320]

Used to define a name as representing a user-written, externally defined, subprogram, procedure, or dummy procedure name. More than one name can be declared external in a single EXTERNAL statement. The purpose of the EXTERNAL statement is to allow the name to appear as an actual argument in an argument list. If an intrinsic function name is entered in an EXTERNAL ⟨proc. name⟩ list, the name then refers to a user-written function and the library function can no longer be referenced.

INTRINSIC ⟨func. name⟩
[pp.318–319]

Similar to EXTERNAL, but applies to library functions. All intrinsic function names that appear as actual arguments in an argument list must be declared INTRINSIC. A function cannot be declared both intrinsic and external. Type conversion functions (e.g., FLOAT, IFIX) and min/max functions (e.g., MIN, MAX) cannot be used as actual arguments.

COMMON [/blockname/] ⟨name list⟩
[pp.326–328]

Stores all the variables in name list together in a block of memory which can be given a block name. These variables may then be accessed by different program units without using argument lists. As with argument lists, elements in the name list are assigned values by position. Two program units that share data via COMMON blocks must have a COMMON statement with the same block name (or unnamed) and be of identical lengths. More than one block name may be defined in a single COMMON statement. The block name may also be used as a variable name without conflict. If any variable in a COMMON block is of type CHARACTER, then *all* variables in the block must be of type CHARACTER. Entries in a labeled COMMON block can be initially defined via a DATA statement in BLOCKDATA subprogram only. The variables in blank COMMON are automatically saved upon return from a subprogram, while those in labeled COMMON *may* not be. (They are saved if the labeled COMMON in the subprogram also appears in the main program.)

(∗)EQUIVALENCE (⟨name list⟩)

Provides for the sharing of the same memory locations by two or more variables, arrays, array elements, or character substrings. When coupled with COMMON statements, the effect of EQUIVALENCE statements can be extremely complex, and use of this statement is not recommended.

SAVE [⟨name list⟩]
[p.321]

Preserves the value of variables in a subprogram after a RETURN has been executed. The value of the variable may then be referenced in a subsequent call to the subprogram. Dummy variable names, names in a COMMON block, and procedure names must not appear in the name list. A SAVE statement with no name list will SAVE *all* allowable variables in the subprogram.

(∗)ENTRY ⟨entry name⟩[(arg. list)]

In addition to the main entry point of a subprogram (the top), the ENTRY statement may be used to begin the subprogram execution anywhere except within a DO loop or an IF-THEN-END IF block. The subprogram is initiated at the alternate entry point by replacing the subprogram name by ⟨entry name⟩ in the referencing line. The argument list in the entry statement should be similar to the argument list in the subprogram definition. ENTRY statements are often used to skip repetitive computations or assignments in a subprogram, such as

```
PROGRAM MANE
  ...      ...
CALL XX(A,M)
  ...      ...
CALL XXMID(B,N)
  ...      ...
END
SUBROUTINE XX(C,L)
DIMENSION C(L)
DO 1 I = 1,L
1 C(I) = 0.0
ENTRY XXMID(C,L)
C(5) = ...
  ...      ...
END
```

A.3 ASSIGNMENT AND PROGRAM CONTROL STATEMENTS

Statement
Assignment Statement:
⟨var. name⟩ = ⟨exp⟩
[p.41]

Comment
Where ⟨var. name⟩ is the name of a variable or array element. If the expression ⟨exp⟩ on the right is arithmetic, it is first evaluated according to the hierarchy rules and the type of the dominant variable type, then converted to the type of ⟨var. name⟩ and then assigned to ⟨var. name⟩. If the expression is of type CHARACTER, the variable name must also be of type CHARACTER and if the string lengths differ the expression is

either padded with blanks to the right or truncated on the right to match the length of ⟨var. name⟩. If ⟨var. name⟩ is of type LOGICAL, then the expression must have a value of either .TRUE. or .FALSE. Multiple assignments of the form A = B = C = D = 5 are not standard Fortran 77.

(∗)ASSIGN ⟨stmt. no.⟩ TO ⟨name⟩

In this statement, ⟨stmt. no.⟩ is the statement number of an executable statement or a FORMAT statement, and ⟨name⟩ is the name of an integer variable. This statement is used in conjunction with the ASSIGNED GO TO statement or with WRITE(5, ⟨name⟩). . . statements and in general is not recommended.

END
[p.49]

Used to mark the end of a compilation unit. The END statement can have a statement number. If during execution of the main program, the program flow branches to, or encounters, an END statement, the program terminates. The same situation in a subprogram will result in a RETURN. Both are considered poor style.

STOP [tag]
[pp.15, 49]

The STOP statement terminates the execution of the program wherever it is encountered in a program or subprogram and the word "STOP" is displayed in the day file or on the terminal screen. The optional [tag] can be a positive integer (of five digits or less) or a character string constant and will be displayed along with STOP. Example,

```
STOP 'SUCCESSFUL RUN -
                JOB TERMINATED'
STOP 97
```

A program may have more than one STOP statement.

(∗)PAUSE [tag]

Similar to the STOP statement. This statement causes the program execution to be interrupted and the word

"PAUSE" followed by the optional [tag] to be displayed. If the program is being run in batch mode, only the operator at the console can cause the program to continue. In interactive mode, the user enters either "DROP" to terminate or "GO" to continue. Use of this statement is not recommended.

A.4 FLOW-CONTROL STATEMENTS

Statement
RETURN [exp.]
[p.296]

Comment
A RETURN statement causes the termination of a subprogram procedure and the return to the referencing program or subprogram. RETURN statements may only appear in subprograms and each subprogram may have more than one RETURN. If the optional alternate return address expression [exp] is omitted, the procedure returns to the next statement in the referencing program unit; the normal situation.

(*)*Alternate RETURN*

An alternate RETURN (from subroutines only) is effected by including an integer or integer expression following RETURN *and* a sequence of asterisks in the defining subroutine argument list which will function as dummy address labels and will be associated with the actual statement numbers in the referencing call. Thus RETURN 3 will cause a return to the statement number in the third position in the actual argument list. The statement numbers in the actual argument list must refer to executable statements and be preceded by single asterisks.

```
PROGRAM MANE
   ...        ...
CALL CAL(A,B,*1,*3,*7)
```

```
                STOP
                  ...          ...
           3  C  =  A  +  B
                  ...          ...
           1  C  =  A  -  B
                  ...          ...
           7  C  =  A/B
                  ...          ...
                END
                SUBROUTINE  CAL(S,T,*,*,*)
                  ...          ...
                RETURN  3      ⟨causes return
                                  to stmt. 7⟩
                  ...          ...
                RETURN         ⟨normal
                                  return,
                                  executes the
                                  STOP⟩
                END
```

CALL ⟨subname⟩ [(arg. list)]
[p.296]
    Initiates a transfer of control to the subroutine named ⟨subname⟩. The argument list contains actual arguments that may be constants, expressions, variable names, array names or elements, procedure names, or an alternate return address in the form *⟨stmt. no.⟩.

CONTINUE
[pp.78, 229]
    The CONTINUE statement is an executable statement that performs no operation and may be placed anywhere among the executable statements. It is most commonly used as the terminus of DO loops. The CONTINUE statement should have a statement number.

GO TO Statements

Statement

GO TO ⟨stmt. no.⟩
[pp.21, 77]

Comment

Unconditional GO TO: Simply transfers control to the statement labeled with ⟨stmt. no.⟩, which must be an integer constant, not a variable name, and correspond to an existing executable statement anywhere in the same program or subprogram. When possible, use of the GO TO should be avoided and replaced with structured Fortran.

GO TO (⟨stmt. no. list⟩), ⟨exp⟩
[p.89]

Computed GO TO: The statement number list contains labels of existing executable statements, and ⟨exp⟩ is an integer or integer expression. If ⟨exp⟩ is 1, control transfers to the statement identified by the first number in the list; if ⟨exp⟩ is 2, to the second; etc. If ⟨exp⟩ is less than 1 or greater than the number of labels in the list, execution continues with the next line of the program after the GO TO.

(∗)GO TO ⟨ivar⟩, [(⟨stmt. no. list⟩)]

Assigned GO TO: The integer variable ⟨ivar⟩ must have been previously assigned a value by an ASSIGN TO statement. The statement then acts much like the unconditional GO TO if the statement number list is not present and like the computed GO TO if it is. Use of this statement should be avoided.

IF Statements

Statement	Comment
IF(⟨arith. exp⟩) s_-, s_0, s_+ [p.90]	*Arithmetic IF:* The arithmetic expression is evaluated and if negative, zero, or positive control is transferred to the executable statement labeled by the statement number s_-, s_0, or s_+, respectively.
IF(⟨log. exp⟩) ⟨exec. stmt.⟩ [p.88]	*Logical IF:* The logical expression ⟨log. exp⟩ is evaluated, and if true, the executable Fortran statement ⟨exec. stmt.⟩ is executed, otherwise the program continues with the next line after the IF. The executable statement ⟨exec. stmt.⟩ cannot be a DO, IF, ELSE, ELSE IF, END, or END IF statement.

Block IF Statements

Statement	Comment
IF(⟨log. exp⟩)THEN [p.62]	If the logical expression ⟨log. exp⟩ is true, then the program execution continues with the next line; otherwise the control branches to the next ELSE or ELSE IF statement if present, or to the END IF, the terminus of the block, if an ELSE or ELSE IF is not present.
ELSE [p.69]	Marks the beginning of the alternate (false) path of an IF(. . .)THEN or an ELSE IF(. . .)THEN statement. The ELSE statement should not have a statement number.
ELSE IF(⟨log. exp⟩)THEN [p.72]	The operation of this statement is the same as the IF(. . .)THEN statement, however, it can only be reached by an evaluation of ⟨false⟩ in a previous IF(. . .)THEN or ELSE IF(. . .)THEN statement.
	ELSE and ELSE IF statements must be placed within a corresponding IF(. . .)THEN–END IF block. Block IF structures can be nested. That is, one block IF structure may contain another only if the entire second block IF is contained within the first. It is permitted to branch to a block IF statement, but not permitted to branch to any statement within a block. Each block IF structure can contain several ELSE IF statements but only a single ELSE statement which must follow all the ELSE IFs in the structure.
END IF [p.62]	Marks the end of an IF(. . .)THEN structure. The END IF statement should not have a statement number. (It is not permitted to GO TO ⟨ENDIF⟩.) Each IF(. . .)THEN must have a corresponding END IF statement. If the program flow arrives at an END IF statement, the program continues with the next line of the program.

DO-loop Structures

Statement	Comment
DO ⟨stmt. no.⟩ $a_c = b_{low}, b_{high} [,b_{step}]$ [p.228ff]	Marks the beginning of a block of statements that are executed with the value of the counter variable, a_c, set equal to the initial limit, b_{low}, and then the entire block is repeated with a_c assigned the value $b_{low} + b_{step}$ (or $b_{low} + 1$ if the optional b_{step} is omitted). This is repeated until the counter variable exceeds the final limit, b_{high}. The program then continues with the line following the loop terminator. The loop terminator is an executable statement with statement number ⟨stmt. no.⟩ which occurs after the related DO statement and which is not an IF, GO TO, RETURN, STOP, END, ELSE, or another DO. CONTINUE statements are always recommended as DO terminators. The DO-loop limits, b_{low}, b_{high}, b_{step}, are numerical constants, variables, or expressions that are converted to the type of the counter a_c before execution of the loop. Generally, a DO loop will execute zero times if $b_{low} > b_{high}$ (if $b_{step} > 0$ or omitted) or if $b_{low} < b_{high}$ and b_{step} is negative. If the loop executes zero times, the counter variable has the value of b_{low}. Transfer out of a loop before completion is permitted and the value of a_c will be its most recent value. Branching into a DO loop that has not yet been initiated is forbidden; however, branching out and then back in is permitted provided the counter variable has not been redefined. This, however, is strongly discouraged.
Nested Loops [p.231]	DO loops can be nested provided one loop is totally contained within the other. Nested loops may share the same terminal line. Branching to a shared terminal statement from an inner loop does not constitute branching out of the inner loop. If a block IF contains a DO

loop, the loop must be completely contained within the block.

A.5   FORTRAN FILE DIRECTIVE STATEMENTS

Statement
OPEN(⟨arg. list⟩)
[pp.168–170]

Comment
The OPEN statement is used to assign a unit number to an already existing file, to a newly created file, or to alter some properties of existing files. The extensive options permitted in the argument list are explained in Section 4.6.2. The most common use is to permit a program to access a separate data file or to write output on a disk-stored file. The shortened form is then

```
OPEN(13,FILE='DATAFL',STATUS='OLD')
OPEN(14,FILE='RESULTS',STATUS='NEW')
```

The unit number can be a positive integer (of three digits or less) or an integer constant name initialized in a parameter statement. When a program begins execution, the input and output files are automatically connected to the program and need not be opened. These files may be specified by using an asterisk in place of the unit number in I/O statements.

CLOSE(⟨arg. list⟩)
[p.172]

The CLOSE statement is used to disconnect a file from a unit number enabling that unit to be connected to a different file. The most common form is

```
    CLOSE(<unit no.>)
or
    CLOSE(UNIT=<unit no.>)
```

Any file that has been previously opened with a status other than "SCRATCH" (i.e., either "OLD" or "NEW") will automatically be retained on the system after a CLOSE. To disconnect *and* delete the file from the system, the option STATUS = "DELETE" is included in the argument list. SCRATCH files are always automatically deleted upon termination of a program. Also, the system will automatically CLOSE all connected files upon program termination, so this statement is often unnecessary.

(∗)INQUIRE(⟨arg. list⟩)

Fortran 77 permits the user to ascertain the present attributes of a file by means of the INQUIRE statement. This statement is generally used to avoid errors in opening a file or in reading or writing a file. The options available in the argument list are very extensive and it is suggested that if you think you have need for this statement, a detailed account can be found in Balfour and Marwich or in Wagener listed in the references.

REWIND([UNIT=]⟨unit no.⟩)
[p.171]

The REWIND statement positions a file to the beginning of the file. The file must have been previously opened and assigned a ⟨unit no.⟩. A shortened form is

```
REWIND <unit no.>
```
⟨no parentheses needed⟩

The ⟨unit no.⟩ is a positive integer (of three digits or less) or an integer constant initialized in a PARAMETER statement.

BACKSPACE ([UNIT=]⟨unit no.⟩)
[p.171]

Similar to REWIND, except the file is only backspaced one record (usually one line). Only sequential files can be backspaced.

ENDFILE([UNIT =]⟨unit no.⟩)
[p.172]

The END FILE statement puts an end-of-file mark on the file connected to ⟨unit no.⟩. This is useful when constructing a data file and the "END =" option in the

---

READ statement is anticipated. Only sequential files can be marked with an end-of-file.

A.6   INPUT/OUTPUT STATEMENTS

Input/Output Lists

The I/O list (designated in what follows as either ⟨in list⟩ or ⟨out list⟩) is that part of an I/O statement in which the elements to be read or written are specified along with their ordering. The elements of an input list may be variables, array names, or elements, or these items enclosed in an implied DO loop. The elements of an output list may additionally include constants, arithmetic expressions, character string constants, or references to functions, provided the functions themselves neither cause I/O operations nor alter other elements of the list. A character string constant is a string enclosed in apostrophes—e.g., 'THE ANSWER IS'. The appearance of the name only of an array in either an input or an output list will cause the entire array to be read or written in the order in which it is stored. An implied DO loop in an I/O list is treated as a single element of the list. (For a more detailed description of implied DO loops see Section 7.5.)

Statement	Comment
READ([UNIT=]⟨unit no.⟩,⟨format⟩[,ERR=⟨err-sl⟩][,END=⟨end-sl⟩])⟨in list⟩ [p.166]	The only essential specifications in the READ statement are the input unit number ⟨unit no.⟩ and the format specification ⟨format⟩ that specifies the arrangement of the items in the input list ⟨in list⟩. The unit number may refer to any opened file or may be replaced by an asterisk, which is the default specification for file INPUT. The optional specifications, ERR = ⟨err-sl⟩ will cause a transfer to the statement labeled by statement number ⟨err-sl⟩ if an error is encountered during the read and END = ⟨end-sl⟩ will do similarly if an end-of-file mark is encountered. Examples,
READ(5,3)X	*Read from file 5 according to format 3, the value of X.*
READ(*,3,ERR=9)Y	*Read X from file INPUT according to format 3; if error, branch to statement 9.*

Additionally, the format specification ⟨format⟩ may be either the statement number of an existing FORMAT statement, a set of format specifications enclosed in parentheses and delimited fore and aft by apostrophes, as

```
'(1X,F5.3,/,10X,I5)'
```

or it may simply be a single asterisk, in which case the variables are read without format as list-directed input. List-directed input data elements are separated by commas. Example,

```
READ(*,*)X
```
Reads X from file INPUT without a format.

WRITE([UNIT=]⟨unit no.⟩,⟨format⟩[,ERR=⟨err-sl⟩])⟨out-list⟩
[p.165]

The meaning of the specifications are the same as in the READ statement except that replacing the ⟨unit no.⟩ by an asterisk will cause the output to be written to the file OUTPUT. A shortened form of this statement is

```
PRINT <format>,<out list>
e.g.
PRINT 6,X    Print X according
to format 6
PRINT *,X Print X without
format specifications
```

If the ⟨format⟩ = '*' option is used, the elements of the ⟨out list⟩ are written in a format controlled by the compiler.

⟨stmt. no.⟩ FORMAT (⟨format spec. list⟩)
[p.187ff]

A FORMAT statement is nonexecutable, must have a statement number, and can appear anywhere within a program unit. The format specification list is a sequence of editing specifications separated by commas and of the form

[n]⟨edit-rep.⟩
⟨edit-nonrep.⟩
[n](format spec. list)

where [n] is an optional positive

(unsigned) integer repeat constant, ⟨edit-rep.⟩ is a repeatable edit specification, ⟨edit-nonrep.⟩ is a nonrepeatable edit specification, and the last form is a multiple of an entire format specification sublist. For a description of the more common edit specifications see Chapter 6.

Fortran 77 References

American National Standards Institute: *American National Standard Fortran X3.9-1978* (this is the standard of Fortran-77.), 1430 Broadway, New York.

Balfour, A., and D. H. Marwick: *Programming in Standard Fortran 77,* North Holland, New York, 1979. A comprehensive listing of all features of Fortran-77.

Kernighan, B., and P. J. Plauger: *The Elements of Programming Style,* McGraw-Hill, New York, 1974.

Merchant, Michael J.: *Fortran-77, Language and Style,* Wadsworth, Belmont, Cal., 1981. A comprehensive treatment of Fortran 77 from a non-engineering perspective.

Wagener, Jerold L.: *Fortran 77,* John Wiley, New York, 1980. Contains a more extensive description of the application of data files.

Numerical Methods References

Acton, F. S.: *Numerical Methods That Work,* Harper & Row, New York, 1970.

Cheney, W., and D. Kincaid: *Numerical Mathematics and Computing,* Brooks/Cole, Monterey, Cal., 1980.

Conte, S. D., and C. deBoor: *Elementary Numerical Analysis,* McGraw-Hill, New York, 1972.

Hornbeck, R. W.: *Numerical Methods,* Quantum Publications, New York, 1975.

Pennington, R. H.: *Introductory Computer Methods and Numerical Analysis,* 2nd ed., Macmillan, London, 1970.

Southworth, R. W., and S. L. DeLeeuw: *Digital Computation and Numerical Methods,* McGraw-Hill, New York, 1965.

ANSWERS AND SOLUTIONS TO ODD-NUMBERED PROBLEMS

Chapter 1

1. a) $(11)_{10} = 2 \times 5 + 1$
$= 2 \times 2 + 1$
$= 2 \times 1 + 0$
$= 2 \times 0 + 1$

$\quad = (1011)_2$

  c) $(100)_{10} = (1100100)_2$

  e) $(12.625)_{10}$: $(12)_{10} = (1100)_2$

$\quad .625 \times 2 = 1 + .25$
$\quad .25 \quad \times 2 = 0 + .5$
$\quad .5 \quad\; \times 2 = 1 + 0$
$\quad (.625)_{10} = (.101)_2$
$(12.625)_{10} = (1100.101)_2$

  f) $(0.1)_{10}$: $.1 \times 2 = 0 + .2$
$\quad .2 \times 2 = 0 + .4$
$\quad .4 \times 2 = 0 + .8$
$\quad .8 \times 2 = 1 + .6$
$\quad .6 \times 2 = 1 + .2$
$\quad .2 \times 2 = 0 + .4$
$\quad .4 \times 2 = 0 + .8$
$\quad .8 \times 2 = 1 + .6$
$\qquad \cdots \qquad \cdots$

$(0.1)_{10} = (.00011001100110011\ldots)_2$

3. a) $\begin{array}{r} 1011 \\ + \; 11 \\ \hline (1110)_2 \end{array}$ $\rightarrow$ $\begin{array}{r} 11 \\ +3 \\ \hline (14)_{10} \end{array}$

  b) $\begin{array}{r} 1010 \\ - \; 11 \\ \hline (111)_2 \end{array}$ $\rightarrow$ $\begin{array}{r} 10 \\ -3 \\ \hline (7)_{10} \end{array}$

5. Faced with an arrangement of two unequal rows of markers, the winning move is to remove markers from the longer row until the remaining two rows are of equal length. The opponent's move must remove this symmetry. Your next move restores it—i.e., results in rows of equal length. Finally, your next to last move results in

$$X$$
$$X$$

which obviously leads to a win.

    The idea is to keep the number of counters in each *column* even. Extending this to more than two rows, the basic idea of preserving a pattern from one move to the next remains. The final winning pattern above can also be described in terms of the base-two number of markers in each row.

$$X \; (0001)_2$$
$$X \; (0001)_2$$

    The signature of this pattern is that the ones or zeros in each *column* of the base-two representation of the numbers add to an even number (here 0 or 2). *Any* alteration of a single row, even empty rows, will result in at least one of the columns being odd.

Chapter 2

1. a) $[(2.^{1/2})^2 - 2.]$ evaluated on a variety of computing devices yields:

CDC-CYBER computer	DEC-20 computer	TI calc.
0.0	2.980E−8	−1.9E−9

while $[(3.^{1/2})^2 - 3.]$ yields:

−1.421E−14	0.0	−2.5E−9

The approximate number of significant figures is then:

14	8	9

  b) Executing the INTEGER program yields the following results:

$$\text{CDC-CYBER} - I_{max} = 100\ 000\ 000\ 000\ 000$$
$$\text{DEC-20} - I_{max} = 1\ 000\ 000\ 000$$

3. a) $\begin{array}{r} 1000.00 \\ + \quad\;\; .00999999 \\ \hline = 1000.00999999 \end{array}$ ⟨Assuming the computer does not round the result.⟩ Thus, EPS = .00999999.

  b) If EPS = 0.0001 and forgetting for a moment that the computer does binary, not decimal arithmetic, the first 100,000,000 terms add to 100.000. Adding .0001 to this number does not change the sum. (100.000 + .0001 = 100.) Thus the 10^8 terms will add to 100.

  c) $10^8 \times 10^{-4} = 10^4 = 10,000$.

5. a) ERROR, no commas   b) ERROR, no decimal point allowed in exponent

c) OK, but base should have decimal point

e) OK

g) OK

i) ERROR, cannot take square root of negative 3.

k) OK, mixed-mode replacement

d) OK, mixed-mode replacement

f) ERROR, 6XA not valid variable name

h) ERROR, two arithmetic operators cannot touch, $\langle **-\rangle$

j) OK, result = 1.

l) OK, but does nothing

i) OK, assuming the variables GE and LE have values.

j) OK, assuming the variable EQ has a value.

k) OK

5. a) True; b) True; c) False

7. Using the features of integer arithmetic, the program can be constructed from

```
INTEGER N
READ *, N
IF((N/2)*2 .EQ. N)THEN
      N is divisible by 2 and is even
ELSE
      N is odd
END IF
```

7. a) 3.   b) 4   c) 0
   d) 4.   e) 1   f) 1.5
   g) 9   h) 5   i) 4.5
   j) 27.   k) 1

9. a) X*Y/(Z + 1)
   b) X**(N + 1)
   c) X**(1./2.) or X**.5
   d) ACOS(ABS(LOG(X))) 〈note three right parentheses〉
   e) (X**A)**B or X**(A*B)

Chapter 3

1. a) THEN must be on same line as IF(. . .)
   b) OK
   c) OK, but note the first two tests are mutually exclusive—i.e., there is no possibility of executing the PRINT.

3. a) ERROR, must GO TO a statement number, not a name.
   b) OK
   c) OK
   d) ERROR, A = 0. does not have a value of true or false. The expression should be A .EQ. 0.
   e) ERROR, both sides of .AND. must be either true or false. Expression should read

```
        X .EQ. 1.5 .AND. Y .EQ. 1.5
```

   f) OK, but odd logic
   g) ERROR, the conditional result of a logical IF must be an executable statement,

```
        IF(A .EQ. 0.)A = 0.
```

   Note, this statement will not alter A.
   h) ERROR, the executable statement following the IF test must not be another IF test.

9. The conditions that must be satisfied if four lengths a, b, c, d are to form a polygon are:

$$a \le b + c + d$$
$$b \le a + c + d$$
$$c \le a + b + d$$
$$d \le a + b + c$$

```
        PROGRAM POLYGON
        REAL A,B,C,D
        READ *,A,B,C,D
*       a) Test four conditions
        IF(A.LE.B+C+D .AND. B.LE.A+C+D .AND.
     +  C.LE.A+B+D .AND. D.LE.A+B+C)THEN
          PRINT *,A,B,C,D,'CAN FORM A POLYGON'
        ELSE
          PRINT *,A,B,C,D,'CAN NOT FORM POLYGON'
        END IF
*       b) Next test for equal sides
        IF(A .EQ. C .AND. B .EQ. D)THEN
          PRINT *,'A RECTANGLE IS POSSIBLE'
          IF(A .EQ. B)THEN
          PRINT *,'WHICH IS A SQUARE'
          END IF
        ELSE
          PRINT *,'RECTANGLE NOT POSSIBLE'
        END IF
        STOP
        END
```

11. The results for the functions given in the problem are:

```
IMAX = 25, DXMAX = 0.8,
EPS = 1.E-5, XO = 0.8
```

a) $x_1 = [\ln(3/x)]^{1/2}$, $x = 1.032683$
      after 14 iterations
   $x_1 = 3e^{-x^2}$, diverges after one iteration
b) $x_1 = [7 - 5x^3]^{1/10}$, diverges after one step
      negative base to real
   $x_1 = [(7 - x^{10})/5]^{1/3}$, excessive iterations
      last $x = 1.0347$

13.

```
PROGRAM PRIME
INTEGER N,P
PRINT *,'ENTER A POTENTIAL PRIME, N'
READ *,N
IF((N/2)*2 .EQ. N)THEN
    PRINT *,N,'IS NOT PRIME'
    PRINT *,'IT IS AN EVEN NUMBER'
END IF
P = 3
1 IF((N/P)*P .EQ. N)THEN
    PRINT *,N,' IS NOT PRIME'
    PRINT *,'IT IS DIVISIBLE BY ',P
    STOP
ELSE
    P = P + 2
    IF(P .GE. SQRT(N + 1.))THEN
       PRINT *,N,' IS PRIME'
       STOP
    ELSE
       GO TO 1
    END IF
END IF
END
```

Chapter 4

1. a) $F(X) = 3.*X*X + X - 1.$
b) $G(X,A,B,C) = A*X*X + B*X + C$
c) $H(X,A) = EXP(-A*X) + LOG(SIN(ACOS(-1.)*X))$
d) $INDEX(I,J) = 3*I + 2*J$

3.

```
PROGRAM SUMS
INTEGER SUMA, SUMB, SUMC,SUMD,
+       TERMA,TERMB,TERMC,TERMD,
+       ANSWA,ANSWB,ANSWC,ANSWD,N,RUN
*  RUN = 1
1 IF(RUN .EQ. 1)THEN
    N = 2
  ELSE IF(RUN .EQ. 2)THEN
    N = 10
  ELSE IF(RUN .EQ. 3)THEN
    N = 25
  ELSE IF(RUN .EQ. 4)THEN
    N = 100
  ELSE
    STOP
  END IF
*
  PRINT *,' '
  PRINT *,' N = ',N,
  PRINT *,'  COMPUTER    THEORY'
  PRINT *,'    SUM        EQ.'
*
  SUMA = 0
  SUMB = 0
  SUMC = 0
  SUMD = 0
  I = 1
2 TERMA = I
  TERMB = I**2
```

```
  TERMC = I**3
  TERMD = I**4
    SUMA = SUMA + TERMA
    SUMB = SUMB + TERMB
    SUMC = SUMC + TERMC
    SUMD = SUMD + TERMD
  I = I + 1
  IF(I .LE. N)THEN
     GO TO 2
  ELSE
     ANSWA = N*(N + 1)/2
     ANSWB = ANSWA*(2*N + 1)/3
     ANSWC = ANSWA**2
     ANSWD = ANSWB*(3*N**2 + 3*N - 1)/5
     PRINT*,'A)    ',SUMA,'          ',ANSWA
     PRINT*,'B)    ',SUMB,'          ',ANSWB
     PRINT*,'C)    ',SUMC,'          ',ANSWC
     PRINT*,'D)    ',SUMD,'          ',ANSWD
  END IF
  RUN = RUN + 1
  GO TO 1
END
```

N = 2

	COMPUTER SUM	THEORY EQ.
A)	3	3
B)	5	5
C)	9	9
D)	17	17

N = 10

	COMPUTER SUM	THEORY EQ.
A)	55	55
B)	385	385
C)	3025	3025
D)	25333	25333

N = 25

	COMPUTER SUM	THEORY EQ.
A)	325	325
B)	5525	5525
C)	105625	105625
D)	2153645	2153645

N = 100

	COMPUTER SUM	THEORY EQ.
A)	5050	5050
B)	338350	338350
C)	25502500	25502500
D)	2050333330	2050333330

5.

```
PROGRAM ETOX
INTEGER I,IMAX,IEXP
REAL    X,XX,SUM,TERM,EPS
PRINT *,'ENTER X'
READ *,X
IMAX = 100
EPS  = 1.E-6
XX   = ABS(X)
```

```
IEXP = 0
IF(XX .GT. 10.)THEN
    IEXP = LOG10(ABS(XX))
    XX   = XX/(10.**IEXP)
END IF
SUM = 1.0
I = 1
TERM = XX
1 SUM = SUM + TERM
  IF(ABS(TERM) .LT. EPS)THEN
    ANSWER = SUM**(10**IEXP)
    IF(X .LT. 0.)ANSWER = 1./ANSWER
    PRINT *,'EXP(',X,') = ',ANSWER
    STOP
  ELSE
    TERM = TERM*XX/(I + 1.)
    I = I + 1
    IF(I .GT. IMAX)THEN
        PRINT *,'DID NOT CONVERGE'
        STOP
    ELSE
        GO TO 1
    END IF
  END IF
END

------------------------------------------

ENTER X
? 50.
 EXP(50.0) = 5.1847047E+21
```

7.
```
PROGRAM GOLDEN
INTEGER I,IMAX
REAL    R0,R1,EPS
IMAX = 100
EPS  = 1.E-6
I = 0
R0 = 1.0
1 R1 = 1./(1. + R0)
  IF(ABS(R1 - R0) .LT. EPS)THEN
    PRINT *,'THE SOLUTION = ',R1
    STOP
  ELSE
    I = I + 1
    IF(I .LT. IMAX)THEN
      GO TO 1
    ELSE
      PRINT *,'DID NOT CONVERGE'
      STOP
    END IF
  END IF
END
```

The equation can be written as

$$r^2 + r - 1 = 0$$

which, from the quadratic equation, has roots $(\sqrt{5} - 1)/2$, $-(\sqrt{5} + 1)/2$.

9. a) $f(x) = x^2 + 2x - 15$, $a = 2.8$, $b = 3.1$

i	x_1	x_3	f_1	f_3	x_2	f_2	L/R
0	2.800	3.100	−1.560	0.810	2.950	−0.398	R
1	2.950	"	−0.398	"	3.025	0.201	L
2	"	3.025	"	0.201	2.9875	−0.099	R
3	2.9875	"	−0.099	"	3.0063	0.050	L
4	"	3.0063	"	0.050	2.9969	−0.025	R
	. . .	. . .		. . .	. . .		

To achieve three significant figures will require n steps, where $n > \ln[(b - a)/\delta]/\ln(2)$ and $\delta = 5E\text{-}3$. That is, $n = 6$ steps.

b) $g(x) = \sin(x)\sinh(x) + 1$, $a = 1$, $b = 4$

i	x_1	x_3	f_1	f_3	x_2	f_2	L/R
0	1.000	4.000	1.989	−19.65	2.500	4.621	R
1	2.500	"	4.621	"	3.250	−.393	L
2	"	3.250	"	−0.393	2.875	3.327	R
3	2.875	"	3.327	"	3.0625	1.84	R
4	3.0625	"	1.843	"	3.1563	0.82	R
5	3.1563	"	0.828	"	3.2031	0.244	R
6	3.2031	"	0.244	"	3.2266	−0.067	L
7	"	3.2266	"	−0.067	3.2148	0.090	R
	. . .	. . .		. . .	. . .		

To achieve three significant figures will require n steps where $n > \ln[(b - a)/\delta]/\ln(2)$ with $\delta = 5E\text{-}3$. Thus $n = 9$ steps are required.

11.
```
PROGRAM MAX
INTEGER I,IMAX
REAL    X0,X1,DX,F0,F1
F(X) = <statement function for f(x)>
PRINT *,'ENTER STARTING X AND STEP SIZE'
READ  *,X,DX
PRINT *,'ENTER MAXIMUM NUMBER OF STEPS'
READ  *,IMAX
I = 0
X0 = X
F0 = F(X0)
1 X1 = X0 + DX
  F1 = F(X1)
  I  = I + 1
  IF(I .GT. IMAX)THEN
    PRINT *,'IN THE RANGE ',X,' TO ',X1
    PRINT *,'THE MAXIMUM OF F(X) IS AT ',
+        'F(',X1,') = ',F1
    STOP
  ELSE
    IF(F1 .GT. F0)THEN
      X0 = X1
      F0 = F1
    ELSE IF(F1 .LT. F0)THEN
      DX = -DX/2.
      X0 = X1
      F0 = F1
```

```
      ELSE
          PRINT*,'THE FUNCTION IS THE SAME',
   +         ' AT X = ',X0,' AND X = ',X1
          PRINT*,'PROGRAM STOPS WITH I = ',I
          STOP
      END IF
      GO TO 1
   END IF
   END
```

13. Finding the inverse of a number C is the same as finding a value of x that satisfies the equation

$$f(x) = \frac{1}{x} - C = 0$$

Using Newton's method and the fact that $f'(x) = -1/x^2$ to find the root of this equation results in

$$x_1 = x_0 + \Delta x, \qquad \Delta x = -f(x)/f'(x)$$

$$= x_0 + \frac{[(1/x_0) - C]}{(1/x_0^2)}$$

which simplifies to

$$x_1 = x_0(2 - Cx_0)$$

If x_1 is to be positive the initial guess must be chosen such that $(2 - Cx_0)$ is greater than zero.

Chapter 5

1.
```
   PROGRAM ZAMGRD
   INTEGER TOTAL,NA,NB,NC,ND,NF,MAX,MIN,
   +        ID,IDMIN,IDMAX,SCORE
   REAL    GPA,SUM
*
   TOTAL = 0
   NA    = 0
   NB    = 0
   NC    = 0
   ND    = 0
   NF    = 0
   MAX   = 0
   MIN   = 100
   SUM   = 0.0
*
   OPEN(21,FILE='GRADES',STATUS=OLD)
   REWIND 21
 1 READ ID,SCORE
   IF(ID .GT. 0)THEN
*
      IF(SCORE .LT. MIN)THEN
         MIN = SCORE
         IDMIN = ID
      END IF
      IF(SCORE .GT. MAX)THEN
         MAX = SCORE
         IDMAX = ID
      END IF
*
```

```
      IF(SCORE .LT. 60)THEN
         PRINT *,'STUDENT WITH ID = ',ID,
   +        'FAILED WITH A SCORE = ',SCORE
         NF = NF + 1
      ELSE IF(SCORE .LT. 70)THEN
         ND = ND + 1
         SUM = SUM + 1.
      ELSE IF(SCORE .LT. 80)THEN
         NC = NC + 1
         SUM = SUM + 2.
      ELSE IF(SCORE .LT. 90)THEN
         NB = NB + 1
         SUM = SUM + 3.
      ELSE
         NA = NA + 1
         SUM = SUM + 4.
      END IF
      GO TO 1
   ELSE
      TOTAL = NA + NB + NC + ND + NF
      GPA = SUM/TOTAL
      PRINT *,'OF ',TOTAL,' EXAMS, THERE ',
   +       'WERE ',NA,'-A ',NB,'-B ',NC,
   +       '-C ',ND,'-D ',NF,'-F GRADES'
      PRINT *,'STUDENT-',IDMAX,' HAD THE ',
   +       'MAXIMUM SCORE = ',MAX
      PRINT *,'STUDENT-',IDMIN,' HAD THE ',
   +       'MINIMUM SCORE = ',MIN
      PRINT *,'OVERALL GPA = ',GPA
   END IF
   STOP
   END
```

3. The tricky part of this problem is to place the revised number of data lines at the top of the updated file. The only way to accomplish this with a Sequential Access file is to read the file twice.

```
   PROGRAM UPDATE
   INTEGER I,ID,CRDTS,NLINES,N,SCAN
*
   OPEN(11,FILE='ROSTER',STATUS='OLD')
   OPEN(12,FILE='NEWLST',STATUS='NEW')
   SCAN = 1
 1 REWIND 11
   N = 0
   I = 0
   READ(11,*)NLINES
*
 2    I = I + 1
      IF(I .LE. NLINES)THEN
         READ(11,*)ID,CRDTS
         IF(CRDTS .LT. 130)THEN
*
            IF(SCAN .EQ. 2)THEN
               WRITE(12,*)ID,',',CRDTS
            ELSE
               N = N + 1
            END IF
*
         END IF
         GO TO 2
```

```
      ELSE
*
         SCAN = SCAN + 1
         IF(SCAN .EQ. 2)THEN
            WRITE(12,*)N
            GO TO 1
         ELSE
            CLOSE(12)
            STOP
         END IF
      END IF
   END
```

5. a) The program to read the file 'GRADES', count the number of students in each college, and to rewrite the file as a direct access file is

```
PROGRAM RERITE
CHARACTER NAME*10
INTEGER CLASS,COLLGE,Q1,Q2,Q3,HW,EXAM,
+       K1,K2,K3,K4,K5,I,RECNO
*
OPEN(22,FILE = 'GRADES')
REWIND 22
OPEN(21,FILE='DIRGRD',ACCESS='DIRECT',
+       RECL = 40)
*
*   K1 is the number in collese 1, etc.
*
K1 = 0
K2 = 0
K3 = 0
K4 = 0
K5 = 0
1  CONTINUE
READ(22,*)NAME,CLASS,COLLGE,Q1,Q2,Q3,HW,
+       EXAM
IF(NAME .EQ. 'END - DATA')GO TO 99
IF(COLLGE .EQ. 1)THEN
   K1 = K1 + 1
   I  = K1
ELSE IF(COLLGE .EQ. 2)THEN
   K2 = K2 + 1
   I  = K2
ELSE IF(COLLGE .EQ. 3)THEN
   K3 = K3 + 1
   I  = K3
ELSE IF(COLLGE .EQ. 4)THEN
   K4 = K4 + 1
   I  = K4
ELSE
   K5 = K5 + 1
   I  = K5
END IF
*
*   Write the student data to the new file
*
RECNO = I + 1000*(COLLGE - 1)
WRITE(23,REC = RECNO)NAME,CLASS,Q1,Q2,
+                     Q3,HW,EXAM
GO TO 1
*
*   Write the number in each collese to the
*   new file.
```

```
*
99 WRITE(23,REC=5001)K1
   WRITE(23,REC=5002)K2
   WRITE(23,REC=5003)K3
   WRITE(23,REC=5004)K4
   WRITE(23,REC=5005)K5
*
   PRINT *,'END OF FILE GRADES'
   STOP
   END
```

b) The program to read the new file and to assign grades is

```
PROGRAM ASSIGN
CHARACTER GRADE*1,NAME*10
INTEGER CLASS,COLLGE,Q1,Q2,Q3,HW,EXAM,
+       N,TOTAL,RECNO
*
OPEN(44,FILE='DIRGRD',ACCESS='DIRECT',
+       RECL = 40)
COLLGE = 1
*
1  RECNO = 5000 + COLLGE
   READ(44,REC=RECNO)N
   WRITE(*,*)' '
   WRITE(*,*)'THERE ARE ',N,' STUDENTS',
+            ' IN COLLEGE ',COLLGE
   WRITE(*,*)'THEIR RECORDS ARE'
   WRITE(*,*)'     NAME CLASS Q1 Q2 Q3',
+            ' HW EXAM   GRADE'
   I = 1
*
2     RECNO = 1000*(COLLGE - 1) + 1
      READ(44,REC=RECNO)NAME,CLASS,Q1,Q2,
+                       Q3,HW,EXAM
      TOTAL = Q1 + Q2 + Q3 + HW + 3*EXAM
*
*   Compute the student's letter srade
*
      IF(TOTAL .GE. 600)THEN
         GRADE = 'A'
      ELSE IF(TOTAL .GE. 540)THEN
         GRADE = 'B'
      ELSE IF(TOTAL .GE. 475)THEN
         GRADE = 'C'
      ELSE IF(TOTAL .GE. 410)THEN
         GRADE = 'D'
      ELSE
         GRADE = 'F'
      END IF
*
      WRITE(*,*)I,'-- ',NAME,CLASS,Q1,Q2,
+                Q3,HW,EXAM,'        ',GRADE
      I = I + 1
      IF(I .LE. N)THEN
         GO TO 2
      ELSE
         COLLGE = COLLGE + 1
         IF(COLLGE .LE. 5)THEN
            GO TO 1
         ELSE
            PRINT *,'END OF GRADE ASSIGN',
+                   'MENTS'
```

```
        STOP
      END IF
    END IF
  END
```
- -

The results of the program are

```
THERE ARE 4 STUDENTS IN COLLEGE 1
THEIR RECORDS ARE
      NAME CLASS Q1 Q2 Q3 HW EXAM  GRADE
1-- GREELEY    2 95 92 91 30 85      B
2-- NOVAK      3 66 50 59 66 62      D
3-- STRAUSS    1 91 96 93 88 94      A
4-- MCDERMITT  2 71 65 66 61 69      D

THERE ARE 2 STUDENTS IN COLLEGE 2
THEIR RECORDS ARE
      NAME CLASS Q1 Q2 Q3 HW EXAM  GRADE
1-- REEVES     2 71 65 80 80 72      C
2-- TAYLOR     3 66 75 77 67 82      C

THERE ARE 6 STUDENTS IN COLLEGE 3
THEIR RECORDS ARE
      NAME CLASS Q1 Q2 Q3 HW EXAM  GRADE
1-- WILSON     2 71 65 82 80 77      C
2-- HUNSICHER  2 82 89 91 75 84      B
3-- LEVY       2 61 68 60 42 67      D
4-- CASSIDY    2 82 71 88 56 71      C
5-- STEPHENSON 2 45 60 62 21 51      F
6-- NELSON     2 91 86 94 92 91      A

THERE ARE 1 STUDENTS IN COLLEGE 4
THEIR RECORDS ARE
      NAME CLASS Q1 Q2 Q3 HW EXAM  GRADE
1-- BROWN      2 71 80 77 65 73      C

THERE ARE 1 STUDENTS IN COLLEGE 5
THEIR RECORDS ARE
      NAME CLASS Q1 Q2 Q3 HW EXAM  GRADE
1-- CHEN       2 71 75 80 86 83      B
END OF GRADE ASSIGNMENTS
```

Chapter 6

1. a)

Method 1: `READ *, . . . <defaults to file INPUT>`

or

Method 2: `OPEN(12,FILE = 'INPUT')`
  `READ(12,*) . . .`

  `READ(*,*) . . .`

b)

Method 1: `READ(*,*,END = . . .)`
Method 2: insert a trailing data line that can be used as a flag.
  `READ(*,*)ID,Q1,Q2, . . .`
  `IF(ID .LT. 0)THEN`
  　　⟨End of data⟩

c) `PRINT '(1X,F9.6)',EXP(ACOS(-1.))`

d) Most compilers will not inform you if a FOR-MAT statement is not referenced in the program.

e) The format specifications T, /, and ' cannot be repeated by preceding with an integer.

f) Examples of when replacing WRITE by READ will result in a compilation error.

`READ(*,*,END = . , .) → WRITE(*,*,END = . , .)`

⟨There is no End-of-File check with a WRITE statement.⟩

`READ 12,A,B,C → WRITE 12,A,B,C`

⟨The correct form of the shortened WRITE statement is PRINT 12, A,B,C.⟩

g) No, FORMAT statements are not executable.

3. a) ERROR. The integer IY is read with an F4.1 format. This will result in an incorrect assignment.

b) ERROR. The REAL variable Y is read with an I5 format and the integer IX is read with an F5.1. If Y contains a decimal point, this will result in an execution time error.

c) No error, but FORMAT 3 is used twice.

d) ERROR. The FORMAT is too wide. This is an execution time error.

e) OK. Note, the entire FORMAT is not used.

f) OK

5. a) OK, results in three lines of output.

b) OK

c) OK

d) ERROR. The FORMAT partially overwrites the values for A and B, and the values for C and D. Three lines are printed.

e) OK

f) OK, however, the zero and the second + sign are not displayed.

g) ERROR. The real quantity A is assigned to an 'A' format.

h) OK. No value for C will be printed. The FORMAT statement is only partially used.

7.
```
col.        1   1   2   2   3
    1...5....0....5....0....5....0
a) ßß1.ßß2.00.3ßßßß2ßßßß3
b) ßßßß    .1E+01    .2E+01
   ß3
c) ßß    .1E+01    2
   ßß    .2E+01    3
   ßß    .3E+00
d)   1.0 = X
e) ßßßßßß
      1.
   ßßßßßß
```

```
        2.
   ЬЬЬЬЬЬ
        0.
f) ЬЬ 0.*****
g) 1.0***    2
   2.0---    3
h) ЬЬЬЬЬ    .333E+00    .300E+01  ⟨the last number
                        will be small and machine dependent⟩
```

9. a) `PRINT '(1X,F10.5)',X,Y,X + Y`
   b) `PRINT '(1X,A,F10.5)','X = ',X,'Y = ',Y`

11.

```
   PROGRAM METALS
   CHARACTER METAL*10
   INTEGER    STEP
   REAL       TEMP
   OPEN(27,FILE = 'MTLDTA')
   REWIND 27
   I = 0
   WRITE(*,10)
 1 READ(27,11,END = 99)METAL,TEMP
   I = I + 1
   WRITE(*,12)I,METAL,TEMP
   IF(TEMP .GT. 1400.)THEN
      WRITE(*,13)'TOO HIGH'
   ELSE IF(TEMP .LT. 600.)THEN
      WRITE(*,13)'TOO LOW'
   END IF
   GO TO 1
99 STOP
10 FORMAT(T4,'        METAL    MELTING',/,
  +         T4,'NUMBER TYPE      TEMP. ',/,
  +         T4,'----    -----    -------')
11 FORMAT(A10,F6.0)
12 FORMAT(T6,I2,3X,A10,1X,F6.1)
13 FORMAT('+',T28,A)
   END
```

13.
```
   PROGRAM PERFEC
   INTEGER I , ITEST
   REAL XI
   I = 2
   WRITE(*,9)
 1 XI = I
   WRITE(*,10)I
   ITEST = SQRT(XI) + .000001
```
   ⟨A small number is added to SQRT to avoid
   truncation of a real number like 1.99999999 . . . .⟩
```
   IF(ITEST**2 .EQ. I)THEN
      WRITE(*,11)ITEST
   END IF
   ITEST = XI**(1./3.) + .000001
   IF(ITEST**3 .EQ. I)THEN
      WRITE(*,12)ITEST
   END IF
   I = I + 1
   IF(I .LE. 100)THEN
      GO TO 1
   ELSE
      STOP
   END IF
 9 FORMAT('1',T10,'LIST OF THE INTEGERS')
```

```
10 FORMAT(T20,I3)
11 FORMAT('+',T25,'IS THE SQUARE OF ',I4)
12 FORMAT('+',T46,'IS THE CUBE OF ',I4)
   END
```

15. a)
```
 1 FORMAT(T10,' MM         MM ',/,
  +        T10,' MMM       MMM ',/,
  +        T10,' MM M     M MM ',/,
  +        T10,' MM  M   M  MM ',/,
  +        T10,' MM   M M   MM ',/,
  +        T10,'MMMM   M   MMMM')
```

c)
```
   PROGRAM ADDER
   CHARACTER RESPNS*3
   INTEGER   I
   REAL      X, SUM
   PRINT *,'DO YOU WISH TO COMPUTE A SUM'
   PRINT *,'ENTER YES OR NO'
   READ'(A)',RESPNS
   IF(RESPNS .EQ. 'YES')THEN
      WRITE(*,*)'ENTREES'
      I = 0
      SUM = 0.0
      PRINT *,'ENTER THE TEN NUMBERS ONE'
  +         ' AT A TIME USING F10.5'
 1    PRINT *,'ENTER NUMBER NOW'
      READ(*,'(F10.5)')X
      WRITE(*,'(T13,F8.4)')X
      I = I + 1
      SUM = SUM + X
      IF(I .LT. 10)THEN
         GO TO 1
      ELSE
         WRITE(*,11)SUM
      END IF
   END IF
   STOP
11 FORMAT('+',8X,'+',//,T12,9('-'),//,1X,
  +        'TOTAL = ',E12.4)
   END
```

Chapter 7

1. a) ERROR. This was probably intended as B(5000). It will be interpreted as B(5,0) and an array cannot be specified as having zero positions for a subscript.

   b) OK, but the array will reserve $9^7 \sim 5$ million computer words. This is very likely beyond the capacity of your computer's main memory.

   c) ERROR. Executable statements may not precede dimension statements.

   d) ERROR. The variable name INTEGERI has 8 characters. This was probably intended as

```
        INTEGER I(50)
        REAL    X(25)
```

   e) ERROR. Only integers may be used to specify subscript bounds.

f) ERROR. Variable names cannot be used to specify subscript bounds.

g) ERROR. Even if A is of type CHARACTER, the expression (1:2) following the specification of the subscript bound has no meaning. This may have been intended as A(5,1:2), which is the same as A(5,2).

h) OK

i) OK

j) OK. That is, A(0) is an integer.

k) OK. However, it is bad style to use the name REAL as a variable.

l) OK. Same comment as above.

m) ERROR. The order must be lower bound:upper bound, i.e., (−7:−5).

3.

```
PROGRAM GRIDS
INTEGER IZ(0:20,0:20),IX,IY
REAL     Z(0:20,0:20),X,Y,ZMAX,ZMIN
F(X,Y) = EXP(X-Y)*SIN(5.*X)*COS(2.*Y)
DO 1 IX = 0,20
DO 1 IY = 0,20
   X = IX/10.
   Y = IY/10.
   Z(IX,IY) = F(X,Y)
1  CONTINUE
ZMAX = Z(0,0)
ZMIN = Z(0,0)
DO 2 IX = 0,20
DO 2 IY = 0,20
   IF(Z(IX,IY) .LT. ZMIN)ZMIN = Z(IX,IY)
   IF(Z(IX,IY) .GT. ZMAX)ZMAX = Z(IX,IY)
2  CONTINUE
DO 3 IX = 0,20
DO 3 IY = 0,20
   IZ(IX,IY) = 10.*(Z(IX,IY) - ZMIN)/
+                  (ZMAX - ZMIN)
3  CONTINUE
DO 4 IY = 20,0,-1
   Y = IY/10.
   WRITE(*,10)Y,(Z(IX,IY),IX = 0,20)
4  CONTINUE
   WRITE(*,11)(IX/10.,IX = 0,20,5)
   STOP
10 FORMAT('1',T3,F3.1,T9,21(1X,I1,1X))
11 FORMAT(T9,5(F3.1,12X))
   END
```

5.

```
PROGRAM SUMS
INTEGER I,IMAX
REAL     ANSWER,TERM,EPS
IMAX = 100
EPS  = 1.E-6
  ANSWER = 0.        ⟨probs.  a–e⟩
         = 1.        ⟨prob.  f⟩
  TERM   = 1.        ⟨probs.  a,b,c,e⟩
         = 0.25      ⟨prob.  d⟩
```

```
         = 4./3.     ⟨prob.  f⟩
DO 1 I = 1,IMAX
  ANSWER = ANSWER + TERM       ⟨probs.  a–e⟩
         = ANSWER * TERM       ⟨prob.  f⟩
  IF(ABS(TERM) .LT. EPS)THEN   ⟨probs.  a–e⟩
    (ABS(TERM-1.)...           ⟨prob.  f⟩
      PRINT *,'ANSWER = ',ANSWER
      STOP
  END IF
  TERM = 1./I**2                            (a)
       = (-1.)**(I + 1)/(2.*I - 1.)         (b)
       = TERM/2.                            (c)
       = TERM/4.                            (d)
       = TERM/I                             (e)
       = (2.*I)**2/(2.*I - 1.)/(2.*I + 1.)  (f)
1 CONTINUE
  PRINT *,'NO RESULT IN ',IMAX,' STEPS'
  STOP
  END
```

7. a)
```
   SUM = 0.0
   DO 1 I = 1,10
       SUM = SUM + A(I)*X(I)
1  CONTINUE
```
  b)
```
   DO 2 I = 1,10
       T(I) = 0.0
       DO 1 J = 1,10
           T(I) = T(I) + A(I,J)*X(J)
1      CONTINUE
2  CONTINUE
```
  c)
```
   DO 2 I = 1,10
   DO 2 J = 1,10
     C(I,J) = 0.0
     DO 1 K = 1,10
       C(I,J) = C(I,J) + A(I,K)*B(K,J)
1    CONTINUE
2  CONTINUE
```

9. The array K_{ij} contains the following integers:

i＼j	0	1	2	3	4
0	0	1	2	3	4
1	10	11	12	13	14
2	20	21	22	23	24
3	30	31	32	33	34
4	40	41	42	43	44

a) ɓ 1 2 3 4

b) ɓɓ 11 12 13 14 21
   ɓɓ 22 23 24 31 32
   ɓɓ 33 34 41 42 43
   ɓɓ 44

c) ɓɓɓɓ   0   1   2   3   4
   ɓɓɓɓ   0   1   2   3   4
   ɓɓ1ɓɓ 10   11   12   13   14
   ɓɓ2ɓɓ 20   21   22   23   24
   ɓɓ3ɓɓ 30   31   32   33   34
   ɓɓ4ɓɓ 40   41   42   43   44

d) ƀƀ 0 1 2 3 4
   ƀƀ 11 12 13 14 22
   ƀƀ 23 24 33 34 44

e) ƀƀ 40 30 20 10 0
   ƀƀ 41 31 21 11 1
   ƀƀ 42 32 22 12 2
   ƀƀ 43 33 23 13 3
   ƀƀ 44 34 24 14 4

f) ƀƀ 0 1 2 3 4
   ƀƀ 11 12 13 14
   ƀƀ 22 23 24
   ƀƀ 33 34
   ƀƀ 44

g) 0 1 2 3 4
    11 12 13 14
       22 23 24
          33 34
             44

Chapter 8

1. a) ƀA
   b) ƀ12345
   c) ƀAƀƀƀƀ
   d) ƀA
   e) ƀ45
   f) ƀAƀƀƀƀ
   g) ƀ12345
   h) ƀA4512
   i) ƀ1
     ƀ2
     ƀ3
     ƀ4
     ƀ5
   j) ƀA1234
   k) ƀ1234512345

3.
```
PROGRAM FOG
CHARACTER LINE*80
INTEGER NSNTC,NWORD,LEFTWD,RGHTWD,
+                    LEFTSN,RGHTSN
NSNTC = 0
NWORD = 0
LEFTSN = 1
1   READ(*,'(A)',END = 99)LINE
LEFTWD = 1
```
⟨LEFTWD is the position of the left end of the current word. LEFTSN is the position of the left end of the current sentence.⟩
```
3   RGHTSN = INDEX(LINE(LEFTSN:),'. ')
    IF(RGHTSN .NE. 0)THEN
        NSNTC = NSNTC + 1
        LEFTSN = RGHTSN + 1
        GO TO 3
    END IF
2   RGHTWD = INDEX(LINE(LEFTWD:),' ')
    IF(RGHTWD .EQ. LEFTWRD + 1)THEN
        GO TO 1
    ELSE
```

```
        NWORD = NWORD + 1
        LEFTWD = RGHTWD + 1
        GO TO 2
    END IF
    GO TO 1
99  PRINT *,'FOG FACTORS'
    PRINT *,'NUMBER OF WORDS = ',NWORD
    PRINT *,'NUMBER OF SENTENCES = ',NSNTC
    STOP
    END
```

5.
```
CHARACTER LINE*50, NUMB*8
1   READ(*,'(A)',END = 99)LINE
    MATCH1 = INDEX(LINE,'JONES, JAMES')
    IF(MATCH1 .NE. 0)THEN
        I = MATCH1 + 12
        MATCH2 = INDEX(LINE(I:),'JENNINGS')
        IF(MATCH2 .NE. 0)THEN
            NPOS = INDEX(LINE,'-')
            NUMB = LINE(NPOS-3:NPOS + 4)
            WRITE(*,'(A,A)')'NO. = ',NUMB
            STOP
        ELSE
            GO TO 1
        END IF
    END IF
    GO TO 1
99  WRITE(*,*)'NOT IN PHONE BOOK'
    STOP
    END
```

7. Simply replace MIN by MAX and .LT. by .GT.

9.
```
    DO 2 I = 1,N - 1    ⟨I is the top element
                          in current comparison
                          set.⟩
        FLAG = 'OFF'
        DO 1 J = N,I + 1,-1    ⟨Compare pairs of
                                remaining set
                                starting at the
                                bottom.⟩
            IF(A(J) .LT. A(J - 1))THEN
                                ⟨If out of order,
                                  exchange.⟩
                TEMP = A(J)
                A(J) = A(J - 1)
                A(J - 1) = TEMP
                FLAG = 'ON '
            END IF
1       CONTINUE
        IF(FLAG .EQ. 'OFF')THEN
            GO TO 3
        END IF
2   CONTINUE
3   CONTINUE
```

11.
```
    PROGRAM BUBBLE
    CHARACTER FLAG*3
    INTEGER   INDX(1000),I,N,J,TEMP
    REAL      A(1000)
    READ(*,*,END = 99)(A(I),I = 1,1000)
99  N = I - 1
```

```
        DO 1 I = 1,N
            INDX(I) = I
1       CONTINUE
        DO 3 I = N,2,-1
            FLAG = 'OFF'
            DO 2 J = 1,I-1
                IF(A(INDX(J)) .GT. A(INDX(J + 1)))THEN
                    TEMP        = INDX(J)
                    INDX(J)     = INDX(J + 1)
                    INDX(J + 1) = TEMP
                    FLAG        = 'ON '
                END IF
2           CONTINUE
            IF(FLAG .EQ. 'OFF')THEN
                GO TO 4
            END IF
3       CONTINUE
4       CONTINUE
```

13.
```
        PROGRAM PLOT2
        CHARACTER 1 LINE(0:60)
        INTEGER     ISTAR,IPLUS,IX,N
        REAL        X(0:25),Y(0:25,2),XSTEP,
      +             YMAX,YMIN,SCALE,Z,F1,F2
        F1(Z) = ...
        F2(Z) = ...
        N = 25
        BLANK = ' '
        STAR  = '*'
        PLUS  = '+'
        READ(*,*)XLO,XHI
        XSTEP = (XHI - XLO)/(N - 1.)
        DO 1 I = 0,N
            X(I) = I*XSTEP
            Y(I,1) = F1(X(I))
            Y(I,2) = F2(X(I))
1       CONTINUE
        YMAX = Y(0,1)
        YMIN = Y(0,1)
        DO 2 I = 0,N
        DO 2 K = 1,2
            IF(Y(I,K) .LT. YMIN)YMIN = Y(I,K)
            IF(Y(I,K) .GT. YMAX)YMAX = Y(I,K)
2       CONTINUE
        SCALE = YMAX - YMIN
        DO 3 I = 0,60
            LINE(I) = ' '
3       CONTINUE
        WRITE(*,12)
        DO 4 IX = 0,25
            ISTAR = (Y(IX,1) - YMIN)/SCALE * 60.
            IPLUS = (Y(IX,2) - YMIN)/SCALE * 60.
            LINE(ISTAR) = '*'
            LINE(IPLUS) = '+'
            WRITE(*,11)X(IX),Y(IX,1),Y(IX,2),
      +               (LINE(I),I = 0,60)
            LINE(ISTAR) = ' '
            LINE(IPLUS) = ' '
4       CONTINUE
        STOP
Formats
        END
```

15. A page plot of the function $e^{-x/3}\sin(\pi x/2)$

Chapter 9

1.
```
READ(*,*)X,Y
CALL LENGTH(X,Y,R)
PRINT *,'THE POINT (',X,',',Y,') IS'
IF((R .LT. 3.) .AND. (R .GT. 1.))THEN
    PRINT *,'BETWEEN THE CIRCLES'
ELSE
    PRINT *,'NOT BETWEEN THE CIRCLES'
END IF
```

3. a) Note: A Fortran function must have an argument list; however, the actual list may be empty; for example,

```
FUNCTION DBUG()
SAVE ICALL
DATA ICALL/0/
PRINT *,'IN THIS DEBUGGING RUN THE PRO',
   +     'GRAM GOT TO THE ',ICALL,'TH CALL'
   +     ' OF THE FUNCTION DBUG'
```

```
ICALL = ICALL + 1
DBUG = 1.
RETURN
END
```

    b) In all reasonable uses of a function subprogram it is expected that the function will either return to the calling program or will stop.

    c) i. yes, ii. yes, iii. no, iv. yes, v. yes.

5.
```
SUBROUTINE MATPRD(A,B,C,N,M)
REAL A(N,N),B(N,N),C(N,N)
DO 7 I = 1,M
DO 7 J = 1,M
  C(I,J) = 0.0
  DO 3 K = 1,M
    C(I,J) = C(I,J) + A(I,K)*B(K,J)
3     CONTINUE
4 CONTINUE
RETURN
END
```

7. a)
```
SUBROUTINE INPUT(FILE,NAME,INIT,ID,
+           CLASS,SEX,SAT,COLL,GPA,N)
CHARACTER NAME(8000)*10,INIT(8000)*1,
+           FILE*6,SEX(8000)*1
INTEGER   ID(8000),CLASS(8000),
+           SAT(2,8000),COLL(8000)
REAL      GPA(2,8000)
OPEN(21,FILE = FILE,STATUS = 'OLD')
REWIND 21
I = 1
1 READ(21,10,END = 99)NAME(I),INIT(I),
+           ID(I),CLASS(I),SEX(I),SAT(1,I),
+           SAT(2,I),COLL(I),(GPA(L,I),L=1,2)
  I = I + 1
  GO TO 1
99 N = I - 1
CLOSE 21
PRINT *,'THE FILE ',FILE,' CONTAINS ',
+   INFORMATION ON ',N,' STUDENTS'
RETURN
10 FORMAT(A10,A1,I9,I2,A1,2I3,I1,2F4.2)
END
```

9. The program solution to this problem is quite long but easy to construct. Each of the constants *a–f* and the variables *x, y* are declared to be integer arrays with the first element representing the numerator and the second the denominator of the rational fraction. For example, if $a = 1/2$, then $A(1) = 1$, $A(2) = 2$. The arithmetic required in the equations for the solutions x, y must be written explicitly in terms of subfunctions to multiply and subtract fractions. Finally, after the numerator and denominator of x and y are obtained, they are reduced to lowest terms by using the function IGCD to divide out the greatest common factor in the numerator and the denominator.

11. Replace the line
```
          IC = 2.*C + 1
```
by
```
          IC = 2.*C - 1
```

13. a) ERROR. An arithmetic operation $(I + 1)$ cannot appear in the definition line of a subroutine.

    b) OK, if the position in which $I + 1$ appears is used only as input to the subroutine.

    c) ERROR. Parentheses are not allowed within the argument list of the definition line of a subroutine or function.

    d) OK

    e) ERROR. The name of the function is not assigned a value before the return.

    f) ERROR. The name of a subroutine cannot be assigned a value.

15. a)
```
FUNCTION POLY(X,N,C)
INTEGER N
REAL    X,C(0:N),SUM,POLY
SUM = C(0)
DO 1 I = 1,N
  SUM = SUM + C(I)*X**I
1 CONTINUE
POLY = SUM
RETURN
END
```
b)
```
FUNCTION POLYB(X,N,C)
INTEGER N
REAL    X,C(0:N),SUM,POLYB
SUM = C(N)
DO 1 I = N,1,-1
  SUM = SUM*X + C(I - 1)
1 CONTINUE
POLYB = SUM
RETURN
END
```

17.
```
FUNCTION CUBRT(X)
REAL CUBRT,C,X
  C = ABS(X)**(1./3.)
  IF(X .LT. 0.)C = -C
CUBRT = C
RETURN
END
```

19. The result of the calculation is:
```
THE MAXIMUM SURFACE SEPARATION IS
            11009.22 MILES
BETWEEN RIO DE JANEIRO AND VLADIVOSTOK
```

21.
```
PROGRAM SOLVE
REAL GUESS,ANSWER,G
EXTERNAL G
GUESS = 1.0
```

```
 CALL ROOT(GUESS,ANSWER,G)
 PRINT *,'THE ROOT IS ',ANSWER
 STOP
END
FUNCTION G(X)
REAL X,ARGMT,G,PI
EXTERNAL ARGMT
 PI = ACOS(-1.)
 G = 5.*EXP(-2.*X**2) - SIN(PI*X/2.)
+                 + XINTGL(0.,X,ARGMT)
RETURN
END
FUNCTION ARGMT(T)
 ARGMT = (4.*T**2 - 5.)*EXP(-T**2)
RETURN
END
SUBROUTINE ROOT(A,ANSWER,F)
   ...          ...
END
FUNCTION XINTGL(A,B,FNC)
   ...          ...
END
```

Chapter 10

1. a) FALSE, but variables assigned values in a PA-RAMETER statement may only be used after the PARAMETER statement. Thus

```
        REAL A(N)
        PARAMETER (N = 12)
```

   would be an error.
   b) False
   c) True
   d) False
   e) False
   f) False

3.
```
PROGRAM CSQRT
COMPLEX GUESS, C
REAL DIFF,EPS
INTEGER I,IMAX
```

```
  IMAX = 30
  EPS = 1.E-6
  PRINT *,'ENTER A COMPLEX NUMBER WHOSE ',
+          'SQUARE ROOT IS DESIRED'
  READ *,C
  I = 0
  PRINT *,'ENTER A GUESS FOR THE SQUARE ',
+         ' ROOT. THE GUESS MUST HAVE ',
+          'A NONZERO IMAGINARY PART'
  READ *,GUESS
1 DIFF = ABS(GUESS*GUESS - C)
  IF(DIFF .LT. EPS)THEN
    PRINT *,'THE SQUARE ROOT OF ',C,' IS'
+        ,X1,' AFTER ',I,' ITERATIONS'
    STOP
  ELSE
    I = I + 1
    IF(I .GT. IMAX)THEN
      PRINT *,'EXCESSIVE ITERATIONS'
      STOP
    END IF
    GUESS = .5*(GUESS + C/GUESS)
    GO TO 1
  END IF
  END
```

5. For $I = -1$ and $I = 0$, D has the value ⟨false⟩ while E is ⟨true⟩. So the statement

```
        THE ORDER OF AND/OR MAKES
        A DIFFERENCE FOR
        F .AND. F .OR. T
```

   is printed twice.

7. a) ERROR. Only arithmetic expressions involving constants, not intrinsic functions, are allowed in a PARAMETER statement.
   b) OK
   c) ERROR. Variables assigned values in a PA-RAMETER statement cannot be used in place of statement numbers.
   d) OK
   e) OK
   f) OK

INDEX

DOUBLE PRECISION
  input/output, 354
  to minimize round-off error, 354
  root solver, programming problem VII-A (373)
  statement, 354–355, 383
Drag, air, 254
Dummy arguments, 111, 296, 298, 381
  protecting, 298

E-format, 187–188
Echo print, 45, 87, 185
Editing programs, 14, 18
Electrical engineering, 153–154, programming problems II-A (154), V-A (286), VII-A (374), VII-B (376)
Electron cloud, 291
ELSE statement, 69, 392
  as terminator of a DO loop, 229
ELSE IF statement, 72, 392
  as terminator of a DO loop, 229
END statement, 49, 296, 381, 388
  as terminator of a DO loop, 229
END option in READ/WRITE statements, 199
End Do in extended Fortran, 73, 75, 363–365
END IF statement, 62, 392
ENDFILE statement, 172, 395
END-OF-FILE FLAGS, 122, 175, 197
ENTRY statement, 386
.EQ. , 63
EQUIVALENCE statement, 386
.EQV. , 357
ERF(X), 157–159, 349
Error
  compilation, 24
    misspelling variable names, 358
    translating algebra to Fortran, 53
  execution, 24–25, 49
    array index out of bounds, 222
    floating point overflow, 326
    in formatted input/output, 186
    noninitialized variables, 49, 116
    using CHARACTER variables in arithmetic expressions, 43
  function, Efr(x), 157–159, 349
  option
    in OPEN statements, 169
    in READ/WRITE statements, 199
  round-off, 354

  syntax, 24
  truncation, 37
Exchange sort, see selection sort
Execution error, see error, execution
EXP(X), 47, 366
Explicit typing, 34
Exponentiation (**), 46–47
  successive, 38
Expression
  arithmetic, 35ff
  character, 262
  mixed mode, 40
  logical, 62
Extended Fortran, 363–365
EXTERNAL statement, 318–320, 385
Extremum of a function, see minimum/maximum

F-format, 183–185
Factorial, 119, 230, 279
  Fortran code for, 230
Failure path, 87
.FALSE. , 63, 357, 383
Fermat's last theorem, extension to complex numbers, 370
file, 16, see also data file
  specifier in OPEN statements, 169
Fixed point iteration, see successive substition
Fixed point numbers, 30, see also INTEGER
Floating point
  numbers, 30, 183, see also REAL
  overflow, 325–326
Flow control statements, 389–394
  CALL, 296, 390
  CONTINUE, 78, 229, 390
  DO, 228ff, 233ff, 393
  ELSE, 69, 392
  ELSE IF, 72, 392
  GO TO, 21, 77, 89, 390–391
  IF, 62ff, 88, 90, 391–392
  RETURN, 296, 389
Flowchart, 58–60
  examples, 60, 71, 74, 76, 81, 85
  symbols, 59
Fluid flow, 94
Fluidized-bed reactor, 148
FORMAT
  of a Fortran line, 15–16, 380
  input/output errors, 186

List-directed input/output, 51–53, 166, 182, 192
 of complex numbers, 356
Load map, 25
Locally defined variables, 321
LOG(X), 47–48, 366
LOG10(X), 311, 366
LOGICAL
 block IF structure, 62ff, 392
 combinational operators, 63
 constants, 357
 expressions, 62
 IF statement, 88, 391
 type statement, 356–357, 383
 variables, 63
Loop structures, 73ff
 iterative, 80
 flowchart example, 76
 repetitive, 78
.LT. , 63

Machine language, 12
Main memory, 7–9
Main program name, 41, see also PROGRAM
Materials engineering 156–157, programming problem II-B (157)
Matrix, 239, see also array
 multiplication, 329
 notation, 235
 printing, 239
MAX( ), 366
Maximum/minimum, see minimum/maximum
Mechanical engineering, 100, sample problem V (282), programming problems I-A (100), IV-A (249)
Memory
 main, 7–9
 preset to zero, 116
Merging of files, programming problem III-A (209)
Metallurgical engineering, see materials engineering
MIN( ), 366
Minimum repair costs, sample problem III (243)
Minimum/maximum
 algorithm, 112
 and horizontal tangent lines, 115
 exam scores, Fortran code, 121

of an array, 238
of a multivariable function, 252
Mixed mode
 expressions, 40
 replacement, 42–43
MOD(X,Y), 366
Mode conversion, 42–43, 355, 367
Multiple roots, 140

Name
 conflict with intrinsic function, 110
 of data file, 169
 of Fortran variables, 33–35
 of FUNCTION, 305
 of main program, 43
 of statement function, 109
 of SUBROUTINE, 295
Named constants, 360
Nested
 block IF structures, 70
 DO loops, 231–232, 393
 loops, 80, 99
Newton-Raphson method, see Newton's method
Newton's algorithm for square roots, 80, 365, 368
 Fortran code for, 86
Newton's method, 138ff
 Fortran code for, 141–142
 problems with, 140–141
Nim, 27
NINT(X), 367
Nonexecutable statement, 32, 183
Nonrepeatable edit descriptors, 191
.NOT. , 63
Nuclear engineering, programming problem III-C (214)
Numerical differentiation, 254
Numerical integration, 123ff

Object code, 12
Ohm's law, 155
OPEN statement, 168–170
.OR. , 64–65
Ordering of Fortran statements, 361–363
Overprinting, 196
Overwriting constants, 301
Oxygen deficiency of a polluted stream, programming problem I-B (103)

Parabola, area beneath, 144
PARAMETER statement, 360–361, 384

Statement   ## Comment

BACKSPACE ⟨unit no.⟩   Backspace a sequential file one record (usually one line).

ENDFILE(⟨unit no.⟩)   Write an END-OF-FILE mark on a sequential file.

(*)INQUIRE(⟨arg. list⟩)   Not used in this text.

INPUT/OUTPUT STATEMENTS

READ(⟨unit no.⟩,⟨format no.⟩,[options],⟨in-list⟩)
READ ⟨format no.⟩,⟨in-list⟩
READ *,⟨in-list⟩   List-directed READ.

WRITE(⟨unit no.⟩,⟨format no.⟩,[options],⟨out-list⟩)
PRINT ⟨format no.⟩,⟨out-list⟩
PRINT *,⟨out-list⟩   List-directed PRINT.

FORMAT(⟨format spec. list⟩)   I/O editing specifications.

THE ORDER OF FORTRAN STATEMENTS

PROGRAM/SUBROUTINE/FUNCTION				Comments
IMPLICIT		PARAMETER	FORMAT	
Type Specifications (REAL, INTEGER, etc.)				
Other Specifications DIMENSION/COMMON/EXTERNAL/INTRINSIC/				
Statement functions				
All executable statements	DATA			
END				